AF292391

Biosensors with Fiberoptics

Contemporary Instrumentation and Analysis

Gary M. Hieftje, Series Editor

Biosensors with Fiberoptics, edited by *Donald L. Wise and Lemuel B. Wingard, Jr.,* 1991

Lasers in Chemical Analysis, edited by *Gary M. Hieftje, John C. Travis, and Fred E. Lytle*

Introduction to Bioinstrumentation, *with Biological, Medical, and Environmental Applications, Clifford D. Ferris*

Biosensors with Fiberoptics

Edited by

Donald L. Wise

Northeastern University, Boston, Massachusetts

and

Lemuel B. Wingard, Jr.

University of Pittsburgh, Pittsburgh, Pennsylvania

 Springer Science+Business Media, LLC

Library of Congress Cataloging-in-Publication Data

Biosensors with fiberoptics / edited by Donald L. Wise and Lemuel B.
   Wingard, Jr.
         p. cm. — (Contemporary neuroscience)
      Includes bibliographical references and index.
      ISBN 978-1-4612-6782-9   ISBN 978-1-4612-0483-1 (eBook)
      DOI 10.1007/978-1-4612-0483-1
      1. Biosensors. 2. Fiber optics. I. Wise, Donald L. (Donald
Lee), 1937– . II. Wingard, Lemuel B. III. Series.
      [DNLM: 1. Biosensors. 2. Fiber Optics. QT 34 B61546]
R857.B54B564 1991
610'.28—dc20
DNLM/DLC
for Library of Congress   91-6350
                                                        CIP

In Memory of

Lemuel B. Wingard, Jr., PhD

(1930–1990)

This is to recognize the memory of "Lem" Wingard, who passed on in August 1990 as this text was nearing completion. Lem was slated to be coeditor of this text, based on his early review of material submitted to him by the Editor. Lem had a long and productive interest in applied biochemistry and the development of biomedical devices and instrumentation and, with his graduate student, completed the first chapter in this text. Most recently, Lem was focusing his interest on biosensors based on fiberoptic techniques.

At the time of his passing, Lem was: Professor of Pharmacology (main appointment), Department of Pharmacology, School of Medicine, University of Pittsburgh; Professor of Anesthesiology (secondary appointment) in the same School of Medicine; and also Adjunct Professor of Chemical Engineering (secondary appointment) in the School of Engineering.

Lem Wingard's major research contributions included important studies on GABA Type A receptor; gene expression in foreign cell lines; large-scale preparation, characterization, and reconstitution; immobilization of cofactors, enzymes, receptors, and drugs, especially for analytical/electrochemical/biosensor applications; mechanisms of action of anticancer agents on surface components of mammalian cells; coupling of drugs to antibodies; and clinical pharmacokinetics.

For those colleagues who wish to remember Lem's scientific lineage, the following is a reminder: He was an undergraduate in 1948–1953 at Cornell University, Ithaca, NY, where he was awarded the BChE in Chemical Engineering. Lem also took his graduate training during 1961–1965 at Cornell University, where he earned his PhD, under Professor R. K. Finn, in biochemical engineering. Subsequently,

Lem did postgraduate work in 1970–1972 at the State University of NY at Buffalo, with Professor G. Levy, in pharmokinetics, and in 1979–1980 at Yale University Medical School, New Haven, CT, with Professor A. C. Sartorelli, where he studied cancer chemotherapeutic mechanisms.

Clearly, Lem Wingard made major contributions in the applications of biotechnology. The world became a better place because of his dedicated work, and he will be missed.

Donald L. Wise

Preface

This reference text on fiberoptic probes as biosensors focuses on the rapidly enlarging intersection of the fields of biotechnology and advanced fiberoptics/electronics. In preparing the chapters, all principal authors were asked to place their emphasis on the three major phases of the developmental process from concept to marketplace, namely, research, development, and applications.

The use of fiberoptic probes for advanced instrumentation is now well-recognized, and the term biosensor is increasingly used to describe the unique joining of forefront electronics and modern biotechnology. Thus, fiberoptic biosensors represents an entirely new and advanced field. The present text consists of contributed chapters prepared by experts directly involved in using key areas of fiberoptic biosensor research, development, and applications. They describe novel biotechnology-based fiberoptic biosensors, such as those used for detection of very low levels of chemical and biological moieties. Rather than present traditional systems in which a fiberoptic instrument is used simply as an observer, the biosensors described herein provide for direct assay and readout of bioinstrumentation information.

The authors also discuss new biotechnology-based fiberoptic biosensors used for direct chemical and biological analysis, as in laboratory or process control instruments, and biosensors used to measure chemical and biological moieties in the body and in the environment. One suggested commercial objective is to use fiberoptic probes, prepared as immunosensors, for direct in-field monitoring of environmental toxins. By placing single bundles of these microsensors around the perimeter of a hazardous waste site—and also having probes from each fiber bundle set a different depths—it should be possible to provide continuous *in situ* surveillance of hazardous leachates. For this case, the use of fiberoptic probes as immunosensors may require preparing biopolymer coatings on the fibers such that standard immunodiagnostics may be used. Such a system should bring the sensitivity and selectivity of laboratory medical diagnostics to environmental situations. With practical fiberoptic biosensors, it may be possible to provide a service business of in-field environmental toxin assays. This work could include the supplying of fiberoptic probes for initial assay, monitoring during bioremediation, and continued surveillance. These fiberoptic biosensors may be used at the surface or placed at various depths for *in situ* assay.

Further applications of fiberoptic biosensors will be to the rapid "doctor's office"-type systems as well as over-the-counter systems; substantial improvements in laboratory and clinical instrumentation are anticipated. An array of biological (and chemical) moieties may be detected (no one disease vector need be focused upon).

It is hoped that this reference text will be of keen interest to a wide audience, including instrument manufacturers, the electronics industry, and many of the increasing number of biotechnology firms, especially those looking for additional applications of their expertise. University professors and government officials, as well as industrial executives, and all who are working in the area where modern biotechnology, advanced electronics, and instrumentation meet should find this text to be extremely valuable.

Donald L. Wise

Contents

Concepts, Biological Components, and Scope of Biosensors
Lemuel B. Wingard, Jr. and Jerome P. Ferrance

Chemical Sensing with Fiberoptic Devices
Carmen Cámara, María Cruz Moreno, and Guillermo Orellana

Fluorescent Labels

Richard P. Haugland

Chemistry and Technology of Evanescent Wave Biosensors

Richard B. Thompson and Frances S. Ligler

Optical Characteristics of Fiberoptic Evanescent Wave Sensors
Theory and Experiment

Walter F. Love, Leslie J. Button, and Rudolf E. Slovacek

Evanescent Wave Immunosensors for Clinical Diagnostics

Barry I. Bluestein, Mary Craig, Rudolf Slovacek, Linda Stundtner, Cynthia Urciuoli, Irene Walczak, and Albert Luderer

Instrumentation for Cylindrical Waveguide Evanescent Fluorosensors

Steve J. Lackie, Thomas R. Glass, and Myron J. Block

Immunoassay Kinetics at Continuous Surfaces

John F. Place, Ranald M. Sutherland, Andrew Riley, and Ciaran Mangan

Luminescence in Biosensor Design

Pierre R. Coulet and Loïc J. Blum

In Vivo Applications of Fiberoptic Chemical Sensors

Amos Gottlieb, Skip Divers, and Henry K. Hui

Contributors

MYRON J. BLOCK • *Ord Inc., North Salem, NH*
BARRY I. BLUESTEIN • *CIBA-Corning Diagnostics Corp., Medfield, MA*
LOÏC J. BLUM • *Universite Lyon, Villeurbanne, France*
LESLIE J. BUTTON • *Corning Inc., Corning, NY*
CARMEN CÁMARA • *Universidad Complutense, Ciudad Universitaria, Madrid, Spain*
PIERRE R. COULET • *Universite Lyon, Villeurbanne, France*
MARY CRAIG • *CIBA-Corning Diagnostics Corp., Medfield, MA*
SKIP DIVERS • *Puritan-Bennett Corp., Carlsbad, California*
JEROME P. FERRANCE • *University of Pittsburgh School of Medicine, Pittsburgh, PA*
THOMAS R. GLASS • *Ord Inc., North Salem, NH*
AMOS GOTTLIEB • *Random Technologies, San Francisco, CA*
RICHARD P. HAUGLAND • *Molecular Probes Inc., Eugene, OR*
HENRY K. HUI • *Puritan-Bennett Corp., Carlsbad, CA*
STEVE J. LACKIE • *Ord Inc., North Salem, NH*
FRANCES S. LIGLER • *Naval Research Laboratory, Washington, DC*
WALTER F. LOVE • *Corning Inc., Corning, NY*
ALBERT LUDERER • *CIBA-Corning Diagnostics Corp., Medfield, MA*
CIARAN MANGAN • *Commission of the European Communities, Brussels, Belgium*
MARIA CRUZ MORENO • *Universidad Complutense, Ciudad Universitaria, Madrid, Spain*
GUILLERMO ORELLANA • *Universidad Complutense, Ciudad Universitaria, Madrid, Spain*
JOHN F. PLACE • *Dakopatts a/s, Glostrup, Denmark*
ANDREW RILEY • *University of Utah, Salt Lake City, UT*
RUDOLF E. SLOVACEK • *CIBA-Corning Diagnostics Corp., Medfield, MA*
RANALD M. SUTHERLAND • *Abbott GmbH Diagnostika, Delkenheim, Germany*
LINDA STUNDTNER • *CIBA-Corning Diagnostics Corp., Medfield, MA*
RICHARD B. THOMPSON • *Naval Research Laboratory, Washington, DC*
CYNTHIA URCIUOLI • *CIBA-Corning Diagnostics Corp., Medfield, MA*
IRENE WALCZAK • *CIBA-Corning Diagnostics Corp., Medfield, MA*
LEMUEL B. WINGARD, JR. • *University of Pittsburgh School of Medicine, Pittsburgh, PA*

Concepts, Biological Components, and Scope of Biosensors

Lemuel B. Wingard, Jr.
and Jerome P. Ferrance

1. Concept of a Biosensor

In the late 1960s, when techniques for immobilizing enzymes on solid supports were being developed, one proposed application was analytical chemistry *(1,2)*. By attaching an appropriate enzyme to an electrode it was possible to generate electrical currents that were related to the concentration of substrate for the enzyme *(2)*. These immobilized enzyme electrodes subsequently were tested as direct readout analytical devices to quantify the concentrations of many analytes.

It soon became apparent that such direct readout analytical devices could be designed using biological materials other than enzymes and readout techniques besides those based on electrochemistry, and these devices have become known as biosensors. The concept of a biosensor is summarized in Fig 1. Two components are needed, one a biological material to provide molecular-level recognition for the analyte (compound being quantified), and a second to convert the molecular level recognition into a readable output signal.

From the great number of papers, books and conferences dealing with biosensors over the last few years, it is easy to see that this is an

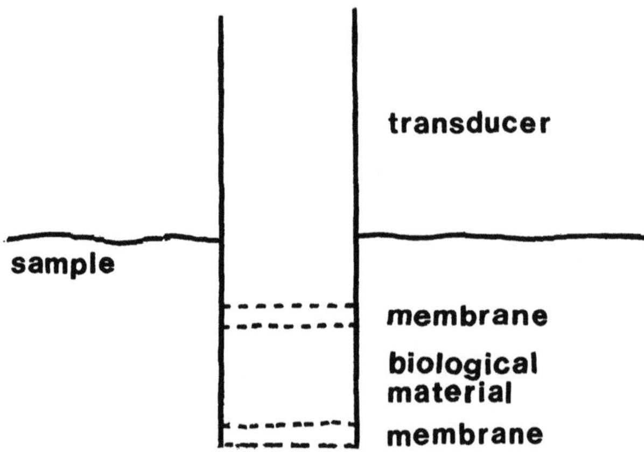

Fig. 1. Components of a biosensor.

exciting and expanding field. A wide variety of readout schemes and biological materials, as well as anticipated applications, have been proposed or demonstrated experimentally thus far *(3–5)*. Investigators, from biologists to chemists to engineers, continue to become involved in perfecting existing devices and developing new ones. As they join the field, they bring with them new ideas and applications for biosensors that are not even imagined today.

2. Biological Components

2.1. Introduction

The primary function of the biological component of a biosensor is to provide selectivity for the analyte of interest. In most instances this is accomplished through the molecular-level selective binding affinity that the biological material displays for the analyte. This selective binding may in turn trigger some other process, such as an enzyme-catalyzed reaction or a receptor modulated transmembrane ion flux that may participate in the transduction process. It is the inherent binding selectivity, however, that justifies most use of biological material in biosensors. Some of the more widely studied biological materials for use as biosensor components are listed in Table 1 *(6,7)*.

Table 1
Biological Materials Used in Biosensors

Enzymes: Purified or crude mixtures;
    single or multiple enzymes
Purified antibodies
Receptor proteins: purified or crude mixtures
Transport proteins
DNA
Lectins
Mammalian tissue slices
Plant tissue slices
Organelles
Microorganisms

In addition to selectivity, several other variables need to be considered in selecting the biological material for a specific biosensor design. These include stability of the component to maintain binding ability and functionality, reversibility of the binding, sensitivity of detection, and availability of the material. Stability and availability are described for each of the biological materials discussed separately, but sensitivity and reversibility of binding are both related to a binding constant of the interaction.

Most of the binding reactions of biological materials are reversible, but the kinetics of dissociation may be so slow that attempts to develop reusable or continuous-reading biosensors becomes impractical. For example, for ligand L binding to site R, the binding constant K_d is defined a follows:

$$L + R \underset{k_{-1}}{\overset{k_1}{\longleftrightarrow}} LR \tag{1}$$

$$k_1 [L] [R] = k_{-1} [LR] \tag{2}$$

$$K_d = [L] [R]/[LR] = k_{-1}/k_1 \tag{3}$$

If Y is the fraction of sites occupied and $[R_T]$ the concentration of total binding sites, then

$$Y = [LR]/[RT] \text{ and } [R] = [R_T] - [LR] \tag{4}$$

From Eqs. (3) and (4) it follows that

$$K_d = ([L] - [L]\, Y)/Y \tag{5}$$

which can be rearranged to give Eq. (6):

$$Y = 1/\{(K_d/[L]) + 1\} \tag{6}$$

Thus, when $[L]$ is equal to K_d, Y is 0.5, and one-half of the sites are occupied. For ligand–binding-site combinations that result in K_d magnitudes of about $10^{-6}M$ or greater, the speed of dissociation is fast enough to make biosensor reuse practical; however, the sensitivity will be limited to about $0.1 \times K_d$ or an analyte concentration of about $0.1\ \mu M$. Such sensitivity restrictions should still enable determination of practical concentrations for many compounds of medical, environmental, or chemical processing interest, but eliminate those compounds that exist at practical concentrations of 10^{-12}–$10^{-14}M$ or less. High-affinity binding systems that have K_d values of $10^{-13}M$ can provide the needed sensitivity, but are restricted to a single use because of the very slow rates of dissociation of bound ligand.

2.2. Enzymes

Enzymes are proteins designed with an active site that binds a specific substrate and then catalyzes the transformation of the substrate into a product. Product is then released from the active site so that the enzyme can continue to catalyze the reaction. Each enzyme is designed to carry out a single reaction on a substrate or group of substrates that binds to the active site on the enzyme. Enzymes may also have additional binding sites for cofactors that they use during the reaction to form product from the substrate. For those enzymes that require a cofactor in order to function, a supply of cofactor or a means of cofactor regeneration must be available. The continuous removal of product from the binding site allows enzyme biosensors to function continuously or to be used for more than a single determination. The need for cofactor regeneration, however, continues as a difficult hurdle to be overcome in the application of some enzymes in multiple-use biosensors.

There are few points in the enzyme's catalytic process at which transduction can be carried out. Binding of the substrate to the active site of the enzyme, like antigen/antibody binding, may be used to produce a measurable response. The formation of product is another step at which transduction may be accomplished. Since the rate of product formation will usually be a function of the concentration of the substrate, a linear response can often be obtained; this will not be true, however, if the enzyme is saturated with substrate. Changes in cofactors that take place during the reaction can also be transduced into a signal.

In the design of biosensors, it is usually necessary to immobilize the biological material in order to prevent diffusion into the sample, as well as to keep the event being detected close to the transducer. A wide variety of methods are available for immobilization of enzymes on solid supports *(8–10)*. They vary from entrapment in small vesicles or behind essentially enzyme-impervious membranes to chemical attachment to a solid surface. To find out if a specific technique will immobilize a particular enzyme without loss of enzyme activity, the system must be tested experimentally; however, valid predictions can often be made by comparison with known results for similar enzymes and similar coupling procedures.

Availability of enzyme is another point in favor of their use in biosensors. Because enzymes that carry out the same reaction are often produced in a number of different organisms, it is often possible to find a source from which large amounts of the enzyme can be obtained. Enzyme production is also possible using gene-expression systems, if the gene for the enzyme is available. Having the gene for the enzyme also establishes the possibility of modifying the enzyme to control both the affinity and specificity of the binding site on the enzyme.

Two other advantages of enzymes are that they are usually stable and that they can be used in organic solvents. Although the stabilities of enzymes vary greatly, many enzymes show sufficient thermal stability to be stored and used in biosensors at room temperature. Where stability at a higher temperature is required, it may be possible to find such variants in organisms that live in boiling mud flats; deep, hot ocean vents; or hot geysers, and to clone the genes for the enzymes that are more thermally stable *(11)*. Other enzymes, especially membrane and intracellular types, are inherently less stable in extremes of temperature, ionic conditions, and other environmental factors, and provide less hope for finding stable variants. Numerous enzymes have been found to retain significant activity in organic solvents, but only when the enzymes retained small trace amounts of water or when the solvents were not anhydrous *(12,13)*. This may enable the development of enzyme-based biosensors that would function in mixed aqueous–organic solvents and thus be capable of determining lipid-soluble analytes. Response curves of the biosensors are likely to be solvent-dependent, however, a consequence of partitioning of the analyte between the solvent and the aqueous binding site.

Enzymes have also been used in biosensors in an amplification capacity *(14–16)*. Since a single enzyme can run a reaction over and over, the sensitivity of a biosensor can be increased using enzyme-linked assays. These are carried out by linking an enzyme with a high activity to a compound that competes with the analyte in the sample for the biosensor's binding sites. In the presence of the substrate of the enzyme, the amount of bound enzyme can be determined. Knowing the binding constants for the analyte and the labeled competitor, the amount of analyte in the sample can be determined. Other types of enzyme-linked assays, such as enzyme-linked antibody assays, have also been used in amplification schemes.

2.3. Immunocompounds

A second group of proteins, widely studied for use in biosensors, are the globulin antibodies produced by the mammalian immune system. The function of these immunoglobulin molecules is to recognize the presence of foreign substances in the organism and to bind them so that they can be eliminated. The foreign invaders to which the antibodies bind are called antigens, and the specific sites on the antigens that are recognized by antibodies are called epitomes or determinants. Each antibody produced by an oraganism will recognize only a single epitome.

Immunoglobulin G (IgG), the most prevalent antibody, as well as immunoglobulins A, M, D, and E, all have a structure similar to that shown in Fig 2 *(17)*. The IgG-type compounds have two light chains

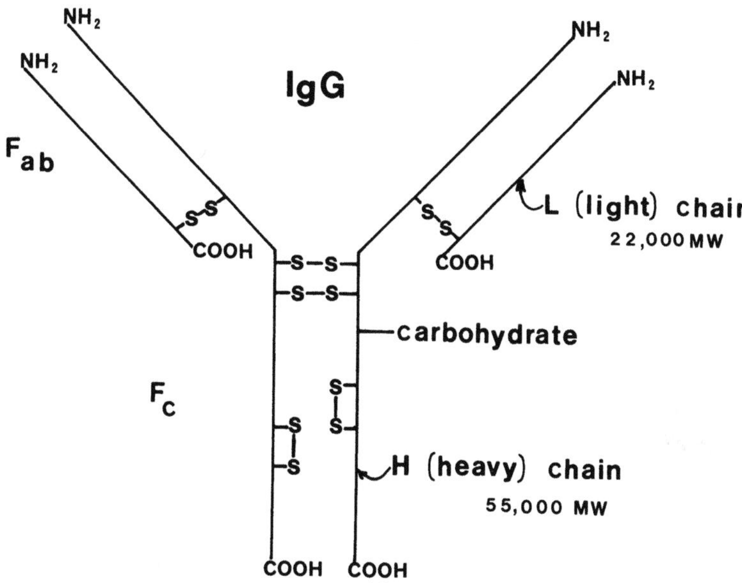

Fig. 2. General structural features of IgG antibody protein.

(each about 25,000 in mol wt) and two heavy chains (each about 55,000 in mol wt), linked together by disulfide bridges. The region where an antibody binds an epitome (F_{ab} portion) varies in composition between different antibodies and is located at the *N*-terminus of each of the four chains, as indicted in Fig 2. The composition of the remaining portions of the IgG molecule is relatively constant, with one or more carbohydrate (glycosylation) groups usually found on the heavy chains.

As with enzymes, for most biosensor applications it is necessary to immobilize the antibody on or near the transducer surface so that antigen–antibody binding can lead to a detectable output signal. Most of the same reactions used to immobilize enzymes on solid supports can be utilized to immobilize antibodies. The main precaution is that the antigen-binding (F_{ab}) portion of the antibody not be involved or blocked by the immobilization procedure. Methods using the carbohydrate and sulfhydryl sites often can be used for covalent attachment of the antibody to a solid support without disruption of F_{ab} geometry and thus with retention of antibody binding *(18–20)*.

If an antibody can be produced, it can usually be generated in gram or larger quantities. Thus, the ability to produce an antibody against a specific antigen and to purify the antibody to homogeneity is the important factor. For the generation of antibodies, large-mol-wt antigens can be used directly; small-mol-wt compounds having at least one determinant site (haptens) must be coupled to a large-mol-wt carrier group. The antigen, or hapten coupled to a carrier, is used to immunize a host animal, which in turn produces antibody-producing spleen cells. The latter cells are fused with myeloma cells to form hybridoma cells, which can be cultured and grown in large quantities to secrete antibodies. By screening the hybridoma cells, antibodies can be obtained that all bind to a single determinant site (monoclonal), or that bind either to a variety of epitomes on a single antigen or to determinant sites on more than one antigen (polyclonal).

Antibodies, like enzymes, show a range of stabilities against thermal and other environmental conditions. In general, antibody stabilities do not preclude their use on biosensors, but the storage conditions need to be evaluated and tested, and the immobilizaton procedures must be tested for each biosensor–antibody combination.

2.4. Receptors

Receptors, especially neuroreceptors, are another class of proteins that are suitable biological components for biosensor recognition components *(6,7,21)*. In effect, receptors are Nature's own biosensors, since they must recognize and bind a specific compound in the environment and signal to the organism that the compound is present. Most receptors contain one or more selective binding sites and produce a signal through a transmembrane ion channel and/or a secondary-messenger enzyme-activation system *(21)*. These systems have the possibility of signal amplification, since the binding of one or two molecules of analyte to a binding site may result in the transmembrane flux of several thousand ions or in the activation of an enzyme to generate several thousand secondary-messenger molecules.

Receptors exist for all substances that act as messengers or signals, including hormones and neurotransmitters. Some of the more common receptors are listed in Table 2. Receptors are being investi-

Table 2
Common Receptor Types That May Be Used in Biosensors

Adenosine	Glutamate
Adrenergic	Glycine
Calcium	Histamine
Cholinergic: muscarinic;	Hormone
nicotinic	Opiate
Dopamine	Serotonin
γ-amino-butyric acid (GABA)	Steroid

gated for use in biosensors because of the great number of drugs that mimic the action of natural compounds by binding at receptors; drugs that bind to receptors and modify the action of the natural messenger system are also available (e.g., benzodiazepine, which binds to the GABA receptor). There are still problems, however, that must be worked out before a large number of receptor biosensors will be seen.

Many receptors are membrane proteins, which means that they have hydrophobic regions interacting with the lipids in the membrane. This interaction may contribute to both the protein's shape and its functioning, and purification of the proteins opens up the possibility that the receptor will not function properly in a completely aqueous environment. This means designing some type of lipid environment near the transducer, in which to place these proteins. Some success at immobilizing protein-containing lipid membranes on surfaces has been shown using Langmuir–Blodgett techniques, but problems with these methods still exist *(22,23)*. Purification, however, may be an even bigger problem if there is not a good source for the receptor. At present, only one receptor, the acetylcholine nicotinic receptor, has been found in large enough quantities to make biosensor production feasible *(24–27)*.

This means that some method for cloning the receptors must be found to produce enough receptor protein for use in biosensors. This also has problems, since many receptors are made up of more than one subunit, each of which may undergo posttranslational processing, including glycosylation, phosphorylation, and disulfide-bond formation. In such a case, the system used for cloning the receptor must be able to produce more than one subunit protein at the same time, process the proteins, and assemble the receptor—all of this and still produce large enough quantities of the receptor so that purification is relatively easy.

2.5. Others

Lectins are recognition proteins that bind to specific carbohydrates on glycoproteins that can also be used as the biological component in biosensors *(28)*. The binding sites in lectins are usually specific for a single sugar residue or a short oligosaccharide structure in the carbohydrate. Concanavalin A, the most widely studied lectin, has binding sites for both mannose and glucose, and it is not unusual for lectins to have more than one binding site. The major problem with lectins is that only a few have been studied; thus, more work will need to be conducted before lectins are routinely used in biosensors.

Transport proteins are another large class of proteins that can be used as the biological component in biosensors. Examples of transport proteins, and the level at which they function, are found in Table 3. Binding sites on transport proteins are not usually as specific as enzymes or antibodies, but often will recognize an entire class of compounds; the indole binding site on human serum albumin is a good example. Transport proteins, including those that transport drugs, often have more than one binding site on them as well, as feature that could be used to some advantage if separate transduced signals for binding to each site could be developed.

Availability and stability of transport proteins will be dependent on the level at which they function. Extracellular transport proteins should be easily obtained, and stability should not be a major problem, because these proteins are normally found in the bloodstream. Proteins that transport things across the membrane will be more difficult to purify and use in biosensors, since they rely on the lipid environment in which they exist for both their structure and functioning. Cytoplasmic transport proteins should not prove difficult to use, as long as cells can be found from which sufficient quantities can be purified.

One special type of membrane-transport protein that has been used in biosensors is the ion channel *(23)*. Ion channels are proteins

Table 3
Transport Proteins

External transport (found in bloodstream)
  Lipoproteins
  Amino acid transport proteins
  Sugar transport proteins
  Anion transport proteins
  Peptide transport proteins
  Albumin
  Transferrin
Membrane transport (found in cell membrane)
  Maltose binding protein
  Na^+/K^+ pumps
  Neurotransmitter reuptake proteins
  Ion-channel proteins
Intracellular transport (found in cytoplasm)
  Protein A
  Ligandin

---

that open and close to allow the flow of specific ions into or out of the cell. The direction of the flow will be determined by the voltage across the membrane and the difference in ion concentration between the inside and the outside of the cell. Besides being able to detect changes in specific ion concentrations, ion channels often have binding sites for one or more toxins that affect the properties of the channel, which could also be detected. Like other membrane-transport proteins, however, these channels require a lipid environment to retain their functioning.

Recently, there has also been interest in the detection of DNA using biosensors *(29)*. There are two materials that have DNA-binding properties that would be of use in these biosensors: DNA-binding proteins and DNA itself. DNA-binding proteins, which recognize and bind to a specific sequence of bases in a DNA chain, function in a variety of capacities in the expression of genes. Specific sequences of DNA can also be recognized and bound by complementary DNA sequences. The degree of specificity of the binding can easily be controlled by the length of the DNA probe and the temperature during measurement. Both DNA and DNA-binding proteins should be stable enough for use in biosensors, and production or purification of these materials is possible.

Until now, we have been discussing the use of purified molecules for use in biosensors. Complete purification of the material is not always necessary, and crude mixtures (including enzyme mixtures; receptors, and ion channels in membrane fragments; whole cells including microorganisms; and tissue slices) have been used in designing biosensors *(30–35)*. When using unpurified materials, the possibility of some contaminant in the material affecting the function of the sensor must be kept in mind. Such an effect can be the result of interference with binding, change in the response vs concentration characteristics attributable to extraneous compounds, other materials that bind the analyte of interest, or interference with the transduction mechanism.

3. Transducers

When a biosensor is placed in the presence of the analyte it is designed to detect, the transducer in the biosensor must detect some change in the environment, or in the biological material itself, and then generate a measurable signal based on the change. Although this is all that is necessary, it is usually desired to have some relation between the concentration of analyte in the sample and the intensity of the generated signal. This allows the biosensor to be used for quantitative as well as qualitative measurements. A list of common types of transducers, along with the mechanisms that are used, is given in Table 4.

3.1. Electrical Transducers

Amperometric transducers consist of an electrode that is held at a constant voltage relative to a reference electrode. The change brought about by the analyte produces a change in the current through the electrode, which is measured. Chemical sensors using this type of transduction have been used for over 30 years, and quite a bit of work has been conducted on the reactions of coupling enzymes to these

Table 4
Transducers for Biosensors

Electrochemical	Optical	Other
Amperometric	Absorbance	Piezoelectric
Potentiometric	Reflectance	Bulk wave
Conductance	Optical rotation	Surface wave
Capacitive	Fluorescence	Calorimetric
	Fluorescence quenching	Thermistor
	Evanescent wave	Pyroelectric
	Surface-plasmon resonance	Optical enthalpimeter
	Luminescence	

chemical sensors *(36)*. By joining a chemical sensor to an enzyme that produces or uses up a molecule that can be measured by that sensor (e.g., O_2, H_2O_2), one can produce a biosensor *(37)*.

A lot of the emphasis has also been placed on using amperometric transducers with oxidoreductase enzymes, in which cofactors used by the enzyme are oxidized or reduced during the enzyme reaction *(38,39)*. By transferring an electron between the electrode and the cofactor, either through a direct transfer or through some type of mediated transfer, the cofactor is regenerated as it is used in the reaction. The amount of analyte present can be determined by the current through the electrode, since this is related to the amount of cofactor used. Work is continuing, but unfortunately this idea has not been applicable to the large number of enzymes that use the cofactor NADH, because of the difficulty of regenerating the NADH cofactor *(40)*.

Ion channels, and receptors that contain ion channels, are also candidates for use with amperometric transducers. Both of these normally function to allow ions to flow (current) only at specific times; the binding of a single molecule to a receptor causes a channel to open, which allows millions of ions to flow through. Using these materials with amperometric transducers, then, allows the tremendous amplification factor of these materials to be exploited. The difficulty in using these materials is the need to immobilize them onto the surface of the electrode in such a way that a potential can be put across them which would cause the current when the channel opened.

The use of other biological materials with amperometric transducers is possible using enzyme-linked analytes or antibodies, but it is not advantageous. Other systems for the detection of analyte through enzyme-linked amplification schemes are much more sensitive. Amperometric transducers also lack the ability to determine multiple analytes at the same time. However, these tranducers can be integrated with other transducer mechanisms to allow multiple-detection schemes *(41)*.

Potentiometric transducers are electrodes to which a constant current is applied, and the change in the potential at the electrode is measured. Potentiometric sensors are in common use for measuring such things as pH and other ions. Like amperometric chemical sensors, these sensors can be coupled to biological materials that produce or consume these ions, to form biosensors. Potential changes at the electrode can also be established by the binding of charged analyte to the biological material. That not only makes this type of transducer applicable to enzymes, but also allows ion channels, antibodies, nerve cells, and tissue slices to be incorporated into this type of sensor *(16,42–45)*.

One special type of potentiometric transducer, which has received a lot of attention, is the field effect transistor (FET). In this device, the conductivity of a semiconductor material is controlled by a potential generated at a gate in the semiconductor, thus modulating the current through the device. Biosensors have been designed by immobilizing the biological materials, which change the potential, in the gate, as shown in Fig. 3 *(46)*. The biggest advantages of FETs are their small size, which causes them to use very little biological material, and the relative ease with which they can be manufactured.

Because of their small size, it is possible to place more than one gate on the same chip, thus allowing multiple analytes to be detected at the same time, using the same sensor. Work on multiple-detection systems has also been carried out on light-addressable potentiometric sensors. These sensors use a set of light-emitting diodes to illuminate selected regions on a semiconductor. This induces a photocurrent in the chip, the magnitude of which will be changed by the potential in the region that was illuminated. This type of device has been used to

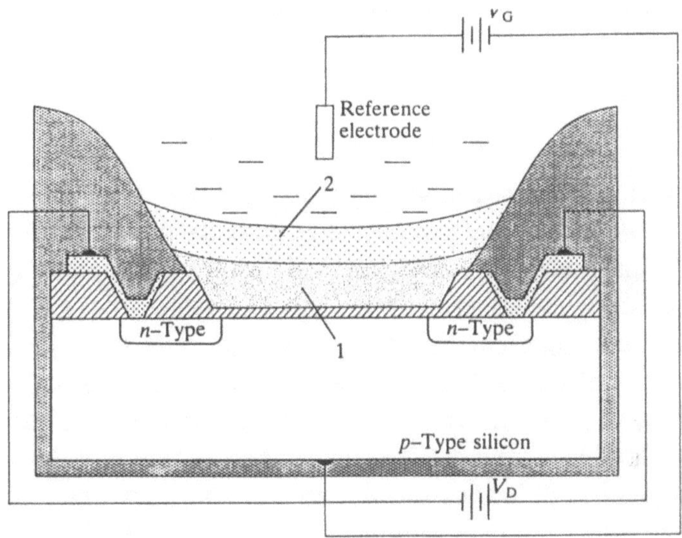

Fig. 3. Diagram of an FET type of biosensor with (1) a chemical sensitive membrane over (2) an immobilized biological material. Reproduced from Figure 26.21, G. F. Blackburn, "Chemically Sensitive Field Effect Transistors," in *Biosensors: Fundamentals and Applications* (Turner, A. P. F., Karube, I., and Wilson, G. S., eds.), by permission of copyright holder, Oxford University Press, Oxford, UK.

detect as many as 23 sites on a single sensor while achieving high sensitivity *(47,48)*.

Other types of electrical transducers, which have been investigated but are not as widely used, are conductometric and capacitance devices. Conductometric or impedance devices measure the current caused by migration of ionic compounds to or from the biological material, or the change in the conductance of the sample as compounds are produced, consumed, or removed from the sample. Use of these transducers with immobilied enzymes has been suggested, but ion channels and other biological materials could also be used in these sensors *(36,49,50)*. Capacitive sensors measure the change in capacitance caused by the analyte binding to the biosensor. This type of transducer has been used with immobilized acetylcholine receptors, but could also be used with any biological material that affects the di-electric constant of the local sensor environment when binding occurs *(51)*.

3.2. Optical Transducers

Tranducers for biosensors have also been designed using optical signals to determine the presence and quantity of a desired analyte. Optical transducers, which use optical fibers to transport the signal from the biological material to the detection system, have certain advantages over electrical transducers. Optical transducers avoid such problems as electrical interference, electrical connections, junction potentials, and the need for a reference electrode. Optical fibers are thin and flexible, allowing access to a variety of remote environments without significant loss of signal, and they can carry multiple wavelengths of light simultaneously, making detection of multiple analytes simple. Real-time and on-site measurements can be made, but interference from ambient light, sample turbidity, or other sample components must be avoided with these devices.

The simplest optical systems use absorbance measurements to determine changes in concentration of species that absorb a given wavelength of light. The light is brought to the sample through an optical fiber, and the amount of light absorbed by the system is detected through the same, or a second, optical fiber. The biological material is immobilized near the end of the optical fibers and either produces or extracts the analyte that absorbs the light. Measurement of compounds that change the optical rotation of plane-polarized light, as well as reflectance caused by agglutination reactions, can also be used to measure analytes in these simple systems.

Fluorescent measurements are another method being used as an optical transduction mechanism. Although there are a few analytes that may naturally be detected in this manner, this method is usually used with artificially labeled compounds. Usually competition for binding sites between the analyte of interest and fluorescent-labeled analyte is carried out. A direct measurement of the amount of bound, labeled analyte is used if the biosensors do not contain the labeled compound. These sensors require the addition of labeled analyte each time a measurement is to be made, however, so they cannot be used for *in situ* measurements.

If the labeled analyte is entrapped within the biosensor, then measuring the amount of unbound labeled analyte will give the concentration of anlyte in the sample. Here, both the biological material and the labeled analyte must be kept near the end of the optical fiber. However, they must not interfere in the measurement of the unbound fluorescent-labeled analyte when unlabeled analyte is present. The general method for doing this is to immobilize the biological material on a dialysis membrane with a molecular-weight cutoff sufficient to keep the labeled analyte trapped, but allowing free exchange of unlabeled analyte with the sample. An example of an immunosensor using this type of transducer is shown in Fig. 4 *(16)*.

Methods have been developed to prevent the interference of the extra labeled analyte in fluorescent biosensors. One idea is to use fluorescent-energy transfer properties by which the fluorescent energy of the labeled compound (donor) can be absorbed by a second compound (acceptor), which then gives off the energy by fluorescing at a longer wavelength *(52)*. The donor and acceptor must be in close proximity, so the acceptor must be bound to the biological material close to the binding site of the analyte of interest. This sensor will then measure the number of sites that are filled with labeled analyte by measuring the fluorescent emissions of the acceptor compound. Another idea is to form membranes using fluorescent-labeled lipids, which change their emission spectra when analyte binds to proteins in the membrane *(53)*.

Fluorescence quenching, in which the fluorescence of a compound bound within the biosensor is affected by the presence of another compound, can also be used as a transducer mechanism. These sensors are usually used with enzymes, where the production of a product, such as H^+, causes quenching of the fluorescence of the bound compound. Chemiluminescence can also be employed in this type of biosensor, in which the formation of products by the enzyme causes a bound compound to give off light energy *(54,55)*. Direct production of a luminescent compound by reaction of an enzyme is also a possible transduction mechanism. All of these mechanisms can be

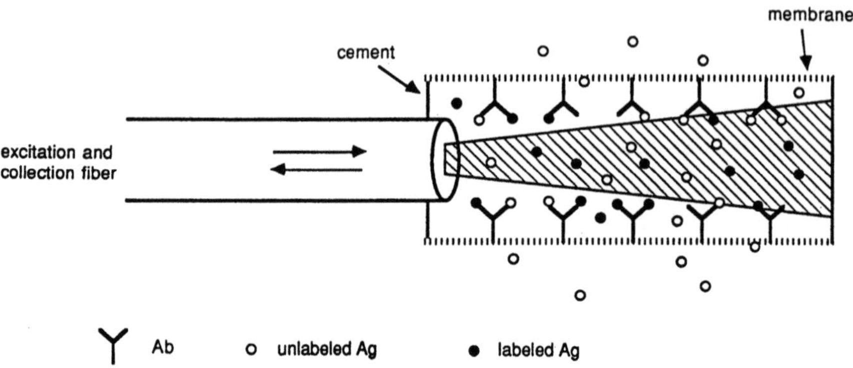

Fig. 4. Diagram of a competition-based immunosensor. (Reprinted from *American Biotechnology Laboratory*, volume 7, number 3, page 18, 1989. Copyright 1989 by International Scientific Communications, Inc.)

applied to other biologial materials through the use of enzyme-linked assays.

Other types of optical transducer mechanisms that are being investigated involve the evanescent field of fiberoptic cables. Since light is propagated through an optical fiber by total internal reflection, the light intensity is not zero at the interface where the reflection takes place. This means that some light penetrates the wall of the fiber and actually extends into the surrounding medium. The intensity of this light decays quickly, but within about 1000 Å an evanescent field is generated. By immobilizing biological materials within this field, one can detect binding of analytes that absorb the light being propagated or fluoresce when excited by this light *(56,57)*.

Plasmon resonance has also been shown to be an effective optical transducer mechanism for biosensor use *(36)*. In this type of transducer, light of a particular angle and wavelength will induce surface plasmons in a metal conductor, which causes a decrease in the reflected light intensity. The critical angle will be sensitive to the electric properties near the surface of the conductor, and this fact can be used to measure the concentration of analyte bound to biological material immobilized near the conductor surface. Recent work using metal-island films in these transducers may make miniaturization of these transducers possible *(58)*.

3.3. Other Transducers

Enzyme reactions present another interesting method for generating a signal as a result of the presence of an analyte: thermometric transducers. During the reactions that enzymes catalyze, heat is often given off, and the amount of heat produced will be proportional to the amount of analyte. By immobilizing the enzyme near a thermistor, the temperature increase can be measured and related to the concentration of analyte in the sample *(49,59)*. Through the use of enzyme-linked antigens, thermometric biosensors that use antibodies have also been developed *(15,36,60,61)*. This allows great amplification of the immunological binding of the antigen, and it greatly increases the sensitivity of these biosensors. This type of enzyme-linked assay can also be applied to other biological materials.

The calorimetric effects of reactions are also being exploited in other types of transducer mechanisms *(62)*. Pyroelectric materials (materials that develop a potential when exposed to a temperature gradient) are being investigated, as are two optical devices. The first is an enthalpimeter, which uses two optical fibers, one of which is coated by the biological material in which the heat-producing reaction will take place. Laser light is launched down both fibers, and the exit beams are superimposed to create an interference pattern. As the reaction takes place, the heat will affect the light-propagation properties of the coated fiber, and the interference pattern will be altered. A second device, which also takes advantage of the change in light-propagation properties, uses only a single birefringent fiber to carry both the reference and the sample light beams. As the reaction proceeds, the two waves will be retarded differently, and an interference pattern will be seen along the longitudinal axis of the fiber.

Piezoelectric devices are other types of transducers that have been investigated for use in biosensors *(63)*. Piezoelectric materials are materials that generate a potential when placed under a mechanical stress. By placing a potential across these materials, resonant waves can be set up in them, with the frequency of the wave being dependent on the mass of the material. Changing the mass by binding analyte to a biological material immobilized on a piezoelectric material causes a change in the frequency of the wave. These devices are called bulk-wave devices because the entire quartz crystal (the most frequently used piezoelectric material) vibrates when the potential is placed across the crystal.

Surface acoustic wave (SAW) devices are another type of piezoelectric transducer. In these devices, waves are set up in only one face of the crystal, using interdigital transducers. Because only a single surface is oscillating, the frequency of the oscillations is much higher in SAW devices than in the bulk wave devices, which allows for a greater sensitivity of detection *(64)*. The biological material in these devices must be immobilized on the oscillating surface. This has caused problems when using these devices in liquid samples, because of both frequency changes and removal of the biological material from the surface, caused by the liquid interface *(65)*. Since piezoelectric devices detect only changes in the mass on the surface, these devices are also prone to error from non—specific binding of sample components to the biological material.

4. Potential Applications

Application of a given biosensor will ultimately determine the choice of biological material and transducer used in construction of the biosensor. Such considerations as where the sensor will be used, the needed sensitivity of the sensor, requirement for quanitative or just qualitative detection, as well as a host of others, will be taken into account in determining the proper biosensor to use in a given situation.

The first thing one has to consider is what biological materials can be used to detect the analyte of interest. Is the analyte acted upon by a specific enzyme; if so, does it produce a product that can be easily monitored, or use a cofactor that can be regenerated? Does the analyte bind tightly to another protein that occurs naturally? Is an antibody that binds the analyte available, or would one have to be produced? Is the analyte something that could be detected by tissue slices or whole cells?

The type of sample, the concentration of analyte in the sample, and the changes in concentration that must be detected also play a part in choosing the biological material to be used. If high sensitivity is needed, then it may be necessary to use amplification schemes to detect the analyte. A large number of similar compounds in the sample will require high selectivity by the biological material, unless one does not need to differentiate among the compounds. If more than one biological material has the proper selectivity, sensitivity, and availability, then the type of transducer to be used may help to determine which biological material will work best for the biosensor.

One of the most important things to consider when selecting a transducer for a biosensor is where the biosensor will be used. This will determine a variety of other factors, including response times, calibration needs, and possibility of reuse, which will help to determine the type of transducer to be selected. The number of analytes that are to be detected, either by a single biosensor or by an array of biosensors, will also help to determine what tranducer should be used.

The ultimate goal of many biosensor researchers is to design a biosensor that can be used at home or in a doctor's office, to determine quickly the presence or absence of a condition. These sensors must be small, cheap, easy to use, and require no external equipment for monitoring the signal from the sensor. Biosenors requiring external equipment, such as a spectrophotometer for optical biosensors or an electrical device for electronic biosensors, may also be used in a doctor's office if a variety of biosensors can be developed to be used with the same piece of equipment. These biosensors would be disposable, require a response time of a few minutes, not need to be calibrated before use, but would often provide only qualitative results. These sensors would not require long-term stability of use, but would need to be stable for long periods of time before use without loss of response.

A second environment in which biosensors will be used is the clinical laboratory, where thousands of samples will be investigated. Here, the demand will be for sensors that can be used over and over again without loss of activity. These sensors must be reversible so that, between uses, the sensor can be regenerated for the next sample. Use in a variety of samples may require calibration of the sensor before each use, or at regular intervals, to determine if changes in the biosensor have occurred. These sensors could require more elaborate external equipment, but will often be required to detect a variety of analytes at the same time. The response time must be within minutes, and the sensors must be stable for repeated use. Cost is not as big a factor, but quantitative results will normally be expected from these sensors.

Biosensors will also be employed in a variety of control and detection processes in manufacturing. Here, in-line sensors that can monitor changes in concentration of products, contaminants, and reactants will be needed, so that changes in the system can be implemented to optimize production. These sensors will have to have long-term stability during use, and they must contain all the materials they need to produce a signal entrapped within the sensor. The response time must be very short, so changes can be detected and responded to quickly. The binding characteristics of the sensor must be such that continuous or short-interval monitoring can be carried out. Cost is not a factor, but placement of the external monitoring and control equipment may be far from the measurement site. This means that the generated signal must be strong enough to allow use in a remote location and still give quantitative results.

Environmental detection of compounds is another area in which biosensors are expected to be used, for detection of compounds in soil samples, water samples, air samples, and residues wiped from surfaces. These detectors must be reusable for a few samples with no loss of activity and must need a minimum of work to regenerate the sensor. The response time should be in the minute range, and calibration of the sensor will be needed because of the wide variety of compounds that may contaminate the sample. Sample preparation should be as simple as possible, and cost per sample will be an important factor. The sensors will often be used qualitatively to eliminate negative samples, with positive samples being brought back to the lab for extensive evaluation. The external equipment required for these sensors must be portable, so that on-site detection of analytes can be carried out.

One last area for which biosensors are being designed is *in vivo* monitoring. Implantable biosensors, which can not only detect the level of an analyte, but also regulate the level of the analyte, are receiving a lot of attention. Glucose levels, and controlling the level of insulin in the bloodstream using a biosensor, have been the major thrust of this research *(66)*. These biosensors would have to be small, and the detection equipment would have to be either self-contained, or small and portable. Long-term stability during use would be necessary so that repeated insertion and removal of the sensor could be avoided. The response time would have to be fast enough to allow corrective actions, and the cost could not be prohibitive. Sensor materials would have to be sterilizable, as well as compatible with *in vivo* use. These sensors must also be nonintrusive enough so that patients will be willing to use them.

Other possible applications of biosensors are limitless. Whatever compound needs to be detected, somewhere there is an organism that has, or can produce, a binding site for it. The types of biological materials, the types of transduction mechanisms, and the amplification mechanisms that can be used in biosensors will continue to increase, as researchers continue to conceive new ideas. Improvements in the known methods and materials will continue as more attempts are made to commercialize biosensors. Someday, an ultimate generic biosensor will be developed, in which a transducer and biological material can be combined in minutes to detect almost any analyte.

References

1. Guilbault, G. G. (1972) Analytical uses of immobilized enzymes, in *Enzyme Engineering* (Wingard, L. B. Jr., ed.), Interscience, New York, pp. 361–376.
2. Clark, L. C. Jr. (1972) A family of polarographic enzyme electrodes and the measurement of alcohol, in *Enzyme Engineering* (Wingard, L. B. Jr., ed.), Interscience, New York, pp. 377–394.
3. Lowe, C. R. (1985) An introduction to the concepts and technology of biosensors. *Biosensors* 1(1), 3–16.
4. Wingard, L. B. Jr., Katchalski-Katzir, E., and Goldstein, L., eds. (1981) *Analytical Applications of Immobilized Enzymes and Cells* (Academic, New York).
5. Turner, A. P. F., Karube, I., and Wilson, G. S., eds. (1987) *Biosensors: Fundamentals and Application* (Oxford University Press, Oxford).
6. Wingard, L. B. Jr. (1990) Amplification possibilities with neuroreceptor-based biosensors, in *Biosensor Technology: Fundamentals and Application* (Bowden, E., Buck, R., Hatfield, W. E., and Umana, M., eds.), Marcel Dekker, New York, in press.
7. Rechnitz, G. A. (1987) Biosensors: Challenges for the 1990s, in *Biosensors International Workshop 1987* (Schmid, R. D., Guilbault, G. G., Karube, I.,

Schmidt, H.-L., and Wingard, L. B., eds.) VCH, Weinheim, Germany, pp. 3–12.

8. Wingard, L. B. Jr., Katchalski-Katzir, E., and Goldstein, L., eds. (1976) *Immobilized Enzyme Principles* (Academic, New York).

9. Mosbach, K., ed. (1976) *Immobilized Enzymes (Methods in Enzymology,* vol. 44), Academic, New York.

10. Mosbach, K., ed. (1988) *Immobilized Enzymes and Cells,* part D (*Methods in Enzymology,* vol. 137), Academic, New York.

11. Coolbear, T., Daniel, R. M., Cowan, D. A., and Morgan, H. W. (1988) Proteases from extreme thermophiles. *Ann. NY Acad. Sci.* **542,** 279–281.

12. Zaks, A. and Klibanov, A. M. (1988) Enzymatic catalysis in nonaqueous solvents. *J. Biol. Chem.* 263, 3194–3201.

13. Laane, C., Boeren, S., Vos, K., and Veeger, C. (1987) Rules for optimization of biocatalysis in organic solvents. *Biotechnol. Bioeng.* **30,** 81–87.

14. Aizawa, M. (1983) Molecular recognition and chemical amplification of biosensors, in *Chemical Sensors, Analytical Chemistry Symposia Series 17* (Seiyama, T., Fueki, K., Shiokawa, J., and Suzuki, S., eds.), Elsevier, Amsterdam, pp. 683–692.

15. Monroe, D. (1989) Thermistor biosensors with TELISA applications. *Am. Biotechnol. Lab.* **Oct.,** 18–23.

16. Borman, S. (1987) Biosensors: Potentiometric and amperometric. *Anal. Chem.* **59,** 1091A–1098A.

17. Roitt, I., Brostoff, J., and Male, D., (1985) *Immunology* (C. V. Mosby, St. Louis).

18. Gameiner, P. and Viskupic, E (1981) Stepwise immobilization of proteins via their glycosylation. *J. Biochem. Biophys. Methods* **4,** 309–319.

19. Narasimhan, K. and Wingard L. B. Jr. (1986) Enhanced direct electron transport with glucose oxidase immobilized on (aminophenyl) boronic acid modified glassy carbon electrode. *Anal. Chem.* **58,** 2984–2987.

20. Ishikawa, E., Yoshitake, S., Imagawa, M., and Sumiyoshi, A. (1983) Preparation of monomeric F_{ab}-horseradish peroxidase conjugate using thiol groups in the hinge and its evaluation in enzyme immunoassay and immunohistochemical staining. *Am. NY Acad. Sci.* **420,** 74–89.

21. Wingard L. B. Jr. (1987) Possibilities for biosensors based on neuroreceptors, in *Biosensors International Workshop 1987* (Schmid, R. D., Guilbault, G. G., Karube, I., Schmidt, H.-L., and Wingard, L. B., eds.) VCH, Weinheim, pp. 133–137.

22. Umezawa, Y. and Sugawara, M. (1989) Langmuir-Blodgett membranes for some biosensor developments, in *Biosensors (Proceedings of the MRS International Meeting on Advanced Materials, 14)* (Doyama, M., Somiya, S., and Chang, R., eds.), MRS, Pittsburgh, pp. 191–204.

23. Ligler, F. S., Fare, T. L., Seib, K. D., Smuda, J. W., Singh, A., Ahl, P.,

Ayers, M. E., Dalziel, A., and Yager, P. (1988) A receptor-based biosensor: 1. Fabrication of key components. *Med. Inst.* **22,** 247–256.

24. Eldefrawi, M. E., Sherby, S. M., Andreou, A. G., Mansour, N. A., Annau, Z., Blum, N. A., and Valdez, J. J. (1988) Acetycholine receptor-based biosensor. *Anal. Lett.* **21,** 1665–1680.

25. Taylor, R. F., Marenchic, I. G., and Cook, E. J. (1988) An acetylcholine receptor-based biosensor for the detection of cholingeric agents. *Anal. Chim. Acta* **213,** 131–138.

26. Dalziel, A. W., Georger, J., Price, R. R., Singh, A., and Yager, P. (1987) Progress report on the fabrication of an acetycholine receptor-based biosensor, in *Membrane Proteins* (Goheen, S. C., ed.), Bio-Rad Laboratories, Richmond, CA, pp. 643–673.

27. Rogers, K. R., Valdes, J. J.,.and Eldefrawi, M. E. (1989) Acetycholine receptor fiber-optic evanescent fluorosensor. *Anal. Biochem.* **182,** 353–359.

28. Schultz, J. S. and Sims, G. (1979) Affinity sensors for individual metabolites, in *Computer Applications in Fermentation Technology* (Armiger, W. B., ed.), John Wiley, New York, pp. 65–72.

29. Downs, M. E. A., Kobayashi, S., and Karube, I. (1987) New DNA technology and the DNA biosensor. *Anal. Lett.* **20,** 1897–1927.

30. Suzuki, H., Tamiya, E., Karube, I., and Oshima, T. (1988) Carbon dioxide sensor using thermophilic bacteria. *Anal. Lett.* **21,** 1323–1336.

31. Karube, I. (1987) Micro-organism based sensors, in *Biosensors: Fundamentals and Application* (Turner, A. P. F., Karube, I., and Wilson, G. S., eds.), Oxford University Press, Oxford, pp. 13–39.

32. Arnold, M. A. and Rechnitz, G. A. (1987) Biosensors based on plant and animal tissues, in *Biosensors: Fundamentals and Application* (Turner, A. P. F., Karube, I., and Wilson, G. S., eds.), Oxford University Press, Oxford, pp. 30–59.

33. Bush, R. M. and Rechnitz, G. A. (1989) Neuronal biosensors. *Anal. Chem.* **61,** 553A–542A.

34. Frew, J. E. and Hill, H. A. O. (1987) Electrochemical biosensors. *Anal. Chem.* **59,** 933A–944A.

35. Matsuoka, H. and Homma, T. (1989) Use of plant leaf as odor sensing material, in *Biosensors (Proceedings of the MRS International Meeting on Advanced Materials, 14)* (Doyama, M., Somiya, S., and Chang, R., eds.), MRS, Pittsburgh, pp. 205–210.

36. Hall, E. A. H. (1986) The developing biosensor arena. *Enzyme Microb. Technol.* **8,** 651–658.

37. Wilson, G. S. and Thevenot, D. R. (1990) Unmediated amperometric enzyme electrodes, in *Biosensors: A Practical Approach* (Cass, A. E. G., ed.), IRL, Oxford, pp. 1–18.

38. Hill, H. A. O. and Sanghera, G. S. (1990) Mediated amperometric enzyme

electrodes, in *Biosensors: A Practical Approach* (Cass, A. E. G., ed.), IRL, Oxford, pp. 19–46.

39. Albery, W. J. and Craston, D. H. Amperometric enzyme electrodes: Theory and experiment, in *Biosensors: Fundamentals and Application* (Turner, A. P. F., Karube, I., and Wilson, G. S., eds.), Oxford University Press, Oxford, pp. 180–210.

40. Bartlett, P. N. (1990) Conducting organic salt electrodes, in *Biosensors: A Practical Approach* (Cass, A. E. G., ed.), IRL, Oxford, pp. 47–96.

41. Kimura, J., Murakami, T., Kuriyama, T., and Karube, I. (1988) An integrated multibiosensor for simultaneous amperometric and potentiometric measurements. *Sensors Actuators* **15,** 435–443.

42. Wingard, L. B. Jr. and Castner, J. (1987) Potentiometric biosensors based on redox electrodes, in *Biosensors: Fundamentals and Application* (Turner, A. P. F., Karube, I., and Wilson, G. S., eds.), Oxford University Press, Oxford, pp. 153–162.

43. Rechnitz, G. A. (1988) Biosensors. *Chem. Eng. News* **Sept. 5,** 24–36.

44. Monroe, D. (1986) Potentiometric immunoassay. *Amer. Biotechnol. Lab.* **Nov/ Dec.,** 28–40.

45. Ohashi, E., Tamiya, E., and Karube, I. (1990) A new enzymatic receptor to be used in a biosensor. *J. Membr. Sci.* **49,** 95–102.

46. Karube, I. (1987) Micro-biosensors based on silicon fabrication technology, in *Biosensors: Fundamentals and Application* (Turner, A. P. F., Karube, I., and Wilson, G. S., eds.), Oxford University Press, Oxford, pp. 471–480.

47. Hafeman, D. G., Parce, J. W., and McConnell, H. M. (1988) Light addressable potentiometric sensors for biochemical systems. *Science* **240,** 1182–1186.

48. Lucas, M. E., Huntington, M. F., Regina, F. J., Bolts, J. M., Alter, S. C., Ballman, M. E., and Jirk, G. L. (1990) Rapid filtration-based immunoassays performed with a silicon biosensor, in *Biosensor Technology: Fundamentals and Application* (Bowden, E., Buck, R., Hatfield, W. E., and Umana, M., eds.), Marcel Dekker, New York, in press.

50. Kell, D. B. and Davey, C. L. (1990) Conductimetric and impedimetric devices, in *Biosensors: A Practical Approach* (Cass, A. E. G., ed.), IRL, Oxford, pp. 125–154.

51. Ives, J. T., Lin, J. N., and Andrade, J. D. (1989) Fiber-optic fluorescence immunosensors. *Am. Biotechnol. Lab.* **March,** 10–18.

52. Meadows, D. and Schultz, J. S. (1988) Fiber-optic biosensors based on fluorescence energy transfer. *Talanta* **35,** 145–150.

53. Brennan, J. D., Brown, R. S., Krull, U. J., and McClintock, C. L. (1989) The potential of fluorescent lipid membranes as transducers in biosensors. (Abstract from *Symposium on Biosensors*), ACS: North Carolina section, p. 12.

54. Abdel-Latif, M. S., Suleiman, A., Guilbault, G. G., Dremel, B. A. A., and Schmid, R. D. (1990) Fiber optic sensors: Recent developments. *Anal. Lett.* **23,** 375–399.

55. Seitz, W. R. (1984) Chemical sensors based on fiber optics. *Anal. Chem.* **56,** 16A–34A.
56. DeGrandpre, M. D., Burgess, L. W. (1988) Long path fiber-optic sensor for evanescent field absorbance measurements. *Anal. Chem.* **60,** 2582–2586.
57. Rogers, K. R., Valdes, J. J., and Eldefrawi, M. E. (1989) Acetylcholine receptor fiber-optic evanescent fluorosensor. *Anal. Biochem.* **182,** 353–359.
58. Debono, R. F., Thompson, M., Mallon, A. L., and Scaini, M. J. (1990) The use of metal island films to support radiative surface plasmons as a method of transducing interfacial events, in *Biosensor Technology: Fundamentals and Application* (Bowden, E., Buck, R., Hatfield, W. E., and Umana, M., eds.), Marcel Dekker, New York, in press.
59. Danielsson, B. and Mosbach, K. (1987) Theory and applications of calorimetric sensors, in *Biosensors: Fundamentals and Application* (Turner, A. P. F., Karube, I., and Wilson, G. S., eds.), Oxford University Press, Oxford, pp. 572–598.
60. Scheller, F., Siegbahn, N., Danielsson, B., and Mosbach, K. (1985) High-sensitivity enzyme thermistor determination of L-lactate by substrate recycling. *Anal. Chem.* **57,** 1740–1743.
61. DeYoung, H. G. (1983) Biosensors: The mating of biology and electronics. *High Technol.* **3,** 41–49.
62. Dessy, R., Arney, L., Bugess, L., and Richmond, E. (1990) A comparison of three thermal sensors based on fiber-optic and polymer films for biosensor applications, in *Biosensor Technology: Fundamentals and Application* (Bowden, E., Buck, R., Hatfield, W. E., and Umana, M., eds.), Marcel Dekker, New York, in press.
63. Clarke, D. J., Blake-Coleman, B. C., and Calder, M. R. (1987) Principles and potential of piezo-electric transducers and acoustical techniques, in *Biosensors: Fundamentals and Application* (Turner, A. P. F., Karube, I., and Wilson, G. S., eds.), Oxford University Press, Oxford, pp. 551–571.
64. Ballantine, D. S. Jr. and Wohltjen, H. (1989) Surface acoustic wave devices for chemical analysis. *Anal. Chem.* **61,** 704A–715A.
65. Thompson, M., Arthur, C. L., and Dhaliwal, G. K. (1986) Liquid-phase piezoelectric and acoustic transmission studies of interfacial immunochemistry. *Anal. Chem.* **58,** 1206–1209.
66. Monroe, D. (1989) Novel implantable glucose sensors. *Am. Clin. Lab.* **8,** 8–16.

Chemical Sensing
with Fiberoptic Devices

Carmen Cámara, María Cruz Moreno,
and Guillermo Orellana

1. Introduction

Fiberoptic sensors are devices that when exposed to a chemical
or physical stimulus, cause alterations in the optical properties of a
reagent and, consequently, changes in the radiation traveling along an
optical fiber connected to a detector. The main attraction of absorbance, fluorescence, and reflectance chemical sensors lies in their
ability to continuously indicate the concentration of a variety of
analytes. These devices usually consist of an optical fiber coupled to a
reagent phase in which the optical properties are modified by the
presence of an analyte. Modified radiation is transmitted to the detector along an optical fiber. Transmission is by total internal reflection,
the light propagating through a *core* material of higher refractive index whereas an enveloping layer, or *cladding,* of lower refractive index, effects the guiding process *(1)*.

Fiberoptic chemical sensors offer several advantages over conventional sensors:

1. Since the signal is optical, they are not subject to electrical interfer-
    ences. The absence of electrical connectors makes them safer than
    electrochemical sensors in certain environments, such as those con-
    taining explosive vapors.

2. Unlike potentiometric sensors, they do not require a "reference signal.
3. Fiberoptic sensors can be miniaturized and used for monitoring medical parameters in vivo (catheterization) *(2–4; see also* Chapter 10).
4. Low-loss optical fibers allow remote monitoring of hazardous environments *(5)*.
5. Fiberoptic sensors may be developed to respond to analytes for which selective electrodes are not available.
6. A single probe may be used to measure different analytes by changing the reagent phase attached to the optical fiber *(6)*.
7. The use of multiwavelength monitoring allows one to obtain sequential information on a number of different analytes *(7)*.

Nevertheless, a variety of drawbacks detract from the utility of fiberoptic sensors *(8,9)*:

1. Ambient light may create interference, this is avoidable by using a dark environment or modulated radiation.
2. Long-term stability is limited because of photobleaching or washout in sensors with reagent phases.
3. When analyte and indicator are in different phases, there is a need for a mass-transfer step before constant response is achieved.
4. A three-way trade-off among amount of reagent phase, intensity of probe radiation, and stability, which acts as a constraint on performance.
5. Sensors based on immobilized pH indicators or chelates have a smaller dynamic range of performance than do some electrodes.
6. Irreversibility may be a problem (or may be insignificant in nonregenerable sensors provided a low amount of reagent is consumed).
7. Need exists for more selective indicators and more reproducible immobilization procedures to enable high sensitivity and long-term stability.

New advances in chemically sensitive electrooptic devices, and recent development of optical fibers for light transmission in a variety of wavelength ranges, have sparked growing interest in optical chemical sensors for continuous monitoring. Several reviews *(1–24)*, patents *(25–27)*, and books *(28–30)* deal with fiberoptic chemical sensors and their applications. Fiberoptic sensors show great potential for rapid *in situ* measurement, being employed for monitoring in industrial processes

(31–34), petrochemistry *(35)*, biotechnology *(2,36–38)*, medicine *(3,4,39)*, immunoassays *(40)*, and the environment *(41–44)*.

This chapter contains a review of the many types of existing chemical sensors, an overview of the instrumentation and a summary of analytical applications.

2. Classification

The enormous variety of fiberoptic sensors described to date makes their classification difficult. In this chapter, only *chemical* sensors are described, which include sensors for gases (e.g., oxygen, carbon dioxide, anesthetics, moisture), ions (e.g., Cl^-, Br^-, I^-, S^{2-}, Be^{3+}, Al^{3+}, Fe^{3+}, Zn^{2+}, Cu^{2+}), pH (H^+), and organic chemicals (e.g., organochlorides, hydrocarbons). *Physical* sensors, which respond to variations in a physical parameter, such as temperature, pressure, velocity, or particle size, are not included.

Fiberoptic sensors can be classified, according to their interaction with the analyte, as *reversible* if the reagent phase is not consumed upon exposure to the analyte, or *irreversible* if the reagent phase is consumed in the measuring process. Irreversible sensors (integrating devices) require steady-state mass transfer to give a constant signal. They can be further divided into *regenerable* sensors that can be readied for reuse after the measurement by treatment with a reagent, and nonregenerable sensors, which have long lifetimes if the rate of reagent consumption is low in comparison to the total amount of reagent available.

Fiberoptic sensors are also referred to as *intrinsic* or *extrinsic*. *Intrinsic* sensors carry an analyte-sensitive reagent at the end of the optical fiber, commonly immobilized on a solid support (sometimes the optical fiber itself). The supported reagent usually is held at the fiber tip by means of a membrane that prevents leaching of the sensitive phase into the sample and is permeable to mass transfer by chemical species. *Extrinsic* sensors consist of bare-ended optical fibers dipped into a sample to which the analyte-sensitive compound has been added, normally as a solute. In the extrinsic case, the sensor is used only to carry light to the sample solution and collect the modified radiation.

Finally, fiberoptic sensors are referred to in terms of the optical property measured. These properties include:

1. *Absorbance or reflectance* changes of the reagent phase upon interaction with the analyte;
2. *Luminescence emission* changes of the sensitive compound (fluorescence- or chemiluminescence-based), or
3. *Transmission* losses along the optical fiber as a consequence of analyte-dependent variations in the refractive index of a sensitive part of the light-guide cladding.

3. Instrumentation

Optical-fiber-based sensors show a wide variety of instrumentation, from sophisticated equipment using laser sources, high-cost monochromation systems, and computer-based signal processing units to inexpensive sensors developed with tungsten lamps, glass filters, and simple photodetectors. Although the fiberoptic sensors described to date are not easily summarized in one representative system, it is commonly accepted that a complete device (Fig. 1) consists of optical fiber, light source, monochromation system, detection system, and readout device.

The communications industry has developed a wide variety of optical-fiber cables. Measurements in the UV region use fused-silica fibers, whereas quartz, glass, or poly(methyl methacrylate) fibers can be used to drive visible radiation. To extend infrared spectroscopy to remote measurements, optical fibers such as special fused silica, zirconium fluoride, or chalcogenide glasses are required. These allow working wavelengths up to 11 μm with low losses (<10 dB/m). The selection of a suitable optical fiber depends on the particular application. In general, highest versatility and lowest losses are achieved with quartz fibers, although their high price and fragility make them unsuitable for inexpensive or flexible devices. Glass fibers are cheaper, but can be used only for wavelengths above 325 nm. Plastic optical fibers are the cheapest and easiest to handle, but applicability is severely restricted by their higher losses, narrower usable range (450–700 nm), and lower stability on prolonged exposure to radiation or chemicals. Fiber core diameters are typically in the range of 50–500 μm.

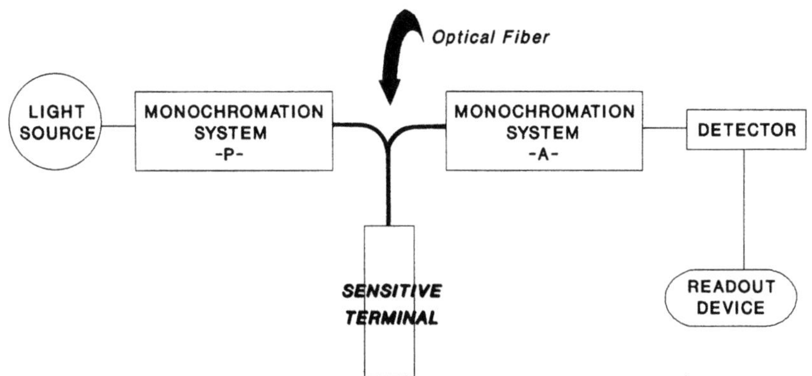

Fig. 1. General device for measurement with fiberoptic chemical sensors.

Low-cost tungsten–halogen lamps are most commonly employed for reflectance and absorbance measurements that do not require high probe radiation intensity. If measurements are to be carried out in the UV range, deuterium lamps are the best choice. Recently, high intensity light-emitting diodes (LEDs) have been coupled to fiberoptic guides to exploit their relatively monochromatic emission and low cost. Nevertheless, the intensity provided by this system is only high enough to perform reflectance and absorbance measurements. Fluorescence sensors require higher probe-radiation intensities, such as those emitted by xenon or low-pressure mercury lamps. To obtain high-intensity light of a single wavelength, helium–neon, helium–cadmium, or argon ion lasers have been used. Although they enhance sensor sensitivity, these light sources have found little acceptance for fiberoptic devices because of their high cost.

Two types of systems are used to select the desired wavelength of light: cutoff or interference filters, and grating monochromators. In most cases, optical filters can be used to meet analytical requirements, thus making the use of expensive grating monochromators unnecessary.

A wide variety of sensitive-terminal probes have been developed that can be classified into three categories according to the arrangement of the optical fibers at the sensitive tip (Fig. 2):

1. *Single strand:* A single optical fiber is used to drive the probe and emerging radiation. This type can be subdivided into one- and two-way types:

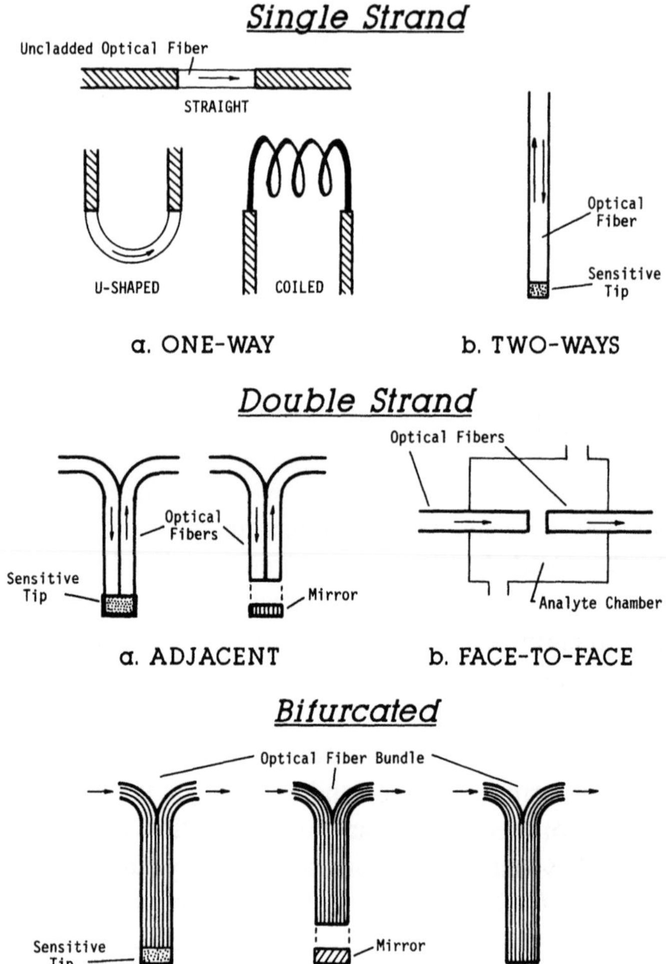

Fig. 2. Types of sensitive terminals according to arrangement of optical fiber used in chemical sensors.

- *One-way terminals* consist of an unclad optical fiber with the surface coated with a sensitive reagent layer. Changes in color or refraction index of this coating caused by varying analyte concentrations can be detected by monitoring transmission losses of radiation passing through the sensitive zone.
- *Two-way terminals* drive incident and emerging radiation along a single fiber. In this case, optical separation, modulated or phase-

shifted radiation is required to avoid interference between probe light and radiation returning after interaction with the sensitive layer placed at the end of the single guide.

2. *Double strand:* In this type of sensor, incident radiation is conducted to the sensitive tip by one optical fiber, and emerging radiation is collected by an adjacent optical fiber and driven to the detector. To perform absorbance measurements, a mirror can be placed in the test solution at a certain distance from the fiber end as an alternative to the sensitive reagent layer. Another arrangement for absorbance measurements uses two single optical fibers positioned *face-to-face* in an optical cell that provides a constant path length.

3. *Bifurcated:* This type of terminal consists of a bundle of optical fibers divided into two equal bundles to drive probe and returning radiation. Many bifurcated optical sensors using a sensitive reagent layer or a mirror, or without any sensitive device in the tip, have been described for fluorescence, absorbance, and reflectance measurements.

In most cases, photomultiplier tubes are used as a detection system. Less routinely, germanium photodiodes (for measurements in the spectral range above 700 nm), photocells (cheapest, but low sensitivity) or silicone p-i-n diodes are also employed.

The detector signal is driven to a readout device, usually a digital millivoltmeter or a strip-chart recorder to continuously monitor the output signal. The detector signal also can be directly input into a computer in order to acquire, control, average, treat, or process resulting data.

4. Analytical Characteristics

The characteristics of analytical reactions in the solid phase differ from those of reactions in aqueous solution *(45)*. These differences depend on whether the reagent has been electrostatically or covalently attached to the solid support. The pK_a of acid–base indicators, and consequently the working range of these indicators, differ in homogeneous and heterogeneous phases *(46)*. The optimum pH, ionic strength, and interferences used in measurements based on the formation of a colorimetric or fluorescent complex, also act differently in solid and liquid phases *(47)*. Thus, these parameters need to be controlled.

In some cases, spectral characteristics also vary, and noticeable wavelength shifts in maximum reflectance intensity or emission fluorescence are found between liquid and solid phases. The expected increase in spectral sensitivity from the sample preconcentration has shown only slight improvement for some elements *(48)*.

The dynamic range in which a linear or sigmoidal response to analyte is obtained depends on the amount of reagent immobilized. Wider dynamic range is usually achieved with higher amounts of immobilized reagent. Nevertheless, it is recommended to work far below the saturation concentration in order to keep a constant concentration of reagent on the solid support. The use of solid supports gives a wider dynamic range with fluorometric than with reflectance-based sensors.

Because the reagent and analyte are usually in different phases, there is necessarily a mass-transfer step before constant sensor response is achieved. This mass-transfer step limits response time and is responsible for the sluggishness of solid-phase sensors as compared to reactions in classical solutions. In sensors having two phases separated by a membrane, pore size also affects response time. Higher pore size provides quicker response, but usually lower selectivity. Thus, a compromise must be sought between selectivity and response time when selecting the membrane. Probe stability depends on the reagent and the way it has been immobilized on the solid support. There are some electrostatically immobilized reagents that have proven to be stable for several months, but more research is required before generalizations can be made on stability and its variation with time. One problem that drastically affects the stability of luminescence sensors is photodegradation of the fluorogenic reagent. When enzymatic reagents are used, their inactivation with time causes serious stability problems. When irreversible sensors based on analytical reactions that consume reagent are used, probe lifetime depends on the rate of reagent consumption. This can be quite long when reagent is consumed slowly.

At present, fiberoptic chemical sensors provide accuracy and precision comparable to that of conventional analytical methods. It is still too soon to predict what course this field will take with continuing research.

5. pH Sensors

A wide variety of fiberoptic sensors have been described for measurement of pH. Although this parameter is quantified satisfactorily by glass electrodes, there are certain situations in which H^+ activity is better determined by fiberoptic sensors than by electric devices. These situations include:

1. When no electric signal can be tolerated (avoiding risk of electric shock, as for catheterized measurements);
2. When a size smaller than that of a glass electrode is needed (as for in vivo measurements);
3. When the possibility of H^+ concentration is within a range in which glass electrodes cannot operate properly;
4. When use of a reference electrode is not desired;
5. For titration of glass-attacking compounds such as hydrofluoric acid; and
6. For use with nonaqueous solvents.

Fiberoptic sensors for pH measurement can be divided into two general types: (a) those using colorimetric acid-base indicators for monitoring changes in absorbance or reflectance at the sensitive end and (b) fluorometric acid-base indicators for detecting changes in fluorescence intensity upon protonation.

Peterson et al. *(49)* were probably the first to report a fiberoptic-based pH sensor. The indicator dye phenol red was copolymerized with acrylamide and *bis*-acrylamide to yield dyed polyacrylamide microspheres 5–10 µm in diameter. The device also contained smaller polystyrene microspheres to provide effective light scattering. Both were packed in cellulosic dialysis tubing at the end of a pair of optical fibers in order to monitor absorbance (or, more accurately, reflectance) changes. The sensor measured in the physiological pH range 7.0–7.4 to ±0.01 pH units. Ionic strength and temperature affected response only slightly. The small size of the probe (0.4 mm in diameter) allowed blood-pH monitoring in vivo.

A detailed in vivo study with this sensor has been reported by Abraham et al. *(50)*, who developed a miniaturized fiberoptic probe for continuous monitoring of intravascular pH in anesthetized dogs. The sensor worked equally well in the presence of normotension or hypotension, as well as respiratory or metabolic acidosis and alkalosis. The difference between fiberoptic and reference-electrode pH determinations averaged 0.060 ± 0.004 pH units. Other pH sensors of this type have been patented *(51–53)*.

The dissociation constant of the covalently bonded dye is of prime importance in defining the working range of a fiberoptic sensor. Accordingly, Kadin *(46)* has described a useful second-derivative spectrophotometric method for determination of the apparent pK_a of an immobilized dye sensor (phenol red on polyacrylamide), based on the absorbance spectra of opaque 50% glycerol suspensions of the phenol red–resin system.

Narayanaswamy et al. *(54)* have recently developed a new method for determining acidity constants of acid–base indicators immobilized on hydrophobic polymer (styrene–divinylbenzene) supports. Their results indicate that adsorption of a dye on a nonpolar surface causes a decrease in the extent of dissociation of the solute thereby altering the dynamic range of bonded dye-based fiberoptic sensors.

Kirkbright and coworkers *(55,56)* have designed and patented another fiberoptic pH sensor based on the immobilization of different acid-base indicators on styrene–divinylbenzene copolymers. Their first device was a flow cell to find the optimal indicator for each pH range. The light coming from the lamp is chopped and guided through a bifurcated optical fiber with the sensing end placed in the flow cell. A solution of known pH is passed through the cell, and the reflectance signal at a given wavelength is continuously monitored.

The same authors later reported *(57)* a fiberoptic pH sensor based on the use of bromothymol blue immobilized on the same copolymer. The sensitive layer was placed at the distal end of a 16-fiber bifurcated bundle, with the supporting polymer retained in position by a polytetrafluoroethylene (PTFE) membrane. On testing turbid solutions, a slight difference from the response in clear solutions was observed.

This was attributed to ambient light reduction at the sensitive tip or absorption of scattered radiation at the membrane/solution interface. The effect of ionic strength was considerable. Probe response to temperature change was nonlinear between 19 and 50°C. The solid-state instrument used for the measurements was later improved *(58)* by using an LED light source, incorporating a photodiode detector, and adding another LED to provide a reference signal. Compared to the conventional instrumentation used in earlier measurements, this smaller, more portable system was less expensive and had less consumed power.

A similar pH-reflectance sensor is described by Moreno et al. *(59)*. They use cresol red as the immobilized dye supported on an anionic Dowex® 1X10 resin. Since the indicator contains two protonation sites, the probe can be used to monitor H^+ concentration in two ranges: 0.025–0.6M HCl and pH 6.1–7.2. Sensor stability was longer than 6 mo and the sensor was accurate to ±0.02 pH units. The effect of ionic strength was not critical unless anionic concentration exceeded $10^{-1}M$. Colored anions (e.g., CrO_4^{2-}) did not affect the response at concentrations lower than $10^{-3}M$.

Saari and Seitz *(60)* have developed a pH sensor based on fluorescence quenching by H^+ of the excited state of immobilized fluoresceinamine. The probe was constructed by immobilizing this reagent on either controlled pore glass or cellulose. Fluorescence response increased significantly from pH 3 to 6. Above pH 8, the glass-bound fluoresceinamine emission continued to increase whereas the response of the cellulose-bound indicator decreased because of hydrolysis. The high background signal from scattered radiation might be reduced by finding a system in which excitation and emission wavelengths are further apart (larger Stokes shift). Another drawback is the low fluorescence intensity from the immobilized fluoresceinamine.

Munkholm et al. *(61)* have recently prepared a pH sensor also based on the use of fluoresceinamine. Here the reagent is incorporated into an acrylamide–methylene–*bis*(acrylamide) copolymer covalently attached to a surface-modified glass fiber by thermal- or photopolymerization. Since the method does not require a membrane, response is rapid. Construction of a serviceable pH sensor demonstrated that polymer immobilization is a valuable means of securing and increasing the signal in fiberoptic analysis. Since polymer modification adds very little to the diameter, the coated fiber may be used for in vivo tests, in which a fast response can be an advantage. A shortcoming is the lack of reproducibility because of the difficulty of removing the fiber from the bulk polymer during the coating process.

Another fluorescence-based fiberoptic pH sensor for operation in the physiological range (6.5–8.5) has been developed by Zhujun and Seitz *(62,63)*. The sensitive end contains the trisodium salt of 8-hydroxy-1,3,5-pyrenetrisulfonic acid (HPTS) immobilized on an anion-exchange membrane. The pH-dependent production of fluorescence in HPTS occurs as shown in Fig. 3. The excited state of HPTS ionizes more rapidly than it returns to the ground state. Consequently, below pH 7.3, the observed fluorescence signal is characteristic of the excited state of PTS^- (PTS^{-*}), even though HPTS is the major ground-state species. However, the excitation wavelengths for HPTS and PTS^- differ considerably. Therefore, a pH-dependent response can be obtained by selectively exciting either HPTS or PTS^-. Alternatively, both species are excited and the ratio of their fluorescences is measured. This ratio is not affected by such variables as temperature, ionic strength, and the presence of anions or cations, nor is membrane-bound HPTS subject to concentration quenching. Precision of the measurements is ±0.02 pH units, but HPTS-treated membranes rapidly photodegrade. Increase in HPTS loading ($>15 \mu g/cm^2$) improves the response, but the inner filter effect may influence the measured pH values.

A new fiberoptic sensor containing a fluorophore (eosin), and an absorber (phenol red), coimmobilized on the distal end of an optical fiber has recently been developed by Jordan et al. *(64)* to measure pH in the 6.0–8.0 range with a precision of ±0.008 pH units. This chemical sensor is based on nonradiative energy transfer from eosin to the basic form of phenol red. With increasing pH, the amount of energy transferred is greater, resulting in diminished fluorescence intensity. Thus, changes in the pH-dependent absorption of phenol red are detected as

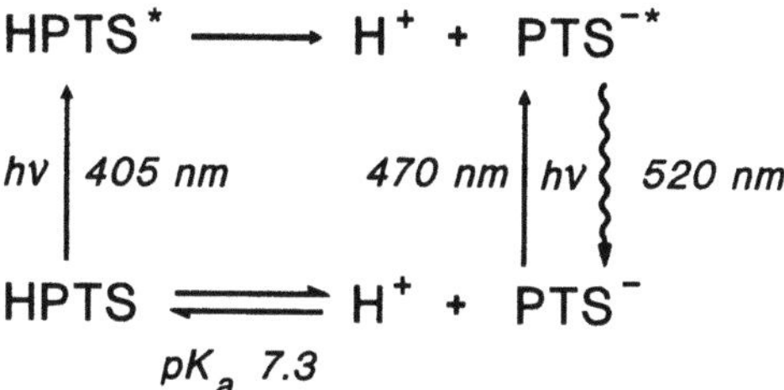

Fig. 3. Acid-base processes of ground-state and photoexcited HPTS (refs. *62, 63*).

changes in the fluorescent signal. The small thickness (<10 μm) and porosity of the polyacrylamide end-coating yield short response times and 100% equilibration times of 10 s or less. Furthermore, the fiber can be repeatedly dried and wetted without loss of signal or sensitivity. Photobleaching of phenol red can be minimized by lessening shutter aperture time of the exciting Ar-ion laser beam.

Two bare-ended fiberoptic devices have been described to aid in liquid titration studies of different acids and bases. The first is the pH sensor developed by Benaim et al. *(65)* based on absorbance measurements with a bifurcated optical fiber. The acid-base indicator dye is added to the solution to be titrated and the color change monitored using a polished stainless steel reflecting cap placed at a known distance from the fiber end to form an *in situ* absorbance "cell." The second device, reported by Wolfbeis et al. *(66)*, is similar, but uses fluorescence as measured signal. It consists of a bifurcated optical fiber and a blue LED (or a Xe lamp) as a light source. A photodiode acts as a detector to monitor changes in the pH-dependent fluorescence of an added indicator (e.g., fluorescein or HPTS) during the course of the titration. Advantages include the possibility of performing microtitrations with sample volumes as small as 0.5 mL, which may be of particular use in the assay of expensive biophosphates (adenosine phosphate and related species).

Problems inherent to optical sensing of pH and other ion activities have been summarized *(67)*. The author concludes that, although neither optical electrodes nor pH electrodes can actually measure pH, the thermodynamic compromises that have to be made in electrochemical estimation of pH are less severe than those that have to be made when using optical sensors, as a result of the fact that the latter are sensitive to hydrogen ion concentrations instead of activities.

Monici et al. *(68)* have developed a sensor for the continuous monitoring of seawater pH, based on the immobilization of phenol red on the nonionic polymer Amberlite® XAD-2 placed at the tip of an optical fiber. A Cuprophane® membrane separates the reactive phase from the surrounding medium, and a Mylar™ film separates it from a graded index rod lens held at the fiber tip to improve sensitivity by beam expansion.

Optosensing flow injection analysis (FIA), a technique in which an indicator immobilized on a cellulose pad is placed in an FIA channel with a dilute buffer as a carrier, has been applied by Woods et al. *(69)* to rainwater pH monitoring. They found that metal-ion interference did not appear until the concentrations were higher than those present in rainwater. The system gives comparable results to a pH glass electrode, but faster since the sampling rate is 60 samples/h, with the potential for even shorter times. The sensor has a working range from pH 3.5 to 5, with a precision of 0.0–0.11 pH units.

The pH sensor developed by Jones and Porter *(70)* is based on the immobilization of the pH indicator congo red on a porous cellulosic polymer film placed at a flow cell used for continuous measurement. The bifurcated-type sensor provides a large dynamic range (>4 pH units), a very short response time (1.26 s) upon a drop in pH from 10.9 to 2.0, and good stability over a pH range of 13–0, although it degrades below pH 0. Mercuric ions are chelated by the indicator in solution, but interference is not observed when the reagent is immobilized on the cellulosic matrix.

Posch et al. *(71)* report an optical-fiber pH sensor able to measure pH in the 0–7 range and suitable for gastric pH determination. Fluorescein and eosin are immobilized separately on aminoethylcellulose particles. Sensing layers are obtained by fixing these particles in the desired proportion onto a polyester film by means of an adhesive. Several layers of an amino-modified hydrogel are applied over the sensing layers and, when dry, placed in a flow-through cell *(72)*. Fluorescence emission (cutoff 540 nm) is monitored via optical fibers. The sensor has an accuracy of ±0.05 pH units, and its characteristic response time is 25–30 s.

Luo and Walt *(73)* have developed a fiberoptic pH sensor based on the controlled release of HPTS from the polymeric matrix ethylene-vinyl acetate (EVA). The 515-nm fluorescence ratio obtained when HPTS is excited at 405 and 450 nm, are related to pH in the range 5.5–8.0 with a precision of ±0.7 pH units. In a second method, HPTS and the pH-insensitive compound sulforhodamine 640 are entrapped in an EVA polymer after dissolving in methylene chloride. In this case, the emission intensity at 530–610 nm when excited at 488 nm is used to measure pH with a higher precision (±0.1 pH units). The dyed polymer is placed in a reservoir-like section above the fiber tip, which is connected to a drilled Teflon™ tube sealed at one end with Parafilm™. Long response times are, nevertheless, a major drawback in this design.

6. Sensors for Anions

The development of sensors for continuous anion monitoring is of great interest for the study of environmental samples and for in vivo anion monitoring. Urbano et al. *(74)* were the first to describe an optical sensor for halides and pseudohalides. This sensor is based on dynamic quenching of fluorescence emission of glass-immobilized acridinium (reagent A) and quinolinium (reagent B) heterocyclic indicators. The fluorometric process involved is known to obey the Stern-Volmer equation *(75,76)* whereby fluorescence decreases with halide concentration. Quenching becomes more efficient in the order $Cl^- < Br^- < I^-$. However, the quenching constants are around 10–20% lower than those of the nonimmobilized indicators in solution *(76)*. Variation of pH in the range 4.2–7.2 has a negligible effect on sensor response; increasing ionic strength leads to negative deviation of the calibration curve. Reagent A gives higher sensitivity than reagent B, and provides detection limits of 0.15 mM for I^-, 0.40 mM for Br^-, and 10 mM for Cl^-. Sulfite, isothiocyanate, and cyanate interfere in halide determination, but phosphate, perchlorate, or nitrate up to 1M does not.

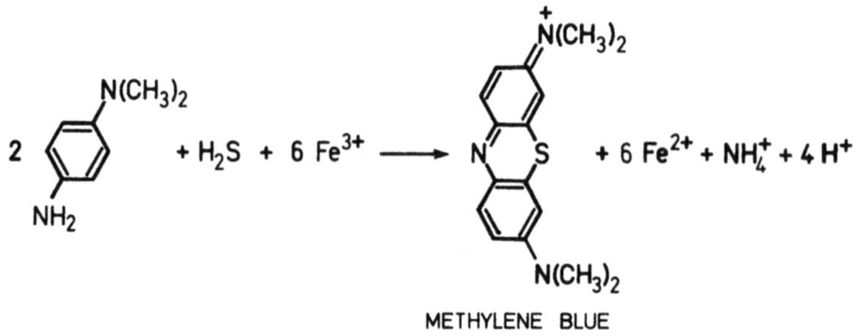

$$2\ \text{(}p\text{-aminophenyl-}N,N\text{-dimethylamine)} + H_2S + 6\ Fe^{3+} \longrightarrow \text{(methylene blue)} + 6\ Fe^{2+} + NH_4^+ + 4\ H^+$$

METHYLENE BLUE

Fig. 4. Analytical reaction for determination of sulfide in aqueous samples.

Using an extrinsic sensor and acid solution, chloride and other halides have been titrated with silver nitrate in the presence of acridine *(77)*. At first, fluorescence is weak by virtue of fluorescence quenching by chloride. As the titration proceeds, the concentration of chloride becomes smaller and reaches a minimum at the end point, when quenching is least efficient. From this point on, the addition of excess silver ion, which acts as a quencher, leads to a decrease in fluorescence.

Cámara and coworkers *(78)* have described a fiberoptic sensor to determine sulfide in aqueous samples. It is based on the change of reflectance intensity with sulfide concentration as a result of formation of methylene blue. This product is generated *in situ* from *N,N*-dimethyl-*p*-phenylenediamine hydrochloride electrostatically immobilized on the cationic exchange resin Dowex® 50 WXP in the presence of acidic Fe(III) (Fig. 4). The authors have designed an optical cell that contains immobilized methylene blue; reflectance is measured at 680 nm. Product formation is pH-dependent and sensitivity varies with reaction time (3 min being selected when working in $0.67M\ H_2SO_4$). The addition of strong electrolytes, such as NaCl, does not modify the slope of the calibration curve, but the linear response range of 0.05–0.15 ppm sulfide in the absence of added electrolyte is extended to 0.05–0.60 ppm by using $1M$ NaCl as a saline background. The system is quite selective for sulfide, the most severe interferences being produced by thiosulfate, sulfite, alkyl mercaptans, and phosphate. The main advantage of this sensor stems from the stability of the methylene blue immobilized on the solid support, which allows resin to be retained for analysis at a later date. Minimal resin and reagent consumption gives low cost, but the system is irreversible and nonregenerable, thus making this sensor unsuitable for continuous sulfide monitoring.

Narayanaswamy et al. *(79)* have described a fluoride reflectance sensor based on the use of an optical cell filled with alizarin blue–Ce(III) or alizarin blue–La(III), immobilized on Amberlite® XAD-2. The response times are 12 and 22 min for Ce and La chelates, respectively. The sensor can be regenerated in 1 h by soaking in 10^{-3}–10^{-4} Al(III), but the unit is not very selective, and phosphate is the main interference.

Another nonreversible but very sensitive sensor for hydrogen cyanide in air has been developed by Bentley and Alder *(80)*. This reflectance sensor employs a two-bed reactor in order to convert HCN to cyanogen chloride, using XAD-7 resin impregnated with sodium *N*-chloro-4-methylbenzene sulfonamide, and to induce color formation through reaction with 4-picoline and barbituric acid on the same support. The fast response of the sensor (<1 min) and low CN^- concentration range (1–10 µg/L) make it very useful in spite of its irreversibility and short lifetime.

7. Sensors for Cations

Sensors to determine several cations have been developed in the last few years, aluminum being the most widely studied. There is great interest in sensors for continuous monitoring of such electrolytes as Fe(III), Na(I), Ca(II), and Mg(II) in blood and other biological fluids. Nevertheless, very few sensors allow cation determination in vivo because of the difficulty in finding reversible and specific reactions.

Labeyrie and Koechlin *(81)* have designed one of the first optrodes for recording Ca(II) transients at a cellular level. It is based on the measurement of light flux originating from the calcium-sensitive protein aqueorine, which emits light of an intensity proportional to the calcium activity of the medium. The optrode has two channels, which are introduced in brain tissue. The first allows aqueorine to be perfused into the extracellular space and the second is a glass-fiber channel to guide emitted light to a photomultiplier tube. The main advantage of this sensor is the very short response time (<10 ms), which allows monitoring of epileptic activity, in which each interictal epileptic spike is accompanied by sharp variations in calcium activity. Nevertheless, its sensitivity is not high enough to quantify emission intensity when slow changes in Ca(II) activity take place, nor could the authors use the device for measurements in central nervous system because the light captor was too large.

Aluminum, the element that has received the most attention in the development of fiberoptic chemical sensors, is most commonly determined fluorometrically using a well-known reagent that is trivially called morin (3,5,7,2',4'-pentahydroxy flavone). The experimental equilibrium constant of the morin–aluminum complex depends on pH, because Al(III) displaces hydrogen ion when binding to morin.

Saari and Seitz *(82)* have immobilized morin on cellulose and placed it at the end of a bifurcated optical fiber to obtain a sensor with a pH-dependent response to Al(III). There is a linear increase in fluorescence intensity from 10^{-6} to $10^{-4}M$ Al(III) at pH 4.8. The sensor is irreversible, but can be regenerated by dipping into EDTA solution for 3 min. The detection limit is $10^{-6}M$ Al(III) (0.027 ppm) with a linear response to Al(III) in the 1–100 μM concentration range. The selectivity of the device is not very high, Be(III), Fe(III), and Ca(II) being the major interferents. The main advantages of this sensor are its quick response (1–2 min) and ease of regeneration.

The fact that Be(III) is one of the main interferents in determinations by the above sensor has led to its being tested as a sensor for this metal *(83)*. The response time is similar to that for aluminum *(82)*, and it is regenerable by immersion in a fresh buffer solution. The detection limit is 9 ppb, lower than that for Al(III). Al(III), Ca(II), and Mg(II) interfere, with Al(III) being the major interference. Al(III) can be removed in two ways: either by adding EDTA to form a complex the aluminum or by measuring at two excitation wavelengths, to exploit the differences in excitation spectra. In addition, Cu(II) and Fe(III) seriously interfere because they quench fluorescence, so their removal is necessary for correct determination of beryllium.

Wolfbeis and Schaffar *(84)* have also taken advantage of the fluorescence properties of morin to determine Al(III) in aqueous solution by direct titration with DCTA (*trans*-1,2-diaminocyclohexane-*N*,*N*,*N'*,*N'*-tetraacetic acid). The sensor consists of a bared-tip bifurcated fiberoptic light guide that is immersed in the solution containing dissolved morin. Addition of DCTA causes a decrease in the fluorescence of the aluminum–morin complex, the slope depending on morin concentration. On reaching the end point, the fluorescence intensity levels off to a low and constant value. Al(III) can be titrated in this way in the 1–800 μg/mL range. The main advantages of this device are that it offers the possibility of performing titrations in strongly colored, as well as turbid, solutions and that the system is suitable for a variety of titrations by appropriately selecting wavelength and indicator.

When developing a sensor for a specific purpose, it is important to know the exact analytical characteristics of the immobilized reagent. Saari and Seitz *(48)* have reported the analytical features of covalently immobilized calcein for transition metal-ion preconcentration and determination. The high binding constants found for supported and dissolved calcein with Cu(II), Co(II), and Ni(II) ions were similar, indicating that covalent immobilization does not interfere with complex formation. Analytical application of calcein to Co(II), Ni(II), and Cu(II) determination is based on the formation of nonfluorescent metal complexes at neutral pH, where strong fluorescence emission is observed for uncomplexed calcein. This sensor *(48)* has the advantages of responding to several transition metals and of being suitable to determine the end point of titrations. One of its shortcomings is irreversibility, mainly as a result of the large values of the conditional binding constants. When using the sensor as a regenerable optical sensor, it is necessary to work at pH values below the optimum for complex formation, with consequent loss of sensitivity.

Zhujun and Seitz *(85)* have reported a fluorescence-based sensor for Al(III), Mg(II), Zn(II), and Cd(II) determination. This sensor uses quinolin-8-ol-sulfonate (QS) electrostatically immobilized on an anion-exchange resin. Fluorescence intensity of complexes varies with pH, the maxima occurring in the order Al (pH 5), Zn (pH 8), Cd (pH 8), and Mg (pH 11) with increasing pH. Thus, pH control provides a valuable means of achieving selectivity. Fluorescence signal diminishes with increasing temperature and ionic strength. Response time depends on the degree of crosslinking of the anionic exchanger, which shows that internal diffusion of the metal ion is the rate-limiting process.

Zhujun et al. *(86)* have also reported the development of a reversible sensor responding selectively to the sodium ion and based on ion-pair extraction and fluorescence. The indicator phase comprises an anionic fluorophore, a Cu(II)–polyethyleneimine complex, and a neutral sodium-selective ionophore immobilized on a solid substrate. The sensor uses a dialysis membrane permeable to alkali metal ions but not to the polyethyleneimine complex. In the absence of sodium, the anionic fluorophore binds to the cationic Cu(II)–polyethyleneimine polyelectrolyte, fluorescence being quenched by the paramagnetic copper. When sodium ion is added, it combines with the neutral ionophore, which becomes cationic and forms an ion pair with the anionic fluorophore, rendering it fluorescent. Time and intensity of the sensor response increase with greater amounts of immobilized ionophore and depend strongly on temperature and pH. The sensor, which was used for only 3 d also responds to potassium and calcium, the selectivity for sodium over these two ions being 1.2 and 1.7, respectively.

Alder et al. *(87)* have developed an optical sensor that is sensitive to potassium ions in the 10^{-3}–$10^{-1}M$ range at pH 8. Based on the use of a reagent synthesized from 2-hydroxy-1,3-xylyl-18-crown-5 by reaction with the diazonium salt derived from 4-nitroaniline, this reversible sensor has a 2–7 min response time and a K^+/Na^+ selectivity ratio of 6.4:1.

Another ion-selective optrode reported by Schaffar et al. *(88)* for continuous determination of potassium is based on the optical measurement of the membrane potential between an aqueous sample solution and a lipid phase incorporating rhodamine B as a potential-sensitive dye. The fluorescence intensity of this sensor decreases logarithmically with increasing potassium concentration, and its selectivity for potassium over sodium is on the order of 1×10^4. Response to potassium is linear in the range 10^{-1}–$10^{-5}M$.

The first absorbance sensor for cations was developed by Freeman et al. *(89)*. Two jacketed fused-silica optical fibers were positioned directly facing each other in a 2-mm-path absorption cell in order to determine copper(II) in electroplating baths. This sensor was based on absorbance measurements obtained by immersing the cell in a stirred solution and comparing the transmitted signal with that from a blank solution. The sensor provided a linear Cu(II) response in the 50–500 mM range. Varying H_2SO_4 concentration present in the baths produced an error in Cu(II) determination, probably because of changes in refractive index *(90)*. Another reversible sensor has been constructed *(17)* to determine UO_2^{2+} cation. Since UO_2^{2+} fluorescence is strongest in acidic solution and in the presence of phosphate, a sensor reservoir was developed to allow continuous addition of 1% phosphoric acid to the sample solution. This device has been utilized for reproducible detection of uranyl ions at concentrations as low as 10 μM in groundwater *(91)*.

Cámara and coworkers *(92)* have developed the first fiberoptic sensor for Fe(III). It measures reflectance intensity upon formation of the blue chromeazurol–Fe complex. This complex is generated from chromeazurol S (immobilized on an anion-exchange resin) and soluble Fe(III), at pH 5. The sensor is irreversible, but is regenerable by submerging the immobilized complex in a saturated pyrophosphate solution. The sensitive tip is similar to that previously described for sulfide *(78)*. Sensitivity to Fe(III), drastically affected by the moisture content of the polymeric matrix, improves when the support is dried. The sensor provides a linear response in the 0.02–0.5 ppm Fe(III) concentration ranges.

An ammonia-sensitive fiberoptic probe utilizing bromothymol blue immobilized on a hydrophilic polymer has been described by Çaglar and Narayanaswamy *(93)*. This reversible reflectance sensor can be used to measure ammonia vapor at concentrations as low as 1.5 $\times$ 10^{-3}M. Dimethylamine is the main source of interference.

8. Sensors for Gases

Fiberoptic sensors have been developed to measure a wide variety of gases (e.g., oxygen, carbon dioxide, ammonia, halothane, and moisture) mainly because these are of enormous interest in environmental, clinical, and in vivo studies. Probably the first fiberoptic sensor of this type to be described was the in vivo oximetric probe of Kapany and Silbertrust *(94)*. It is based on the shift in the Soret absorption band of hemoglobin upon association with O_2. Responses in animals (fiberoptic catheters inserted in the hearts of dogs) were very fast and sensitive, but calibrated oximeter readings were not the same for every animal, probably because the blood of each individual has different scattering properties.

The same principle applies for the oxygen sensor developed by Zhujun and Seitz *(95)*. In this case, deoxyhemoglobin is immobilized on preswollen CM-Sephadex®-C-50-120 cation-exchange resin and placed at the common end of a bifurcated optical fiber. An O_2-permeable PTFE membrane separates the sensitive layer from the in vivo samples. Determination of the ratio of reflected intensities at 435 and 405 nm is used to measure O_2 partial pressure from 20 to 100 bars. Optimum reagent layer thickness and hemoglobin loading allow a fast response (ca. 3 min). The oxygen sensor's main inconvenience arises from its short lifetime (<2 d at room temperature) and nonlinear response. Responses are profoundly affected by CO_2, pH, and oxidizing gases (acidic and basic).

Another type of reversible O_2 sensor makes use of dynamic quenching of fluorescence from substituted polynuclear hydrocarbons. Thus, Peterson et al. *(96)* based their fiberoptic sensor on the emission quenching of perylene dibutyrate adsorbed onto a polystyrene–divinylbenzene polymeric support. The sensitive layer is fixed at the distal end of a double adjacent plastic optical fiber by means of a highly gas-permeable polypropylene membrane. This sensor allows measurement of partial O_2 pressure ranging from 0 to 150 bars, with response times of <2 min (in aqueous solution) and accuracy of ±1 bar. In vivo tests have been performed with encouraging results, provided that halocarbon anesthetics are not present and clotting around the sensor tip is avoided. Some O_2 sensors have been patented by Peterson and Fitzgerald *(97)* and Fitzgerald *(98)*.

The fact that fluorescence emission is also quenched by halocarbon anesthetics has led Wolfbeis et al. *(99)* to construct a sensor for simultaneous determination of halothane and O_2. In this case, deca-

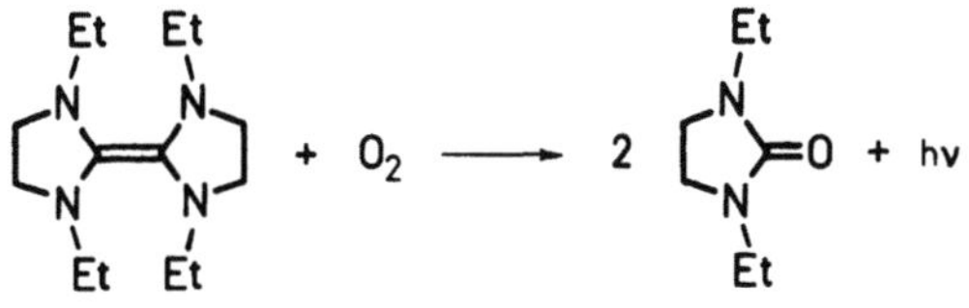

Fig. 5. Chemiluminescent oxidation of 1,1',3,3'-tetraethyl-$\Delta^{2,2'}$-bi(imidazolidone) (EIA) (ref. *101*).

cyclene is dissolved in silicone rubber or polyisoprene/dioctyl phthalate membranes. The O_2-sensitive layer is covered with an additional halothane-impermeable PTFE membrane. Both layers separate the terminal of a bifurcated quartz optical fiber from the gas sample. Concentrations between 0.1 and 4% halothane and between 0 and 100% oxygen can be measured simultaneously. The wide dynamic range, fast response time (10–20 s), and lack of interference from CO_2, N_2O, and flurans make this sensor a promising candidate for in vivo measurements. A third O_2 sensor based on polynuclear aromatic hydrocarbons has also been developed by Wolfbeis et al. *(100)*. It uses pyrene butyric acid covalently attached to controlled-pore glass via an amide bond. Incident radiation is not driven by an optical-fiber system, but such a system could easily be adapted to do so. It provides high selectivity to O_2 and very fast response time (20–50 ms). However, it is useful only for measuring O_2 concentrations in gaseous mixtures, (e.g., clinical inhalation mixtures).

Freeman and Seitz *(101)* described an irreversible oxygen probe based on the chemiluminescence reaction of this analyte with 1,1',3,3'-tetraethyl- $\Delta^{2,2'}$-bi(imidazolidone) [EIA] (Fig. 5). A hexane solution of the reagent is contained in a cell separated from the gaseous or aqueous sample by a PTFE membrane. The emitted light is not driven to the photomultiplier tube by a fiberoptic guide, but the authors state that such a guide can be easily fitted. Sensor lifetime (up to 12 h, limited by the consumption of EIA) could be improved by using thicker membranes and larger reagent reservoirs. Measures are affected by EIA concentration, O_2 flow, temperature, and use of aqueous solutions. Its main value arises from a very low detection limit (1% O_2 in gas samples) and great selectivity vs O_2 (compared to conventional O_2 electrodes, which are subject to interferences from H_2S and oxidizing gases). Irreversible character, limited lifetime, slow response time with decreased slope for aqueous samples (5 h), and ambient light interference detract from its worth.

Zhujun and Seitz *(102)* also have developed a CO_2 sensor for determination of up to $10^{-3}M$ CO_2 in aqueous samples. On an ion-exchange membrane, HPTS is immobilized and covered with a sili-cone rubber membrane, which also separates an internal chamber containing $1 \times 10^{-3}M$ HCO_3^- solution. The diffusion of CO_2 across the membranes causes a change in the pH of the internal solution, which is detected by measuring the variation in fluorescence from the base form of the HPTS (Fig. 3). The resulting bifurcated fiberoptic sensor is therefore completely analogous to the pH sensor previously described by these authors *(62)*. Dynamic range depends on the concentration of HCO_3^- in the internal solution. Response time (roughly 3 min) is limited by the rate of CO_2 diffusion through silicone and pH-sensitive membranes. Major interference arises from SO_3^{2-} and S^{2-} ions. Compared to CO_2 electrodes, this optrode possesses a faster response rate and a lower cost and is not affected by the presence of dissolved ionic constituents. However, its limited dynamic range and stability—intense probe radiation affects optical fibers and causes progressive bleaching of the pH-sensitive dye—are factors that diminish its versatility.

Relative humidity of air (between 40 and 80%) can be measured with the moisture sensor developed by Russell and Fletcher *(103)*. A U-shaped uncladded silica optical fiber is covered with a sensitive layer by immersing in an aqueous solution of gelatin and $CoCl_2 \cdot 6H_2O$ and subsequent drying. Hydration of the cobalt chloride causes variation in transmitted light at 680 nm, yielding a nonlinear, roughly sigmoidal, response at the photodetector. At typical room humidities, sensor stability is >4 mo with high reproducibility (±2%) and fairly short response time (<1 min).

A fiberoptic-based ammonia gas sensor recently has been described by Arnold and Ostler *(104)*. A double, adjacent, sensitive tip is immersed in an aqueous solution of NH_4Cl that contains *p*-nitrophenol as a pH indicator, and it is separated from the aqueous sample by a microporous gas-permeable PTFE membrane. The incoming NH_3 causes a change in the pH of the internal solution, and the resulting color change is detected through a plastic optical fiber. A linear response is observed between 5 μM and 1.0 mM NH_3. Response time (<4 min) and dynamic range are a function of dye concentration, so the optimum conditions strike a compromise between these two factors. Nevertheless, the use of a single pH indicator restricts sensor response to a narrow dynamic range. This shortcoming is remedied *(105)* by using a combination of the indicators chlorophenol red and bromothymol blue that allows an optimum steady-state optrode response. Such a fiberoptic sensor provides the same sensitivity, similar response times, and shorter recovery times than the potentiometric ammonia sensor and has been successfully applied in wastewater analysis.

Over the past few years, transition metal complexes have found increasing use as oxygen-sensitive dyes, because oxygen can dynamically quench their luminescence with high efficiency. Wolfbeis et al. *(106)* have designed a new oxygen-sensing probe containing $Ru(bpy)_3^{2+}$ [tris(2,2'-bipyridine)ruthenium(II)] embedded in a silica gel matrix that is suspended on a silicone sheet and manufactured in thin layers on a polyester support. Silicone gives a very O_2-permeable membrane allowing fast response times. This sensing-layer arrangement has been used *(107)* subsequently to develop a dual sensor for oxygen and carbon dioxide in which HPTS is covalently immobilized on cellulose granules and embedded in hydrogel. The latter is then soaked in a bicarbonate buffer solution, and a thin layer of the oxygen-sensitive indicator, trapped in silica gel and suspended in silicone, is deposited over it. Response intervals are 0–200 torr for oxygen and 0–150 torr for carbon dioxide, with ±1 torr accuracy; detection limits are 0.5 torr in both cases. The sensing setup for CO_2 is based on the same principle as the device described by Zhujun and Seitz *(102)*. Response times for the oxygen sensor are about 40 s, whereas 3.8–5.2 min are required for CO_2. The indicators used in this sensor show no cross-sensitivity, since they are highly O_2 or CO_2 specific, exhibiting very similar excitation maxima but very different emission maxima.

A study of several ruthenium dyes as potentially suitable reagents for use in decay-time sensors *(108)* has led to the development of an oxygen sensor based on lifetime rather than on fluorescence intensity measurements. Lippitsch et al. *(109)* determined the lifetime of $Ru(bpy)_3^{2+}$ immobilized, as described above *(106)*, by means of a fiberoptic system. The lightsource is a frequency-modulated blue LED, the pulses of which are compared with the red fluorescence of the sensor material at the fiber tip. The detection limit is about 2 torr of oxygen. Although the Stern-Volmer plot for lifetime sensors has a negative slope above 35% O_2, the linear range is wider than that of luminescence-intensity sensors. The latter show higher quenching efficiency. One of the main advantages of lifetime-based sensors is that, since the decay time is independent of the fluorophore concentration, they show no signal drift attributable to reagent leaching and bleaching.

The need to improve the stability of optical sensors in terms of calibration has led Lee et al. *(110)* to develop oxygen indicator systems with two luminescence bands, only one of which is quenched by oxygen. Approximately 2 mg of solid-phase indicator is placed in a sensor cap located at the common end of a bifurcated fiberoptic bundle. An oxygen-permeable membrane (polypropylene or porous PTFE) is pressed over the end of the optical fiber and onto the solid material. When supported on γ-cyclodextrin bonded to cellulose, the indicator (5-(4-bromo-1-naphthoyl)pentyltrimethylammonium bromide) shows intense oxygen-dependent phosphorescence and oxygen-independent fluorescence of sufficient stability and sensitivity to determine gas-phase oxygen concentrations in atmospheres of low relative humidity. The results obtained in aqueous samples, however, are not satisfactory, since a dramatic decrease in phosphorescence takes place.

Sharma and Wolfbeis *(111)* have developed a fiberoptic sensor based on electronic energy transfer from pyrene, the fluorescence of which is efficiently quenched by oxygen, to a less oxygen-sensitive acceptor, perylene. Both reagents are immobilized on a silicone matrix and placed at the tip of a bifurcated fiberoptic bundle. The membrane is excited at 320 nm and fluorescence emission is measured at 476 nm (at which level pyrene does not emit fluorescence) and related to oxygen concentration. As the oxygen concentration increases, pyrene fluorescence is quenched and the spectral overlap between the pyrene emission and perylene absorption decreases, resulting in less intense fluorescence at 476 nm. Response time is 1–2 s, and O_2 can be detected in the range 0–150 kPa with a precision of ± 0.3 kPa. The detection limit is 60 Pa oxygen.

The same principle has been applied to the determination of sulfur dioxide, since pyrene emission can be efficiently quenched by this molecule, although perylene is not affected *(112)*. This system yields a quenching constant of 7.0 ($\%^{-1}$) for SO_2 compared with 0.154 ($\%^{-1}$) for O_2. The limit of detection is 10 ppm SO_2.

9. Sensors for Biochemicals

In fiberoptic sensing, use is made of a number of specific colorimetric or fluorometric reactions available for the detection and measurement of metabolites and biochemicals. The particular advantages of these probes may make them suitable for the determination of different medical parameters *(39)*. Because of their small size and flexibility, fiberoptic probes can be easily inserted into catheters for multiple sensing, or into hypodermic needles, where they offer adequate accuracy and stability for in vivo measurements. The absence of electrical connectors to the body adds safety to these relatively biocompatible devices.

The simplest sensor for biochemical monitoring is that developed by Milano and Kim *(113)* for determination of the relative amounts of oxyhemoglobin, carboxyhemoglobin, and reduced hemoglobin in blood, making use of their different absorption spectra. The probe consists of a bare-ended, double, face-to-face optical fiber curved to form a loop at the tip and fitted within a hypodermic needle or catheter. The relative concentration of each of the hemoglobin components can be calculated from the derivative spectra by a least-squares method in 30 s, with a precision of about 1%, which compares favorably to commercial in vivo oximeters. Moreover, standardization of a probe with a solution of known concentration allows semiquantitative determination of total hemoglobin concentration in hemolyzed solutions (unhemolyzed samples give large random errors).

Shultz and Sims *(114)* have developed an affinity fluorescence sensor for determination of glucose concentration. It is based on the displacement of a known labeled molecule from a specific binding site by unlabeled analyte. Concanavalin-A (Con-A) is immobilized on the surface of a fiberoptic bundle carrying the excitation light beam. The coating is thin enough to allow the light to penetrate into the transducer chamber, which contains a small amount of ligand, a dextran tagged with fluorescein isothiocyanate (FITC-dextran), which competes with glucose for Con-A binding sites. The sensor element communicates with the external solution through a dialysis membrane of selected porosity to allow free diffusion of low-mol-wt species, such as glucose, yet to prevent leakage of high mol wt ligand. When the probe is dipped into solutions with higher concentrations of glucose, incoming glucose molecules replace bound FITC-dextran, giving an increase in free FITC-dextran and, therefore, a larger fluorescence signal. This type of sensor has several advantages:

1. An equilibrium basis leaves it insensitive to artifacts that may affect enzyme activity.
2. The rate of glucose transport across the membrane is not critical to the final response.
3. It may be modified to sense several metabolites simultaneously by selecting assay ligands with sufficiently optical characteristics.

Among the main drawbacks are:

1. Possible interference of other sugars, e.g., maltose or fructose, which also displace dextran, and
2. Response time limited by metabolite transport across the membrane.

More recently, Mansouri and Shultz *(115)* have constructed an optical glucose sensor that senses via a hollow dialysis fiber connected to the fluorometer by a single optical fiber. Con-A is immobilized on the inner surface of the dialysis fiber, where the competing ligand,

FITC-dextran, is also placed. This device has been applied in vivo to determination of glucose concentration in blood, with a response time short enough to detect the usual fluctuations and an accuracy of ±0.13 mg/mL. Response is affected by hydrogen ions, which cause quenching at pH values <7, and by temperature, which affects the binding affinities of glucose and dextran with Con-A.

In order to avoid degradation of optical or hollow dialysis fibers, Meadows and Schultz *(116)* have recently proposed maintaining all sensor reagents in the same phase. Their study was performed in solution, but their findings might be used to design an improved sensor. In the absence of glucose, FITC-dextran binds to rhodamine-labeled Con-A; the two chromophores are close enough and fluorescein emission is quenched by rhodamine. In the presence of glucose, the fluorescence intensity signal increases as FITC-dextran is liberated. Rhodamine emission is not significantly affected by the process and can be used as an internal reference to correct for light-source and detector-sensitivity fluctuations.

A different approach was used by Mansouri et al. *(117)* to develop a portable glucose sensor that uses the colorimetric indicator procion green covalently bound to dextran. Their sensor consists of two optical fibers (transmission and reception) with a 1-mm offset between the input and output fibers. Con-A is immobilized in a hollow dialysis fiber as before. The calibration plot is linear in the range 0–400 mg (%) glucose, and response times for this absorbance sensor are longer than those obtained for the fluorescence device (20–30 min vs 10 min).

The probe developed by Ross and Mbanu *(118)* for glucose is based on the critical-angle principle. The light transmitted along an uncladd optical fiber can be used to measure the refractive index of the surrounding liquid, provided the internal rays incident at the fiber surface are within the critical angle. This device has been used to monitor glucose concentration with a sensivity better than 0.1%. Nevertheless, since the refractive index is greatly affected by temperature, this variable must be closely watched, the degree ofsensitivity attained depending on how carefully temperature is controlled. The main advantages are simplicity and compactness. However, the strict temperature control required, along with the low critical incident angle of the laser beam at the fiber surface, makes this sensor unattractive for practical purposes.

Ackerman and Kelley *(119)* have reported a fiberoptic probe to quantify enzymatic or colorimetric assays. The chromophore solution is formed by *p*-nitrophenyl phosphate, alkaline phosphatase, and 3*M* NaOH. The probe is a flexible rubber tube encasing both transmitting and receiving fibers. It ends in a stainless steel sleeve whose distance from the probe end can be adjusted to make possible various immersion depths *(120)*. This colorimeter is inexpensive and saves laborious product transfer in enzyme-linked immunosorbent assays (ELISA) because it allows the use of a through-the-plate reader after addition of chromophore. Careful immersion of the tip is needed in order to avoid air-bubble formation under the sensitive end, and contamination associated with the use of an immersed probe may be a severe drawback.

Coleman et al. *(121)* described a fiberoptic device suitable for certain in vivo true-absorbance measurements in body fluids. The sensor consists of an optical fiber inserted into a hypodermic needle drilled with four holes and cut at 90° below the holes. At the end of the needle, a piece of aluminum foil reflects probe radiation. Sample fluids are aspirated into the irradiation cavity located in the needle, between the fiber tip and the aluminum reflector. When employed to determine bilirubin in human blood serum, the detection limit is 0.005 absorbance units at a signal-to-noise ratio of 2. The response is linear up to 1.5*M*.

Arnold *(122)* has reported a novel type of fiberoptic chemical sensor, in which an isolated enzyme is immobilized on the surface of a bifurcated fiberoptic bundle. This biosensor directly measures the enzymatic generation of a spectrophotometrically detectable product. The feasibility of the probe has been demonstrated with a model system that employs immobilized alkaline phosphatase as the enzyme, *p*-nitrophenylphosphate as substrate, and *p*-nitrophenoxide as detected product. The sensor is composed of two sheets of nylon mesh held at the common end of the fiber bundle. The enzyme is covalently attached to the inner nylon membrane, and the outer membrane scatters the incident radiation.

In the same context, Vo-Dinh et al. *(123)* have developed a simple and portable fiberoptic instrument for fluorometric bioassays. The instrument is designed to be readily adaptable to commercially available biotesting wells in microplates, which can be used as interchangeable sensor heads. The instrument is based on a bifurcated fiberoptic wave guide attached to a sensor probe coated with the appropriate immobilized bioreagents. The device is designed to use a variety of interchangeable source/probe combinations. As little as 5 µL of sample is required to measure concentrations from 10 pg to 1 µg of immunoglobulin G (IgG). The analyte, e.g., rabbit IgG, is first bound to microwells that serve as sensor heads. Antirabbit IgG antibody conjugated to alkaline phosphatase (E-Ab) is added to the microwell. Excess E-Ab is washed off, and the species E-Ab-IgG determined by adding 4-methylumbelliferylphosphate, which is cleaved by the enzyme producing the highly fluorescent 4-methylumbelliferone (MUB). The detection limit with this fiberoptic device is three orders of magnitude lower than that obtained for ELISA assays with a normal spectrophotometer.

Tromberg et al. *(124)* have recently designed a selective fiberoptic sensor for use in competitive-binding solid-phase fluorescence immunoassay. Receptor molecules (antibody or antigen) are covalently immobilized on the distal face of a single-strand quartz optical fiber. Hence, sensor response time is not limited by mass transport across membrane or gel-entrapped reagent phases. In this device, the analyte (IgG) is covalently immobilized on the distal sensing tip of the fiber. The sensor is then exposed to both fluorescein isothiocyanate (FITC) labeled antirabbit IgG and unlabeled antirabbit IgG, which results in competition for the sensor binding sites. Signal intensity is inversely proportional to analyte (unlabeled ligand) concentration. Quenching matrix interferences, self-absorption, and other factors that alter fluorescence are almost negligible for this sensor. Alternatively, rabbit IgG can be attached to a cellulose membrane or directly to the fiber surface via silanization with (β-glycidoxypropyl)trimethoxysilane (GOPS). The

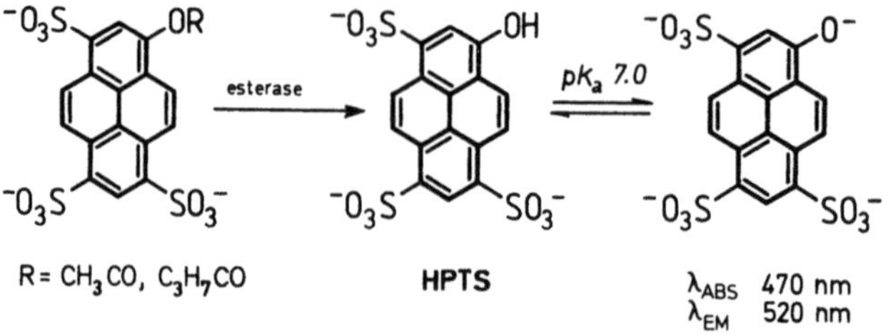

                                            Cámara, Moreno, and Orellana

Fig. 6. Enzymatic generation of HPTS used for fiberoptic kinetic enzyme assays (ref. *125*).

authors conclude that the membrane-drum sensors are slightly less stable, more cumbersome to construct, and yield lower protein loadings than GOPS-derivatized fibers; hence, they use the latter for their assays.

Wolfbeis *(125)* has described a continuous kinetic-enzyme assay by aptical fiber. In contrast to the method of Arnold *(122)*, the substrate is immobilized and the formation of the product, which also remains immobilized, is monitored. The support, a strong anion-exchange membrane with pendant substrates for cholinesterase and related carboxylesterases, is prepared by electrostatic fixation of HPTS, which is subsequently acylated. The enzymatic action results in the cleavage of the substrate to give immobilized HPTS, which, because of its low pK_a in phosphate buffer, is more than 50% dissociated (Fig. 6). The sensing scheme has been depicted in Fig. 3. The sensitivity of the assay is limited by the rates of the enzymatic and nonenzymatic reactions. The detection limit for carboxylesterase is about 20 µg/mL protein. The method potentially can be performed in vivo, because the fluorescent dye, which could be physiologically harmful, remains in the immobilized phase. The problem associated with this probe is that the substrate is irreversibly consumed, and it therefore has a limited lifetime. Calibration, more difficult than in reversible sensors, is performed by using an enzyme solution of defined activity.

Abdel-Latif and Guilbault *(126)* have applied cetyltrimethylammonium bromide to enhance the chemiluminiscence originated in the reaction of *bis*(2,4,6-trichlorophenyl)oxalate and hydrogen peroxide in the presence of perylene. They use optical fibers to follow the process and the measured intensity is linearly proportional to the concentration of hydrogen peroxide in the range 8×10^{-4} to $8 \times 10^{-9} M$, the detection limit being $2.5 \times 10^{-9} M$. The system is suitable for determining glucose concentration by using the enzyme glucose oxidase (GOD) immobilized on an immunoaffinity membrane. Enzymatically generated hydrogen peroxide is related to glucose concentration. The calibration curve for glucose is linear in the range 3×10^{-2} to $3 \times 10^{-6} M$, and the detection limit is $6 \times 10^{-7} M$ glucose. Response times are in the range 4–10 s. The sensor has been successfully applied to determine glucose in orange drink.

Trettnak et al. *(72)* have determined glucose by the use of GOD in combination with an oxygen transducer. Here, too, the enzyme is immobilized on an immunoaffinity membrane, which is held on a silicone membrane in which the oxygen-sensitive reagent decacyclene is trapped. Glucose consumption is accompanied by a decrease in oxygen concentration as a result of the enzymatic reaction and a resulting increase in the emission intensity of the oxygen-quenchable dye. A flow-through cell is used for all continuous measurements.

The same authors *(127)* have developed a new glucose biosensor based on the change in the intrinsic fluorescence of GOD as the enzyme interacts with glucose. Enzyme solutions are placed in the cavity of a methacrylate disk at the end of a fiber and covered with a dialysis membrane that is impermeable to the enzyme, but allows the substrates glucose and oxygen to diffuse into the solution. Fluorescence changes occur over a small glucose-concentration range (1.5–2 mM). Covalent immobilization of the enzyme in a gel gives similar results. To decrease response time and widen the analytical range, glucose solutions can be pumped through the flow-through cell for a fixed time and the peak height or peak integral of the resulting signal measured.

Shah et al. *(128)* have designed a glucose biosensor in which a polymer, poly(2-hydroxyethylmethacrylate), or PHEMA, containing the oxygen-quenchable dye 9,10-diphenylanthracene and the enzyme GOD, is grafted to 1-mm fused-silica fibers using 3-(methacryl) oxypropyltrimethoxysilane as a coupling agent. Different parameters, such as thickness of polymer layer, enzyme loading, and polymerization conditions, are optimized.

Vo-Dinh et al. *(129)* have designed an antibody-based fiberoptic biosensor for the determination of the carcinogenic compound benzo(a)pyrene (BaP). Antibodies to BaP produced by polyclonal techniques are covalently bound to the sensing tip of the optical fiber. BaP molecules present in the sample bind to them, and the measured fluorescence intensity is related to their concentration. Measuring times of about 12 min are needed, and the detection limit obtained is 1 fmol of BaP in a 5-µL drop, which is the sample volume required for the measurements.

In order to avoid the problems associated with preparing, handling, and regenerating sensors that have reactive phases covalently bound to optical fibers, Tromberg et al. *(130)* developed another antibody-based fiberoptic sensor, in which the reactive phase is placed in a reusable sensing tip, for the determination of a benzopyrene derivative. This can consist of either a membrane-trapped liquid antibody or an antibody immobilized on silica beads. When the highly fluorescent compound r-7,t-8,9,c-10-tetrahydroxy-7,8,9,10-tetrahydrobenzo[a]pyrene (BPT) diffuses into the reactive layer, it forms a complex with the monoclonal anti-BPT antibodies and emits fluorescence at intensities proportional to the BPT concentration in the sample. Several factors affecting sensor performance have been studied, such as sensor time-dependent concentration effect, detection limit (about $5 \times 10^{-10}M$ BPT for 15 min of incubation time for the liquid-phase sensor), precision of measurement, varying sensor response caused by changes in BPT and antibody concentration, and interference by polynuclear aromatic compounds (which seem to have little effect on chemical sensitivity). In spite of the advantages that the probe offers, it needs improvement to avoid memory effects and response-limiting trans-membrane diffusion.

Optical fibers have been widely used in the design of new biosensors for several bioanalytes of clinical and biomedical importance, such as urea, which Arnold *(131,132)* measures by immobilizing a deaminating enzyme at the tip of a fiberoptic ammonia sensor, and nicotinamide adenine dinucleotide (NAD^+), which is quantified by direct fluorometric detection. Lactate and pyruvate have also been measured via their equilibrium in the presence of NAD⁺, which is catalyzed by the enzyme lactate dehydrogenase covalently immobilized on a nylon support. If the solution pH is high, pyruvate is the favored product; however, when an appropriate amount of NAD⁺ is added (3 mM), lactate measurements (2–50 μM) can be performed. The optimum NADH concentration for the pyruvate sensor is 1 mM.

The same group *(133)* has applied fluorescence measurements of the NAD⁺/NADH system to develop the first internal enzyme fiberoptic biosensor. The internal solution components are (a) the enzyme alcohol dehydrogenase and (b) NAD⁺, separated from the sample solution by a Teflon™ membrane and held at the tip of a bifurcated fiberoptic bundle. Ethanol in the sample diffuses into the reagent phase and, with NADH formation, is oxidized to acetaldehyde. The rate of NADH formation is measured and related to the ethanol content of the sample. Since reagents are consumed during the process, this is an irreversible and unrecoverable sensor. Parameters studied include the effect of NAD⁺ concentration on the rate of NADH formation, pH, and buffer composition.

Kulp et al. *(134)* have developed penicillin and glucose sensors using fluorescein to follow the pH change produced when penicillinase or GOD catalyze the conversion to penicilloic or gluconic acid, respectively, originating a pH decrease and, consequently, a decrease in indicator fluorescence intensity. Their immobilization procedure involves physical entrapment of the enzyme on a polyacrylamide membrane directly bound to the end of a glass optical fiber.

Fuh et al. *(135)* describe another penicillin biosensor (based on a previously reported fiberoptic pH sensor *[136]*), in which the fluorescent indicator FITC is immobilized on porous glass beads and held at the tip of a single optical fiber by means of a UV-curable polymer. The enzyme penicillinase is immobilized by glutaraldehyde crosslinking in a thin membrane covering the porous glass beads. Buffer capacity affects enzyme activity as well as detection limit (approx 0.1 mM penicillin) and working range of the sensor (approx 0.1–10 mM).

In order to apply fiberoptic devices in medicine, Klainer and Harris *(137)* have developed a 3 α-hydroxyesteroid dehydrogenase based sensor to analyze bile acid concentration in the gastroduodenal tract. This probe should allow study of metabolite concentrations in the intestinal lumen in pathological and normal states, as well as characterization and diagnosis of hepatobiliary disorders.

A different kind of device has been developed by Wolfbeis and Posch *(138)* based on the enzymatic oxidation of ethanol to acetaldehyde with simultaneous production of hydrogen peroxide, followed by decomposition of the latter intermediate to yield H_2O and O_2 (catalyzed by alcohol oxidase and catalase, respectively). The reaction is followed with $Ru(bpy)_3^{2+}$, which is dynamically quenched by oxygen. The two enzymes are immobilized, together with the oxygen indicator on silica gel, and mixed with a silicone prepolymer. The mixture is spread on a glass slide, left to cure, and placed in a flow-through cell for measurement. Sensor response time is about 2 min and working range is 50–500 mmol/L ethanol. Sensor lifetime is limited to 2 wk, mainly because of poor enzyme stability.

10. Sensors for Organic Compounds

Only a few sensors have been reported for measuring small amounts of organic chemicals. Perhaps the main reason for this has been the difficulty in finding specific reactions for selective detection of the wide variety of organic contaminants present in natural samples.

Kawahara and coworkers *(139)* have described an interesting sensor for monitoring hydrocarbons in flowing water. An uncladd, fused-silica optical fiber is covered with an organophilic compound (octadecyltrichlorosilane [OTCS] or other silanes) and dipped into the water to be sampled. Contaminant hydrocarbons adsorb preferentially onto the coating, causing transmission losses of the driven radiation by changing the refractive index at the surface of the optical fiber. The variation in the output signal is related to the concentration of hydrocarbons in water. Detection limits as low as 17 mg/L and 3 mg/L for diesel oil and crude oil, respectively, have been reached by using OTCS, although threshold and dynamic range for different contaminants depend on the silane coating used and the type of hydrocarbon. These high performances require the use of a laser source and a coiled capillary tube containing the optical fiber to increase the contact surface between sample and active layer.

A sensor for detecting volatile chloroorganics in ground water has been constructed by Milanovich et al. *(140,141)*. It is based on quantitative measurement of the fluorescence of the chromophore formed upon interaction of chloroorganics with pyridine in alkaline medium (Fujiwara reaction). A two-phase system composed of pyridine and 10*M* sodium hydroxide, or a single-phase system containing pyridine and tetra(*n*-propyl)ammonium hydroxide, is retained at the end of a single-strand optical fiber by means of a membrane. Because it is an integrating device, the slope of the response curve is directly related to analyte concentration. Among its main advantages are the sensor specificity for organochlorides, fairly good precision (±3%) and suitability for *in situ* monitoring (portable system). However, it is limited by a slow response when using a membrane (response time is shorter without a membrane, but sensor lifetime is reduced to a few hours) and ability to measure only volatile chloroorganics.

Bare-ended fibers have been used by Chudyk et al. *(142)* to design a device for the remote fluorescence analysis of groundwater contaminants using a UV laser as a light source. At instrument-to-analyte distances of 25 m, detection limits for phenol (1-m fibers), *o*-cresol, and humic acid are 10, 1, and 0.1 ppb, respectively. The probe consists of a small, grooved aluminum clamp to keep excitation and emission fibers at a 22° angle when dipped into the sample cuvet.

The same principle was later applied by Kenny et al. *(143)* to develop a portable fiberoptic system. The portable device allows *in situ* detection and monitoring of aromatic compounds, such as phenol and gasoline, with detection limits in the ppb range, as against values in the parts-per-trillion range obtained with laboratory instruments. The use of portable sensors with bare-ended fibers offers multiple advantages over conventional methods, in particular rapid, *in situ* measurement.

Posch et al. *(144)* have designed an optical sensor and a fiberoptic sensor to continuously monitor the concentration of vapors of polar solvents, such as alcohols, ethers, and esters. No response to hydrocarbons and chlorinated hydrocarbons is obtained. Their method is based on reversible decolorization of the blue thermal-printer paper used in graphic plotters. In the optical-fiber approach, the sensing membrane is placed at the distal end of the fiber and changes in its diffuse reflectance are measured. Detection limits vary from 10 to 1000 ppm, depending on the solvent (10 ppm for *n*-butanol, 1000 ppm for diethyl ether).

11. Miscellaneous Sensors

Freeman and Seitz *(145)* have developed a chemiluminescence fiberoptic sensor for hydrogen peroxide determination that uses the well-known luminol reaction. The probe consists of peroxide entrapped in a polyacrylamide gel that also contains luminol. When luminol is in excess, chemiluminescence intensity is proportional to the amount of peroxide present. The reaction is dependent on pH and luminol concentration, optimum conditions being pH 9 and $10^{-3}M$ luminol. The detection limit is close to $10^{-6}M$ peroxide, and the time required to reach a steady response depends on hydrogen peroxide concentration (e.g., at $10^{-5}M$ H_2O_2, steady state is reached in about 4 s). Interestingly enough, steady-state chemiluminescence intensity increases as the rate of stirring increases. This indicates that the rate of mass transfer from the solution to the surface of the enzyme phase is the primary factor limiting the response. The authors draw attention to the feasibility of coupling chemiluminescence-based sensors of this type to a number of other enzymatic processes that yield hydrogen peroxide, e.g., glucose analysis using GOD.

12. Conclusions

The potential for future development that is shown by chemical sensors based on the use of fiberoptics, has awakened great interest in such fields as chemistry, biomedicine, and the environmental sciences. In spite of this promise, fiberoptic sensing is still a young technology that has to surmount serious problems in the use and construction of probes. It is commonly found that lack of reversibility and stability are the greatest obstacles, because they impede the application of fiberoptic sensors in continuous sample monitoring, and these problems are consequently the main focus of current research. A summary of sensors is contained in Table 1.

Table. Summary of the Analytical Features of Fiber-Optic Chemical Sensors.

	ANALYTE	CHARACTER[a]	PROPERTY	REAGENT[b]	IMMOBILIZATION	SUPPORT[c]	MEMBRANE[d]	OPTICAL FIBER	ARRANGEMENT[e]
1	pH	Int. Rev.	Absorbance	Phenol Red	Covalent	Polyacrylamide	Cellulose	Plastic	Adjacent
2	pH	Int. Rev.	Reflectance	Bromophenol blue Bromothymol blue Thymol blue Phenolphthalein Thymolphtalein	Adsorption	Styrene/dvb		Plastic	Bifurcated
3	pH	Int. Rev.	Reflectance	Bromothymol blue	Adsorption	Styrene/dvb	PTFE	Plastic	Bifurcated
4	pH	Int. Rev.	Reflectance	Cresol red	Electrostatic	Anion exchanger	PTFE	Plastic	Bifurcated
5	pH	Int. Rev.	Fluorescence	Fluoresceinamine	Covalent	Cellulose or glass		Plastic	Bifurcated
6	pH	Int. Rev.	Fluorescence	Fluoresceinamine	Covalent	Polyacrylamide/glass		Glass	Single
7	pH	Int. Rev.	Fluorescence	HPTS	Electrostatic	Anion exchanger	Anion exch.	Plastic?	Bifurcated
8	pH	Int. Rev.	Fluorescence	5-aminoeosin + phenol red	Covalent	Polyacrylamide		Glass	Single (two-ways)
9	pH	Ext. Rev.	Absorbance	(Phenol red)					Bifurcated (two-ways)
10	pH	Ext. Rev.	Fluorescence	(Fluorescein) or (HPTS)			Parafilm	Plastic	Bifurcated
11	pH	Int. Rev.	Reflectance	Phenol Red	Adsorption	Styrene/dvb	Cuprophane	Not specified	Single (two-ways)
12	pH	Int. Rev.	Reflectance	Not specified	Covalent	Cellulose		Plastic	Bifurcated
13	pH	Int. Rev.	Absorbance	Congo Red	Adsorption	Cellulosic polymer		Quartz	Bundle (f-to-f)
14	pH	Int. Rev.	Fluorescence	Fluorescein/Eosin	Covalent	Aminoethyl cellulose	Hydrogel	Plastic	Bifurcated
15	pH	Ext.	Fluorescence	HPTS and HPTS/SR 640	Entrapment	Ethylene/vinyl acetate		Glass	Single (two-ways)

	LIGHT SOURCE	MONOCHROMATION SYSTEMS[f]		DETECTOR[a]	DYNAMIC RANGE	RESPONSE TIME[b]	SAMPLES	REFERENCE
		λ_{PROBE}, nm	$\lambda_{ANALYSIS}$, nm					
1	W		560/760 (filt.)	PD	pH 7.0–7.4	40 s (63%)	Aqueous, blood	48,49
2	W-halogen	593 (monochr.)		PMT	pH 3.0–4.6	Not specified	Aqueous	54,55
					5.2–6.8			
					6.0–7.6			
					8.0–9.6			
					8.3–10.0			
					8.3–10.5			
3	W-halogen		593 (monochr.)	PMT	pH 7.0–9.0	5 min (60%)	Aqueous	56
4	W-halogen	570 (monochr.)		PMT	(H+) 0.02–0.6 M		Aqueous	58,59
					pH 6.1–7.2			
5	W-halogen	480 (if. filt.)	520 (if. filt.	PMT	pH 3–8	15–30 s	Aqueous	60
6	Ar-laser	488 (---)	530 (monochr.)	PMT	pH 4.8–8.0	9 s	Aqueous	61
7	W-halogen	405/470 (if. filt.)	520 (if. filt.)	PMT	pH 6–8	70–120 s	Aqueous	62,63
8	Ar-laser	488 or 514.5 (---)	547 (monochr.)	PC	pH 6.0–8.0	0.07 min (63%)	Aqueous	64
9	LED		565 (---)	PD	pH 7–9	<1 µs	Aqueous	65
10	LED or	490 (---) or	530 (monochr.)	PMT	titrations	Not specified	Aqueous	66
	Xe	465 (monochr.)	515 (monochr.)	PD				
11	W-halogen	Not specified (monochr.)		PMT	6<pH<9	1 to 7 min (90%)	Seawater	68
12	W-halogen		580	PMT?	pH 3–5	60 samples/h	Rainwater	69
13	Xe	400–800		PMT	>4 pH units	1.22 s	Aqueous	70
14	Xe-Flash	490 (if. filt.)	>570 (long pass)	PMT	pH 5–7	25 to 30 s	Aqueous	71
15	Xe	405/450 (monochr.)	Ratio at 515	PMT	pH 5.5–8.0	300–10 min pH⁻¹	Aqueous	73
	Ar-laser	488	Ratio 530/610	PMT	5.5–8.0	unit		

	ANALYTE	CHARACTER[a]	PROPERTY	REAGENT[b]	IMMOBILIZATION	SUPPORT[c]	MEMBRANE[d]	OPTICAL FIBER	ARRANGEMENT[e]
16	Cl^-, Br^-, I^-	Int. Rev.	Fluorescence	Acridinium or Quinolinium dye	Covalent	Glass		Not used	
17	S^{2-}	Int. Rev.	Reflectance	PFDA	Electrostatic	Cation exchanger		Plastic	Bifurcated
18	F^-	Int. Irrev.	Reflectance	Alizarin	Adsorption	Styrene/dvb		Plastic	Bifurcated
19	CN^-	Int. Irrev.	Reflectance	4-Picoline/ Barbituric acid	Adsorption	Styrene/dvb		Plastic	Bifurcated
20	Ca(II)	Int. Irrev.	Chemilumines.	Aequorin				Glass	Single
21	Al(III)	Int. Irrev.[1]	Fluorescence	Morin	Covalent	Cellulose	Cellophane	Glass	Bifurcated
22	Co(II),Ni(II) Cu(II),Zn(II)	Int. Irrev.[1]	Fluorescence	Calcein	Covalent	Cellulose		Glass	Bifurcated
23	Be(III)	Int. Irrev.[1]	Fluorescence	Morin	Covalent	Cellulose	Cellophane	Glass	Bifurcated
24	Al(III)	Ext. Rev.	Fluorescence	(Morin)				Plastic	Bifurcated
25	Al(III),Mg(II)	Int. Irrev.[1]	Fluorescence	QS	Electrostatic	Anion exchanger	Cellophane	Quartz	Bifurcated
26	Na(I)	Int. Rev.	Fluorescence	ANS + Cu(II)- polyethyleneimine + Na-ionophore II	Adsorption	Silica gel	Cellophane	Quartz	Bifurcated
27	K(I)	Int. Rev.	Reflectance	Crown ether	Adsorption	Styrene/dvb	PTFE	Plastic	Bifurcated
28	K(I)	Int. Rev.	Fluorescence	Rhodamine B	Adsorption	OVM		Plastic	Bifurcated
29	Cu(II)	Ext. Rev.	Absorbance					Quartz	Double (f-to-f)
30	Fe(III)	Int. Irrev.[1]	Reflectance	Chromeazurol S	Electrostatic	Anion exchanger	PTFE	Plastic	Bifurcated
31	NH_4^+	Int. Rev.	Reflectance	Bromothymol blue	Adsorption	Styrene-dvb	PTFE	Plastic	Bifurcated
32	O_2	Ext. Rev.	Reflectance					Glass	Bifurcated
33	O_2	Int. Rev.	Reflectance	Deoxyhemoglobine	Electrostatic	Cation exchanger	PTFE	Plastic?	Bifurcated
34	O_2	Int. Rev.	Fluorescence	Perylene dibutyrate	Adsorption	Styrene-dvb	Polypropylene	Plastic	Adjacent
35	Halothane and/or O_2	Int. Rev.	Fluorescence	Decacylene	Solution	Silicone rubber or polyisoprene/ dioctyl phthalate	None/PTFE	Quartz	Bifurcated

	LIGHT SOURCE	MONOCHROMATION SYSTEMS[f) λ_{PROBE}, nm	$\lambda_{ANALYSIS}$, nm	DETECTOR[g)	DYNAMIC RANGE	RESPONSE TIME[h)	SAMPLES	REFERENCE
16	Xe	375 (c.-off fil.)	488 (c.-off fil.)	PMT	I^-, Br^- $10^{-3}-1$ M	40 s (95%)	Aqueous	74
		350 (c.-off fil)	460 (c.-off fil.)		Cl^- $10^{-2}-1$ M		Blood	
17	W-halogen	688 (monochr.)		PMT	Fe^{3+} 0.02-0.5 ppm	1 min	Aqueous	78
18	W-halogen	660 (monochr.)		PMT	F^- 0.16-0.95 mM	12 min	Aqueous	79
19	W-halogen	540 (monochr.)		PD	CN^- 1-10 ppb	<1 min	Air	80
20				PMT	Not specified	<10 ms	In vivo	81
21	Xe	420 (if. filt.)	488 (if. filt.)	PMT	Al^{3+} $10^{-6}-10^{-4}$ M	1-2 min	Aqueous	82
22	Xe	Not specified	Not specified	PMT	Not specified	Not specified	Aqueous	47
23	Xe	420 (if. filt.)	488 (if. filt.)	PMT	Be^{3+} $10^{-6}-10^{-5}$ M	1-2 min	Aqueous	83
24	LED	420 (---)	500 (if. filt.)	PMT	Al^{3+} 1-800 ppm	Instantaneous	Aqueous	84
25	W-halogen	363 (if. filt.)	520 (if. filt.)	PMT	Not specified	1 min	Aqueous	85
26	W-halogen	380 (if. filt.)	460 (if. filt.)	PMT	Na^+ 20-200 mM	1-3 min	Aqueous/Serum	86
27	W-halogen	Not specified		PMT	K^+ $10^{-3}-10^{-1}$ M	2-7 min	Aqueous	87
28	Xe	540	620nm	PMT	K^+ $10^{-1}-10^{-5}$ M	2-7 min	Aqueous	88
29	LED	820 (---)		PMT	Cu^{2+} 50-500 mM	10 s	Electroplating baths	89
30	W-halogen	688 (monochr.)		PMT	Fe^{3+} 0.02-0.5 ppm	1 min	Aqueous	92
31	W-halogen	530		PMT	NH_4^+ $1.5.10^{-3}$ M	Not specified	Aqueous	93
32	Not spec.	640/805 (filt.)		PD	Not specified	<1 s	Blood	94a
33	W	433/405 (if. fil.)		PMT	O_2 20-100 torr	3 min	Aqueous	94b
34	D_2	480 (if. fil)	500 (dich. fil.)	PMT	O_2 0-150 torr	<1	Gaseous	95
						0.5-2 min	Aqueous	
35	Xe		510?	PMT	Halothane 0.1-4 %	10-20 s	Gaseous	98
					O_2 0-100 %			

	ANALYTE	CHARACTER[a]	PROPERTY	REAGENT[b]	IMMOBILIZATION	SUPPORT[c]	MEMBRANE[d]	OPTICAL FIBER	ARRANGEMENT[e]
36	O_2	Int. Rev.	Fluorescence	Pyrene butyric ac.	Covalent	Glass		Not used	
37	O_2	Int. Irrev.	Chemilumines.	EIA	Solution		PTFE	Not used	
38	CO_2	Int. Rev.	Fluorescence	HPTS	Electrostatic	Anion exchanger	Silic. rubber	Plastic?	Bifurcated
39	H_2O (g)	Int. Rev.	Absorbance	$CoCl_2.6H_2O$	Suspension	Gelatin		Quartz	Single (U-shaped)
40	NH_3	Int. Rev.	Absorbance	p-Nitrophenol + NH_4Cl (aq)	Solution		PTFE	Plastic	Adjacent
41	NH_3	Int. Rev.	Absorbance	Chlorophenol red/ Bromothymol blue			PTFE	Plastic	Bifurcated
42	O_2	Int. Rev.	Fluorescence	$Ru(bpy)_3^{2+}$	Adsortion	Silica gel	Silicone	Not specified	Bifurcated
	CO_2			HPTS	Covalent	Cellulose	Hydrogel		
43	O_2	Int. Rev.	Emission lifetime	$Ru(bpy)_3^{2+}$	Adsortion	Silica gel	Silicone	Glass	Bifurcated
44	O_2	Int. Rev.	Fluorescence	Pyrene/Perylene	Entrapment	Silicone		Quartz	Bifurcated
45	SO_2	Int. Rev.	Fluorescence	Pyrene/Perylene	Entrapment	Silicone		Quartz	Bifurcated
46	Hemoglobins	Ext. Rev.	Absorbance					Plastic	Double (f-to-f)
47	Glucose	Int. Rev.	Fluorescence	Concanavalin-A	Suspension?	Sepharose gel	Dialysis	Glass	Bifurcated
48	Glucose	Int. Rev.	Fluorescence	Concanavalin-A	Covalent	Cellulose	Dialysis	Glass	Single (two-ways)
49	Glucose	Int. Rev.	Absorbance	Procion green	Covalent	Cellulose	Dialysis	Glass	Adjacent
50	Glusose	Ext. Rev.	Transmission					Quartz	Single (straight)
51	p-Nitrophenyl phosphate	Ext. Irrev.	Absorbance	(Alkaline Phosphatase)				Glass	Adjacent
52	Bilirubin	Ext. Rev.	Absorbance					Quartz	Single (two-ways)

71

	LIGHT SOURCE	MONOCHROMATION SYSTEMS[f] λ_{PROBE}, nm	$\lambda_{ANALYSIS}$, nm	DETECTOR[g]	DYNAMIC RANGE	RESPONSE TIME[h]	SAMPLES	REFERENCE
36	Xe?	360 (monochr.)	380 (c.-off fil.)	PMT	O_2 0-100 %	20-50 m	Gaseous	99
37				PMT	O_2 1 ppm-100 %	20-40 s 5 h	Gaseous	100
38	W-halogen	470 (if. filt.)	520 (if. filt.)	PMT	CO_2 (1-6).10^{-3} M	3 min	Aqueous	101
39	W?	680 (monochr.)		PMT?	$H_2O(g)$ 40-80 %	< 1 min	Gaseous	102
40	W		404.7 (if. filt.)	PMT	NH_3 5 μM-1 mM	< 4 min	Aqueous	103
41	W-halogen	578/618 (monochr.)		PMT	NH_3 10-300 μM?	6-15 min	Wastewater	104
42	W-Halogen	460 (if. filt.)	630 (if. filt.)	PD	O_2 0-200 torr	40 s	Gaseous	105,106
		460 (if. filt.)	520 (if. filt.)	PD	CO_2 0-150 torr	3.8-5.2 min	Gaseous	
43	LED	460 (if. filt.)	570 (c.-off fil.)	PMT	O_2 0-150 torr	30-40 s	Gaseous	108
44	Xe	320 (monochr.)	476 (monochr.)	PMT	O_2 0-150 KPa	1-2 s	Gaseous	110
45	Xe	330 (monochr.)	470 (monochr.)	PMT	SO_2 <10 ppm (det. limit)	< 30 s	Gaseous	111
46	W-halogen		512-621 (polychr.)	Diode array	0-100 %	20-40 ms	Blood	112
47	Xe	Not specified	Not specified	PC	0.25-1.75 g L^{-1}	Not specified	Aqueous	113
48	Xe	488 (if. filt.)	>515 (c.-off fil.)	PMT	0.5-4.0 g L^{-1}	5-7 min	Aqueous Plasma, blood	114a
49	LED	650		PD	0-400 mg (%)	20-30 min (90 %)	Aqueous	114c
50	He-Ne las.	632.8 (---)		Not spec.	0-1 %	Instantaneous	Aqueous	115
51	W?		420 (if. filt.)	PD	0.05-1-10 (absorbance units)	Not specified	Aqueous	116
52	Ar laser	457.8 (---)		PMT	11-30 ppm	Not specified	Blood serum	118

	ANALYTE	CHARACTER a)	PROPERTY	REAGENT b)	IMMOBILIZATION	SUPPORT c)	MEMBRANE d)	OPTICAL FIBER	ARRANGEMENT e)
53	p-Nitrophenyl phosphate	Int. Irrev.	Absorbance	Alkaline phosphatase	Covalent	Nylon	Nylon	Glass	Bifurcated
54	Immunoglobulin G	Ext. Irrev.	Fluorescence	(Alkaline phosphatase + conjugated Ig G + MUB phosphate)			Not specified	Bifurcated	
55	Unlabeled antirabbit IgG	Int. Irrev.	Fluorescence	Rabbit Ig G + FITC antirabbit IgG	Covalent	Quartz (O.F.)	None or cellulose	Quartz	Single (two-ways)
56	Cholinesterease Lipase Acylase I	Int. Irrev.	Fluorescence	HPTS	Electrostatic	Anion exch.	Anion exch.	Glass?	Single (two-ways)
57	Glucose	Ext.	Chemilumines.	bis-(2,4,6-Trichlo-phenyl)oxalate Perylene Glucose oxidase	Covalent	Nylon (Glucose oxidase)		Not specified	Bifurcated
58	Glucose	Int. Rev.	Fluorescence	Glucose oxidase Decacyclene	Covalent Entrapment	Nylon Silicone		Plastic	Bifurcated
59	Glucose	Int. Rev.	Fluorescence	Glucose oxidase			Dialysis	Plastic	Bifurcated
60	Glucose	Int. Rev.	Fluorescence	9,10-diphenylan-tracene/ Glucose oxidase	Entrapment	PHEMA		Quartz	Single (two-ways)
61	BaP j)	Int. Irrev. 1)	Fluorescence	Anti-BaP antibodies	Covalent	Quartz (O.F.)		Quartz	Single (two-ways)
62	BPT k)	Int. Irrev.	Fluorescence	Anti-BPT antibodies	Entrapment	None or Silica beads	Dialysis	Quartz	Single (two-ways)
63	Urea	Int. Rev.	Absorbance	Nitrazine yellow/ Bromothymol blue	Not specified	Not specified	PTFE	Plastic	Not specified

	LIGHT SOURCE	λ_{PROBE}, nm	MONOCHROMATION SYSTEMS[f] $\lambda_{ANALYSIS}$, nm	DETECTOR[a]	DYNAMIC RANGE	RESPONSE TIME[h]	SAMPLES	REFERENCE
64	W-halogen	360 (band pass)	450 (c.-off fil.) PMT		Lactate 2-50 µM Piruvate 0.2 mM (det. lim.)	5-12 min	Aqueous	129
65	W-halogen	350 (band pass)	450 (c.-off fil.) PMT		Ethanol 0.9-9 mM	Not specified	Aqueous	130
66	Not specif.	Not specified	Not specified	Not specif.	0.1-100 mM	Glucose 5-12 min Penicillin 40-60 min	Aqueous	131
67	Ar-laser	488	530 (monochr.)	PD/PMT	0.1-10 mM pH 3-7	20-45 s (90 %) 20-35 s	Aqueous	132 133
68	W/Xe or laser	340 (monochr.)	480 (monochro.)	PMT	2-200 µM (Cholic acid)	Not specified	Aqueous	134
69	W-halogen	460 (monochr.)	610 (monochro.)	PMT	50-500 mmol L^{-1}	ca. 2 min	Aqueous	135
70	He-Ne las.	632.8 (---)		PD	3-1000 mg L^{-1}	Not specified	Aqueous	136
71	Ar laser	488 (---)		PMT	CHCl$_3$ 1-100 ppm	3-5 h	Aqueous	137,138
72	N$_2$/dye las. 270 Nd-YAG las. 266		Filters Total emission	PMT	Not specified	Few min	Groundwater	139
73	Nd-YAG las. 266		Total emission	PMT	ppb (det. limit)	Not specified	Groundwater	140
74	W-halogen	Not specified		PMT	n-Butanol 10 ppm Ethanol 250 ppm (det. limits)	15 s (acetone) 2 min (ethanol) (90 %)	Gaseous	141
75				PMT	H$_2$O$_2$ 10^{-6} M (det.lim.)	4 s	Aqueous	142

[a]Int.: intrinsical; ext.: extrinsical; rev.: reversible; irrev.: irreversible (for definitions see text). [b]HPTS: 8-hydroxy-1,3,6-pyrenetrisulfonic acid; SR 640: sulforhodamine 640; acridinium dye: 3-(10-methylacridinium-9-yl)propionic acid; quinolinium dye: 3-(6-methoxyquinolinium-1-yl)propanesulfonic acid; PFDA: N,N-dimethyl-p-phenylenediamine dihydrochloride; QS: quinoline-8-ol sulfonate; ANS: ammonium salt of 8-anilino-1-naphthalenesulfonic acid; Na-ionophore II: $N,N´,N´´$-triheptyl-$N,N´,N´´$-trimethyl-4,4´,4´´-propylidyne-tris(3-oxabutyramide); EIA: 1,1´,3,3´-tetraethyl- 2,2´-bi(imidazolidine); Ru(bpy)$_3$$^{2+}$: tris(2,2´-bipyridine)ruthenium(II); Ig G: immunoglobulin G; MUB: 4-methylumbelliferyl; FITC: fluorescein isothiocyanate; OCTS: octadecyltrichlorosilane; luminol: 3-amino-phthalhydrazide (reagents used extrinsically appear in parenthesis). [c]Dvb: divinylbenzene; OVM: octadecan-1-ol-valinomycyl; O.F.: optical fiber; PHEMA: poly(2-hydroxyethylmethacrylate). [d]PTFE: poly(tetrafluorethylene). [e]See Figure 2 for fiber-optic arrangements in the sensitive terminal. [f] If. filt.: interference filter; c.-off fil.: cut-off filter; dich. filt.: dichroic filter; monochr.: grating monochromator (see Figure 1). [g]PD: photodiode; PMT: photomultiplier tube; PC: photon counting system. [h]In parenthesis, response level reached. [i]Regenerable. [j]BaP:. Benzo(a)pyrene. [k]BPT: r-7,t-8,9,c-10-tetrahydroxy-7,8,9,10-tetrahydrobenzo(a)pyrene.

	ANALYTE	CHARACTER[a)	PROPERTY	REAGENT[b)	IMMOBILIZATION	SUPPORT[c)	MEMBRANE[d)	OPTICAL FIBER	ARRANGEMENT[e)
64	Lactate	Int. Rev.	Fluorescence	Lactate dehydro-genase/NAD+	Covalent	Nylon		Quartz	Bifurcated
	Piruvate			Lactate dehydro-genase/NADH					
65	Ethanol	Int. Irrev.	Fluorescence	Alcohol dehydro-genase/NAD+/buffer NH4Cl/NaCl			PTFE	Quartz	Bifurcated
66	Penicillin	Int. Rev.	Fluorescence	Fluorescein/ Penicillinase	Covalent (fluorescein)/	Glass (O.F.)	Polyacrylamide	Glass	Single?
	Glucose			Fluorescein/Glucose oxidase/Catalase	Entrapment (enzymes)				
67	Penicillin	Int. Rev.	Fluorescence	FITC/Penicillinase	Covalent	Porous glass		Quartz	Single (two-ways)
	(pH)			FITC	Covalent	Porous glass		Quartz	Single (two-ways)
68	Bile acids	Int. Irrev.	Fluorescence	3α-hydroxysteroid dehydrogenase/NAD+	Covalent	Quartz (O.F.)/ Polyethylene glycol		Quartz	Single (two-ways)
69	Ethanol	Int. Rev.	Fluorescence	Ru(bpy)3²+/ Alcohol oxidase/ Catalase	Adsorption	Silica gel	Silicone	Not specified	Bifurcated
70	Hydrocarbons	Int. Rev.	Transmission	OTCS	Covalent	Quartz (O.F.)		Quartz	Single (coiled)
71	Chloroorganics	Int. Irrev.	Fluorescence	Pyridine + KOH	Reservoir		Mylar	Glass	Single (two-ways)
72	Aromatic Hydrocarbons	Ext. Rev.	Fluorescence					Quartz	Double strand (adjacent)
73	Aromatic Hydrocarbons	Ext. Rev.	Fluorescence					Quartz	Double strand (adjacent)
74	Polar solvents vapours	Int. Rev.	Reflectance	Blue thermal printer paper				Plastic	Bifurcated
75	H2O2	Int. Irrev.	Chemilumines.	Peroxidase/(luminol)	Entrapment	Polyacrylamide	Dialysis	Glass	Single

	LIGHT SOURCE	MONOCHROMATION SYSTEMS[f) λ_{PROBE}, nm	$\lambda_{ANALYSIS}$, nm	DETECTOR[g)	DYNAMIC RANGE	RESPONSE TIME[h)	SAMPLES	REFERENCE
53	W-halogen		404.7 (if. filt.)	PMT	$2.10^{-5}-4.10^{-4}$ M	Not specified	Aqueous	119
54	Hg	360 (filter)	450 (monochr.)	PC	10 pg-1 µg	Not specified	Aqueous	120
	He-Cd las. 325 (---)							
55	Ar laser	488 (---)	520 (monochr.)	PMT	2.5 fmol µL^{-1} (det. limit)	3 min	Aqueous	121
56	Xe	465 (filter)	530 (filter)	PMT	> 20 µg mL^{-1} (det. limit)	Not specified	Aqueous	122
57				Not spec.	$3.10^{-2}-3.10^{-6}$ M	4-10 s	Orange drink	123
58	Xe	410 (band pass)	450 (c.-off fil.)	PMT	0.1-20 mM	1-6 min (90%)	Aqueous	72
59	Xe	450 (band pass)	>500 (c.off fil.)	PMT	1.5-2 mM	2-30 min	Aqueous	124
60	Xe or	360	433 (monochr.)	PMT	Not specified	Not specified	Aqueous	125
	N$_2$-dye las.							
61	He-Cd las. 325		Not specified	PMT	1 fmol (BaP) (det. limit)	ca. 12 min	Aqueous	126
62	He-Cd las. 325		400 (band pass)	PMT	5.10^{-5} M 2.10^{-5} M (det. limits)	15 min 45 min (Incubation times)	Aqueous	127
63	Not specif.	Not specified	Not specif.	Not specif.	Urea 0.4-1.2 mM	Not specified	Aqueous	128

References

1. Wolfbeis, O. S. (1986) Analytical chemistry with optical sensors. *Fresenius Z. Anal. Chem.* **325,** 387–392.
2. Vurek, G. G. (1984) In vivo optical chemical sensors. *Proc. SPIE – Int. Soc. Opt. Eng.* **494,** 2–6.
3. Peterson, J. I. and Vurek, G. G. (1984) Fiberoptic sensors for biomedical applications. *Science* **224,** 123–127.
4. Smith, A. M. (1985) Biomedical sensing using optical fibers. *Anal. Proc.* **22,** 212.
5. Hilliard, L. A. (1985) Application of single optical fibers to remote absorption measurements. *Anal. Proc.* **22,** 210,211.
6. Edmonds, T. E. and Ross, I. D. (1985) Low–cost optic chemical sensors. *Anal. Proc.* **22,** 206,207.
7. Seitz, W. R. (1984) Chemical sensors based on fiber optics. *Anal. Chem.* **56,** 16A–34A.
8. Wolfbeis, O. S. (1988) Fiber optical fluorosensors in analytical and clinical chemistry, in *Molecular Luminescence Spectroscopy: Methods and Applications* (Shulmann, S. G., ed.), Wiley, New York, pp. 129–281.
9. Seitz, W. R. (1987) Optical sensors based on immobilized reagents, in *Biosensors: Fundamentals and Applications* (Turner, A.P.F., ed.), Oxford University Press, Oxford, pp. 559–654.
10. Wolfbeis, O. S. (1985) Fluorescence optical sensors in analytical chemistry. *Trends Anal. Chem.* **7,** 184–188.
11. Narayanaswamy, R. (1985) Fibre optics for chemical sensing. *Anal. Proc.* **22,** 204–206.
12. Alder, J. F. (1986) Optical fiber chemical sensors. *Fresenius Z. Anal. Chem.* **324,** 372–375.
13. Arditty, H. J. and Jeunhomme, L. B. (1986) A fine future for optical fiber sensors. *Spectra 2000* **112,** 42–44.
14. Nylander, C. (1985) Chemical and biological sensors. *J. Phys. [E]* **18,** 736–749.
15. Busuriu, V. I., Semanov, A. S., and Udalov, N. P. (1985) Optical and fiberoptic sensors. *Kuantovaya Elektron.* **12,** 901–944.
16. Weiss, M. D. (1985) Chemically-sensitive transducers: A new breed of miniature sensors. *Intech.* **32,** 45–49.
17. Borman, S. A. (1981) Optrodes. *Anal. Chem.* **53,** 1616A–1618A.
18. Sai, Y. (1984) Application of optical fiber sensor in process. *Keiso* **27,** 10–16.
19. Shirnoishizaka, M. (1985) Optical fiber sensors. *Keiso* **28,** 37–41.
20. Zhujun, Z. (1986) Fiber optics chemical sensors. *Huaxue Tongbao* **1,** 9–16.
21. Stuart, A. D. (1986) Remote determination of chemical concentrations using optical fibers and optical sensors. *Chem. Aust.* **53,** 350–353.
22. Minami, S. (1986) Optical fiber chemical sensor. *Oyo Butsuri* **55,** 56–62.

23. Zemel, J. N., Van der Spiegel, J., Fare, T., and Young, J. C. (1986) Recent advances in chemically sensitive electronic devices, in *Fundamentals and Applications of Chemical Sensors*, (American Chemical Society, Washington DC), pp. 2–38.

24. Harmer, A. L. (1986) Optical fiber sensors. *Cesk. Cas. Fyz.* **36**, 1–34.

25. Cramp, J. H. W., Mortier, R. M., Murray, R. T., and Reid, R. F. (1983) UK Patents 2,103,786 A; 8,124,939.

26. Furukawa Electric Co. Ltd. (1985) Japanese Patents 85,115,901; 60,115,901.

27. Welti, N. A. (1986) Australian Patent 8,605,589 Al.

28. Arditty, H. J. and Jeunhomme, L. B. (eds.) (1986) *Fiber Optic Sensors*, (SPIE — The International Society for Optical Engineering, Bellingham, WA).

29. Kersten, R. T. and Kist, R. (eds.) (1984) Proceedings 2nd International Conference on Optical Fiber Sensors, in *Proc. SPIE.—Int. Soc. Opt. Eng.* vol. 514, VDE-Verlag, Berlin.

30. Turner, A. P. F. and Wilson, G. S. (1987) *Biosensors: Fundamentals and Applications*, Oxford, Oxford.

31. Jones, B. E. (1985) Optical fibre sensors and systems for industry. *J. Phys. [E]* **18**, 770–781.

32. Harmer, A. L. (1984) Optical fiber sensors for industrial applications. *Proc. SPIE—Int. Soc. Opt. Eng.* **514**, 17–22.

33. Philip, G. S. (1985) Fiber optic sensors for process control — a review. *Symp. Pap.-Inst. Chem. Eng.* **2**, 1.1–1.13.

34. Kist, R. (1986) Fiber optic sensors for industrial applications. *NTG—Fachber* **93**, 206–214.

35. Maggioli, V. J. (1985) Fiber optic sensors in petrochemical plants. *IEEE Trans. Ind. Appl.* **6**, 1491–1495.

36. Giallorenzi, T. G., Bucaro, J. A., Dandri, A., Sigel, G. H., Cole, J. H., Rashleigh, C., and Priest, R. G. (1982) Optical fiber sensor technology. *IEEE J. Quantum Electron.* **18**, 626–665.

37. Jordan, G. R. (1985) Sensor technologies of the future. *J. Phys. [E]* **18**, 729–736.

38. Lubbers, D. W. and Opitz, N. (1983) Optical fluorescence sensors for continuous measurement of chemical concentrations in biological systems. *Sensors and Actuators.* **4**, 641–654.

39. Scheggi, A. M. (1984) Optical fiber sensors in medicine. *Proc. SPIE–Int. Soc. Opt. Eng.* **514**, 93–104.

40. Tromberg, B. J., Sepaniak, M., Vo-Dinh, T., and Griffin, G. D. (1987) Fiber optic chemical sensors for competitive binding fluoroimmunoassay. *Anal. Chem.* **59**, 1226–1230.

41. Harada, S. (1986) Japanese Patents 8,640,543; 6,140,543.

42. Gaebler, W. and Sehring, G. (1984) Fiberoptic sensors for environmental surveillance. *Radiat. Risk. Prot., 6th Int. Congr.* **3**, 1235–1238.

43. Hale, P. G. and Jones, R. E. (1986) Optical fiber for offshore monitoring and control. *Proc. SPIE — Int. Soc. Opt. Eng.* **630**, 239–244.
44. Hatano, H. and Ishida, M. (1982) Continuous measurement of local concentration by an optical fiber probe. *Kagaku Kogaku Roubunshu* **8**, 219–224.
45. Klafter, J. and Drake, J. M. (eds.) (1989) *Molecular Dynamics in Restricted Geometries* (Wiley, New York), pp. 1–22.
46. Kadin, H. (1984) Second derivative spectrophotometric determination of the dissociation constant of covalently-bound phenol red–polyacrylamide. *Anal. Lett.* **17**, 1245–1257.
47. Wyatt, W. A., Poirier, G. E., Bright, F. V., and Hieftje, G. M. (1987) Fluorescence spectra and lifetimes of several fluorophores imobilised on nonionic resins for use in fiberoptic sensors. *Anal. Chem.* **59**, 572–576.
48. Saari L. A. and Seitz, W. R. (1984) Immobilized calcein for metal ion preconcentration. *Anal. Chem.* **56**, 810–813.
49. Peterson, J. I., Goldstein, S. R., Fitzgerald, R. V., and Buckhold, D. K. (1980) Fiberoptic pH probe for physiological use. *Anal. Chem.* **52**, 864–869.
50. Abraham, E., Markle, D. R., Fink, S., Ehrlich, H., Tsang, M., Smith, M., and Meyer, A. (1985) Continuous measurement of intravascular pH with a fiber optic sensor. *Anaesth. Analg.* **64**, 731–736.
51. Goldstein, S. R. and Peterson, J. I. (1980) US Patent 4,200,110.
52. Peterson, J. I. (1978) US Patent 855,384.
53. Peterson, J. I. (1978) US Patent 855,397.
54. Narayanaswamy, R. and Sevilla, F. III. (1986) Reflectometric study of the acid-base equilibria of indicators immobilized on a styrene/divinylbenzene copolymer. *Anal. Chim. Acta* **189**, 365–369.
55. Kirkbright, G. F., Narayanaswamy, R., and Welti, N. A. (1984) Studies with immobilized chemical reagents using a flow-cell for the development of chemically sensitive fiberoptic devices. *Analyst* **109**,12–19.
56. Welti N. A. and Kirkbright, G. F. (1984) European Patent EP 126,600.
57. Kirkbright, G. F., Narayanaswamy, R., and Welti, N. A. (1984) Fiberoptic pH probe based on the use of an immobilized colorimetric indicator. *Analyst* **109**, 1025–1029.
58. Guthrie, A. J., Narayanaswamy, R., and Welti, N. A. (1988) Solid-State instrumentation for use with optical-fiber chemical sensors. *Talanta* **35**, 157–159.
59. Moreno, M. C., Martinez, A., Millan, P., and Cámara, C. (1986) Study of a pH sensitive optical fiber sensor based on the use of cresol red. *J. Mol. Struct.* **143**, 553–557.
60. Saari, L. A. and Seitz, W. R. (1982) pH sensor based on immobilized fluoresceinamine. *Anal. Chem.* **54**, 821–823.
61. Munkholm, C., Walt, D. R., Milanovich, F. P., and Klainer, S. M. (1986) Polymer modification of fiberoptic chemical sensors as a method of enhancing fluorescence signal for pH measurement. *Anal. Chem.* **58**, 1427–1430.

62. Zhujun, Z. and Seitz, W. R. (1984) A fluorescence sensor for quantifying pH in the range from 6.5 to 8.5. *Anal. Chim. Acta* **160,** 47–55.

63. Seitz, W. R. and Zhujun, Z.(1984) US Patent 108,531.

64. Jordan, D. M., Walt, D. R., and Milanovich, F. P. (1987) Physiological pH fiberoptic chemical sensor based on energy transfer. *Anal. Chem.* **59,** 437–440.

65. Benaim, N., Grattan, K. T. V., and Palmer, A. W. (1986) Simple fibreoptic pH sensor for use in liquid titrations. *Analyst* **111,** 1095–1097.

66. Wolfbeis, O. S., Schaffar, B. P. H., and Kaschnitz, E. (1986) Optical fibre titrations. Part 3. Construction and performance of a fluorimetric acid-base titrator with a blue LED as a light source. *Analyst* **111,** 1331–1334.

67. Janata, J. (1987) Do optical sensors really measure pH? *Anal. Chem.* **59,** 1351–1356.

68. Monici, M., Boniforti, R., Buzzigoli, G., de Rossi, D., and Nannini, A. (1987) Fibre optic pH sensor for seawater monitoring. *Proc. SPIE—Int. Soc. Opt. Eng.* **798,** 294–300.

69. Woods, B. A., Ruzicka, J., Christian, G. D., Rose, N. J., and Charlson, R. J. (1988) Measurements of rainwater pH by optosensing flow injection analysis. *Analyst* **113,** 301–306.

70. Jones, T. P. and Porter, M. D. (1988) Optical pH-sensor based on the chemical modification of a porous polymer film. *Anal. Chem.* **60,** 404–406.

71. Posch, H. E., Leiner, M. J. P., and Wolfbeis, O. S. (1989) Towards a gastric pH-sensor: An optrode for the 0–7 range. *Fresenius Z. Anal. Chem.* **334,** 162–165.

72. Trettnak, W., Leiner, M. J. P., and Wolfbeis, O. S. (1988) Fiberoptic glucose biosensor with an oxygen optrode as the transducer. *Analyst* **113,** 1519–1523.

73. Luo, S. and Walt, D. R. (1989) Fiberoptic sensors based on reagent delivery with controlled-release polymers. *Anal. Chem.* **61,** 174–177.

74. Urbano, E., Offenbacher, M., and Wolfbeis, O. (1984) Optical sensor for continuous determination of halides. *Anal. Chem.* **56,** 427–430.

75. Wolfbeis, O. S. and Urbano, E. (1983) Fluorescence quenching method for determination of two or three components in solution. *Anal. Chem.* **55,** 1904–1907.

76. Wolfbeis, O. S. and Urbano, E. (1983) Fluorometric analysis. 3. Fluorometric, heavy- metal-free method for the determination of chlorine, bromine and iodine in organic materials. *Fresenius Z. Anal. Chem.* **314,** 577–581.

77. Wolfbeis, O. S. and Hochmuth, P. (1984) A new method for the end-point determination in argentometry using halide-sensitive fluorescent indicators and fiber optical light guides. *Mikrochim. Acta III,* 129–138.

78. Martinez, A., Moreno, M. C., and Cámara, C. (1986) Sulfide determination by *N,N*-dimethyl-*p*-phenylenediamine immobilization in cationic exchange resin using an optical fiber system. *Anal. Chem.* **58,** 1877–1881.

79. Narayanaswamy, R., Russel, D. A., and Sevilla, F. III (1988) Optical-fibre sensing of fluoride ions in a flow stream. *Talanta* **35,** 83–88.

80. Bentley, A. E. and Alder, J. F. (1989) Optical fiber sensor for the detection of hydrogen cyanide in air. Part 1. Reagent characterization and impregnated bead detector performance. *Anal. Chim. Acta* **222,** 63–73.

81. Labeyrie, E. and Koechlin, Y. (1979) Photoelectrodes with a very short time-constant for recording intracerebrally Ca^{++} transients at a cellular level. *J. Neurosci. Methods* **1,** 35–39.

82. Saari, L. A. and Seitz, W. R. (1983) Immobilized morin as fluorescence sensor for determination of aluminum(III). *Anal. Chem.* **55,** 667–670.

83. Saari, L. A. and Seitz, W. R. (1984) Optical sensor for beryllium based on immobilized morin fluorescence. *Analyst* **109,** 655–657.

84. Wolfbeis, O. S. and Schaffar, B. P. H. (1986) Fiberoptic titrations. IV. Direct complexometric titration of Al(III) with DCTA. *Talanta* **33,** 867–870.

85. Zhujun, Z. and Seitz, W. R. (1985) A fluorescent sensor for aluminum(III), magnesium(II), zinc(II) and cadmium(II) based on electrostatically immobilised quinolin-8-ol-sulfonate. *Anal. Chim. Acta* **171,** 251–258.

86. Zhujun, Z., Mullin, J. L., and Seitz, W. R. (1986) Optical sensor for sodium based on ion-pair extraction and fluorescence. *Anal.Chim. Acta* **184,** 251–258.

87. Alder, J. F., Ashworth, D. C., Narayanawasmy, R., Moss, R. E., and Sutherland, I. O. (1987) An optical potassium ion sensor. *Analyst* **112,** 1191–1192.

88. Schaffar, P. H., Wolfbeis, O. S., and Leitner, A. (1988) Effect of Langmuir-Blodgett layer composition on the response of ion-selective optrodes for potassium, based on the fluorimetric measurement of membrane potential. *Analyst* **113,** 693–697.

89. Freeman, J. E., Childers, A. G., Steele, A. W., and Hieftje, G. H. (1985) A fiber optic absorption cell for remote determination of copper in industrial electroplating bath. *Anal. Chim. Acta* **177,** 121–128.

90. Strobel, H. A. (1973) *Chemical Instrumentation: A Systematic Approach,* 2nd Ed., Addison-Wesley, Reading, MA, pp. 432–436.

91. Hirschfeld, T., Deaton, T., Milanovich, F., and Klainer, S. (1983) Feasibility of using fiberoptics for monitoring groundwater contaminants. *Opt. Eng.* **22,** 527–531.

92. Cámara, C. Moreno, M. C., and Faraldos, M. (1988) Reversible fiberoptic sensor to determine iron. *Analysis* **16,** 87–91.

93. Çaglar, P. and Narayanaswamy, R. (1987) Ammonia sensitive fiberoptic probe utilizing an immobilized spectrophotometric indicator. *Analyst* **112,** 1285–1288.

94. Kapany, N. S. and Silbertrust, N. (1964) Fiberoptics spectrophotometer for in vivo oximetry. *Nature* **204,** 138–142.

95. Zhujun, Z. and Seitz, W. R. (1986) Optical sensor for oxygen based on immobilised hemoglobin. *Anal. Chem.* **58,** 220–222.

96. Peterson, J. I., Fitzgerald, R. V. and Buckhold, D. K. (1984) Fiberoptic probe for in vivo measurement of oxygen partial pressure. *Anal. Chem.* **56,** 62–67.

97. Peterson, J. I. and Fitzgerald, R. V. (1983) US Patent 396,055.

98. Fitzgerald, R. V. (1982) US Patent 363,425.
99. Wolfbeis, O. S., Posch, H. E., and Kroneis, H. W. (1985) Fiber optical fluorosensor for determination of halothane and/or oxygen. *Anal. Chem.* **57,** 2556–2561.
100. Wolfbeis, O. S., Offenbacher, H., Kroneis, H., and Marsoner, H. (1984) A fast responding fluorescence sensor for oxygen. *Mikrochim. Acta I.* 153–158.
101. Freeman, T. M. and Seitz, W. R. (1981) Oxygen probe based on tetrakis-(alkylamino)ethylene chemiluminescence. *Anal. Chem.* **53,** 98–102.
102. Zhujun, Z. and Seitz, W. R. (1984) A carbon dioxide sensor based on fluorescence. *Anal. Chim. Acta* **160,** 305–309.
103. Russell, A. P. and Fletcher, K. S. (1985) Optical sensor for the determination of moisture. *Anal. Chim. Acta* **170,** 209–216.
104. Arnold, M. A. and Ostler, T. J. (1986) Fiberoptic ammonia gas sensing probe. *Anal. Chem.* **58,** 1137–1140.
105. Rhines, T. D. and Arnold, M. A. (1988) Simplex optimization of a fiberoptic ammonia sensor based on multiple indicator. *Anal. Chem.* **60,** 76–81.
106. Wolfbeis, O. S., Leiner, M. J. P., and Posch, H. E. (1986) A new sensing material for optical oxygen measurement, with the indicator embedded in an aqueous phase. *Mikrochim. Acta III*, 359–366.
107. Wolfbeis, O. S., Weis, L. J., Leiner, M. J. P., and Ziegler, W. E. (1988) Fiberoptic fluorosensor for oxygen and carbon dioxide. *Anal. Chem.* **60,** 2028–2030.
108. Bacon, J. R. and Demas, J. N. (1987) Determining of oxygen concentrations by luminescence quenching of a polymer-immobilized transition metal complex. *Anal. Chem.* **59,** 2780–2785.
109. Lippitsch, M. E., Pusterhofer, J., Leiner, M. J. P., and Wolfbeis, O. S. (1988) Fiberoptic oxygen sensor with the fluorescence decay time as the information carrier. *Anal. Chim. Acta* **206,** 1–6.
110. Lee, E., Werner, T. C., and Seitz, W. R. (1987) Luminescence ratio indicators for oxygen. *Anal. Chem.* **59,** 279–283.
111. Sharma, A. and Wolfbeis, O. S. (1988) Fiberoptic oxygen sensor based on fluorescence quenching and energy transfer. *Appl. Spectrosc.* **42,** 1009–1011.
112. Sharma, A. and Wolfbeis, O. S. (1989) Fiberoptic fluorosensor for sulfur dioxide based on energy transfer and exciplex quenching. *Proc. SPIE—Int. Soc. Opt. Eng.* **990,** 116–120.
113. Milano, M. J. and Kim, K.-Y. (1977) Diode array spectrometer for the simultaneous determination of hemoglobins in whole blood. *Anal. Chem.* **49,** 555–559.
114. Schultz, J. S. and Sims, G. (1979) Affinity sensors for individual metabolites. *Biotechnol. Bioeng. Symp.* **9,** 65–71.
115. Mansouri, S. and Schultz, J.S. (1984) A miniature optical glucose sensor based on affinity binding. *Biotechnology* **2,** 885–890.

116. Meadows, R. and Schultz, J. S. (1988) Fiberoptic biosensors based on fluorescence energy transfer. *Talanta* **35,**145–150.
117. Mansouri, S., Jones, C., Martin, R., and Heil, R. (1988) Absorbance-based affinity glucose sensor. *Proc. SPIE—Int. Soc. Opt. Eng.* **906,** 57–59.
118. Ross, I. N. and Mbanu, A. (1985) Optical monitoring of glucose concentration. *Optics Laser Technol.* **17,** 31–35.
119. Ackerman, S. B. and Kelley, E. A. (1983) Probe colorimeter quantitating enzyme-linked immunosorbent assays and other colorimetric assays performed with microplate. *J. Clin. Microbiol.* **17,** 410–413.
120. Munsinger, R. A. (1981) Fiberoptic colorimetry. *Electro-Opt. Syst. Des.* **13,** 1–3.
121. Coleman, J. T., Eastham, J. F., and Sepaniak, M. J. (1984) Fiberoptic based sensor for bioanalytical absorbance measurements. *Anal. Chem.* **56,** 2246–2249.
122. Arnold, M. A.(1985) Enzyme-based fiber optic sensor. *Anal. Chem.* **57,** 565,566.
123. Vo-Dinh, T., Griffin, G. D., and Ambrose, K. R. (1986) A portable fiberoptic monitor for fluorimetric bioassays. *Appl. Spectrosc.* **40,** 696–700.
124. Tromberg, B. J., Sepaniak, M. J., Vo-Dinh, T., and Griffin, G. D. (1987) Fiberoptic probe for competitive binding fluoroimmunoassay. *Anal. Chem.* **59,** 1226–1230.
125. Wolfbeis, O. S. (1986) Fiberoptic probe for kinetic determination of enzyme activities. *Anal. Chem.* **58,** 2874–2876.
126. Abdel-Latif, M. S. and Guilbault, G. G. (1988) Fiberoptic sensor for the determination of glucose using micellar enhanced chemiluminescence of the peroxyoxalate reaction. *Anal. Chem.* **60,** 2271–2674.
127. Trettnak, W. and Wolfbeis, O. S. (1989) Fully reversible fibre-optic glucose biosensor based on the intrinsic fluorescence of glucose oxidase. *Anal. Chim. Acta* **221,** 195–203.
128. Shah, R., Call, S., and Gold, M. (1988) Grafted hydrophilic polymer as optical sensor substrate. *Proc. SPIE—Int. Soc. Opt. Eng.* **906,** 65–73.
129. Vo-Dinh, T., Tromberg, B. J., Griffin, G. D., Ambrose, K. R., Sepaniak, M. J., and Gardenhire, E. M. (1987) Antibody based fiberoptics biosensor for the carcinogen benzo(a)pyrene. *Appl. Spectrosc.* **41,** 735–738.
130. Tromberg, B. J., Sepaniak, M. J., Alaire, J. P., Vo-Dinh, T., and Santanella, R. M. (1988) Development of antibody-based fiberoptic sensors for detection of a benzo[a]pyrene metabolite. *Anal. Chem.* **60,** 1901–1908.
131. Arnold, M. A. (1988) Fiberoptic biosensing probes for biomedically important compounds. *Proc. SPIE—Int. Soc. Opt. Eng.* **906,** 128–133.
132. Wangsa, J. and Arnold, M. A. (1988) Fiberoptic biosensors based on the fluorometric detection of reduced nicotinamide adenine dinucleotide. *Anal. Chem.* **60,**1080–1082.

133. Walters, B. S., Nielsen, T. J., and Arnold, M. A. (1988) Fiberoptic biosensor for ethanol based on an internal enzyme concept. *Talanta* **35,** 151–155.

134. Kulp, T. J., Camins, I., and Angel, S. M. (1988) Enzyme–based fiberoptic sensors. *Proc. SPIE.—Int. Soc. Opt. Eng.* **906,** 134–138.

135. Fuh, M. S., Burgess, L. W., and Christian, G. D. (1988) Single fiberoptic fluorescence enzyme-based sensor. *Anal. Chem.* **60,** 433–435.

136. Fuh, M. S., Burgess, L. W., Hirsfeld, T., and Christian, G. D. (1987) Single fiberoptic fluorescence pH probe. *Analyst* **112,** 1159–1163.

137. Klainer, S. M. and Harris, J. M. (1988) The use of fiberoptic chemical sensors (FOCS) in medical applications: enzyme-based systems. *Proc. SPIE—Int. Soc. Opt. Eng.* **960,** 139–147.

138. Wolfbeis, O. S. and Posch, H. E. (1988) A fibre–optic ethanol biosensor. *Fresenius Z. Anal. Chem.* **332,** 255–257.

139. Kawahara, F. K., Fiutem, R. A., Silvus, H. S., Newman, F. M., and Frazar, J.H. (1983) Development of a novel method for monitoring oils in water. *Anal. Chim. Acta* **151,** 315–327.

140. Milanovich, F. P. (1986) Detecting chloroorganics in groundwater. *Environ. Sci. Technol.* **20,** 441,442.

141. Milanovich, F. P., Garvis, D. G., Angel, S. M., Klainer, S. M. and Eccles, L. (1986) Remote detection of organochlorides with a fiberoptic based sensor. *Anal. Instrum.* **15,**137–147.

142. Chudyk, W. A., Carrabba, M. M., and Kenny, J. E. (1985) Remote detection of groundwater contaminants using far-ultraviolet laser-induced fluorescence. *Anal. Chem.* **57,** 1237–1242.

143. Kenny, J. E., Jaruis, G. B., Chudyk, W. A., and Pohlig, K. (1989) Groundwater monitoring using remote laser-induced fluorescence, in *Luminescence Applications in Biological, Chemical, Environmental, and Hydrological Sciences* (Goldberd, M. G., ed.), ACS Symp. Series #383, American Chemical Society,Washington DC, pp. 233–239.

144. Posch, H. E., Wolfbeis, O. S., and Pusterhofer, J. (1988) Optical and fiberoptic sensors for vapor of polar solvents. *Talanta* **35,** 89–94.

145. Freeman, T. M. and Seitz, W. R. (1978) Chemiluminescence fiberoptic sensor for hydrogen peroxide based on the luminol reaction. *Anal. Chem.* **50,** 1242–1246.

Fluorescent Labels

Richard P. Haugland

1. Introduction

It is fortunate that from most biomolecules there is relatively low intrinsic background fluorescence in the visible spectrum, since the low background emission permits the selective detection of extrinsic fluorescent probes to trace and quantitate biomolecules and to elucidate many features of biological structure and function.

Fluorescence is the property of a relatively few molecules, mostly aromatic. A number of fluorescent dyes have been used to probe biomolecules or as labels for biosensor assays. This chapter reviews the properties and uses of those dyes whose principal application has been to form covalent conjugates of biomolecules, including proteins, nucleic acids, and drug analogs. This report focuses on the properties of fluorescent dyes that limit detection, permit multiple color applications, and affect the conjugation of dyes to biomolecules.

2. Fluorescence Excitation and Detection

Fluorescence is unique in being optically detectable at concentrations as low as single molecules in extremely favorable circumstances *(1,2)*. Several factors, only some of which are dye-dependent, determine the detectability of fluorescence. In obtaining fluorescence, the first process is illumination and fluorophore excitation. Commonly used excitation sources for both fluorescence spectrometers and microscopes include tungsten, mercury arc, and xenon lamps. Each of

Biosensors with Fiberoptics Eds.: Wise and Wingard ©1991 The Humana Press Inc.

these has both its characteristic intensity and its spectral output, which is not uniform with wavelength. Illumination with laser sources is commonly used for the high-intensity or focused illumination that is required for the most sensitive detection of fluorescence. A major factor in the ultrasensitive measurement of fluorescence, especially in single-molecule detection, is the ability to focus on an extremely small volume of sample, usually through air, as in the case of flow cytometers and most laser scanning microscopes. Lasers are also used for excitation through fiberoptics.

A common feature of laser illumination is the presence of characteristic lines of excitation at discrete wavelengths. By far the most common laser excitation source suitable for use with several available fluorophores is the argon laser. This laser has particularly intense output at 488 and 514 nm, but also, in some argon lasers of higher power, a relatively small UV output between 350 and 360 nm. In fiberoptic applications of fluorescence, there is the additional potential limitation of the reduced transmittance of optic fibers in the UV range. The argon laser is the laser of choice to pump dye lasers that have broad band output between 550 and 600 nm. Several other continuous-wave lasers have been used to excite fluorescent dyes in the visible wavelengths. The most common are the helium–cadmium (HeCd) laser with maximum output at 325 and 441 nm, and a variety of helium–neon (HeNe) lasers, with output usually near 633 nm. Newer versions of the HeNe laser that have output at 543 or 594 nm have been developed. HeNe lasers are favored for applications that do not require the highest intensity. Because of their relatively low cost, light-emitting diodes are available mostly in infrared wavelengths; however, there are few reactive fluorophores for preparation of bioconjugates that can take advantage of these excitation sources. However, light-emitting diodes that have increased output in the visible part of the spectrum are being developed.

Use of laser excitation introduces major limitations on the selection of available fluorophores, the principal of which is that the discrete excitation wavelengths available must be absorbed by the dye. The ability of dyes to absorb light is determined by the extinction coefficient at the given excitation wavelength. This factor is too frequently ignored in the selection and comparison of dyes. If the dye does not effectively absorb the light at the excitation wavelength, then fluorescence will not result, no matter how high the quantum yield. The strength of the absorbance is measured by the molar absorptivity (extinction coefficient) determined by Beer's law. A useful comparative standard for fluorescent molecules is fluorescein, whose extinction coefficient at its peak absorption wavelength of 492 nm is typically about 80,000 cm^{-1} M^{-1} in aqueous solution at pH 8.0. This is a very high probability of absorbance for a dye. Relatively few dyes have higher absorptivity, with the maximum value for this number being about 400,000 cm^{-1} M^{-1} for a single dye. The phycobiliproteins discussed below have molar extinction coefficients up to 2,400,000 cm^{-1} M^{-1}, however, since they are composed of 34 chromophores, their extinction *per chromophore* still falls in the range of 75,000 cm^{-1}M^{-1}. Many dyes have much lower absorptivities. A more common, but significant, feature is that the absorption peak of the dye does not coincide with the lines available for optimal excitation by the laser or other excitation source. Exciting a dye off the peak of absorbance has the same effect as using a dye that is more weakly absorbent. In most circumstances, the fluorescence that is ultimately collected is linearly related to the absorptivity of the dye. This absorbance can be increased only by use of more dye (with significant limitations described below) or by use of more absorbent dyes with absorbances that better match the laser output.

The ability to fluoresce is the property of relatively few dyes. Even if a molecule has high absorbance of the excitation light, the absorption will not result in fluorescence unless the dye emits the light in a single transition from the excited singlet state back to the ground state. The fluorescence obtained, as measured by the quantum yield, is the ratio of the photons absorbed to the photons emitted (almost always <1, but more commonly <0.1). Unlike absorbance measurements, by which the dye properties can be related to the spectral wavelengths and extinction coefficient, fluorescence measurements are strongly dependent on the instrumentation, and quantum-yield determinations are usually made by comparison of the observed intensity with some quantum-yield standard.

Fluorescence is an intrinsic molecular property, the intensity of which typically shows much more variation with the environment than does absorbance. Sensitivity of fluorescence to the environment is not an unexpected property of dyes. Even though it typically occurs in 1–10 ns, fluorescence is relatively slow compared to absorbance, which occurs in subpicoseconds. Most molecules that absorb light lose their excess energy as heat. This may occur through intramolecular rotation, isomerization or electronic relocalization, or through collisions with the solvent; all of these serve as alternative excited-state deactivation routes that may kinetically dominate photon emission. Since the overall rotational and translational properties of the molecule and the number of collisional encounters with the solvent are dependent on both the viscosity and the rigidity of the environment, it is common for fluorescent dyes to become more fluorescent in viscous solvents or polymers, or in crystals (if other quenching interactions in crystals can be avoided). As expected, increasing the temperature usually decreases the fluorescence by increasing the rates of alternative nonradiative deactivation processes. The variation in dye fluorescence with environment and the effects of quenching processes result in difficulty in the quantitative correlation between fluorescence intensity and dye concentration. This environmental sensitivity, however, is an extremely useful attribute in designing fluorescent probes that can be perturbed by the properties of the environment sensed by the fluorophore, such as ionic composition, analytes, and physical properties of the medium.

The detectability of dyes by absorbance measurements is typically limited to dye concentrations above $10^{-6}M$. In optimal circumstances, however, fluorescent dyes can be detected at concentrations as low as $10^{-16}M$, although concentrations of $10^{-6}M$–$10^{-10}M$ are more common. In absorbance measurements, the optical signal that is related to dye concentration is determined by the diminution of the excitation intensity of the lamp. Since the detection is made in an absorption spectrometer by a photomultiplier that is mounted in line with the strong excitation source, a large signal is subtracted from a very slightly larger signal to obtain the limiting sensitivity. The fluorescence emission is measured by a detector mounted at an angle to the exciting light. Photomultipliers are available that are capable of routinely detecting the emission of even a single photon.

Another important property of fluorescence is that it occurs at a wavelength that is almost always longer than the absorbance (or excitation) wavelength. The loss of energy in the excited state through vibrational energy transfer to the environment before emission is termed the Stokes shift. The magnitude of the actual Stokes shift for a dye is an intrinsic property of the dye. For strongly absorbing fluorophores, such as fluorescein, the Stokes shift is typically 15–30 nm. For most dyes used to prepare biomolecular conjugates, this property is relatively independent of the environment. However, some dyes, such as dansyl and prodan *(3)*, undergo large changes in their Stokes shift related to their environment. Shift of the emission away from the excitation peak is extremely important in reducing the background signal, since detection of low levels of fluorescence at the same wavelength as the excitation is difficult because of scattering of excitation light into the photomultiplier. Light-scattering is particularly apparent in larger biomolecules, cells, and tissues. It is possible to screen out almost all of the exciting light by using monochromators in spectrofluorimeters and interference filters in microscopes and related instrumentation. Reduction of background signal from scattering is facilitated by the use of dyes that have reasonable Stokes shifts (>30 nm). Alternatively, the effect of scattering can be reduced by exciting the dye on the high-energy side of the absorption transition, rather than at the absorption maximum, although this results in less absorbance by the dye. The Stokes shift enhances fiberoptic applications of fluorescent dyes as sensors. Since the light emitted by a fluorophore in the illumination beam occurs at a longer wavelength than the excitation light, fluorescence can be detected through the same fiber following appropriate filtration of any reflected light.

The factor that ultimately limits the sensitivity of detection of fluorescence in most systems is the Raman scattering that occurs at wavelengths longer than the excitation wavelength. In water, excitation at 488 nm results in a broad Raman peak with a maximum at 584 nm.

A factor that is becoming increasingly recognized in the most sensitive detection of fluorescence is the rate of photobleaching of the dye. Photobleaching is a physical property of dyes that, fortunately, occurs with slower kinetics than fluorescence. In this process, the fluorophore is destroyed by light and new chemical products are formed. In most cases the products are nonfluorescent. The total number of photons that can ultimately be detected from a dye depends on the number of times the dye can pass through the excited state and have its emission collected. This, in turn, depends on several factors, the most important of which is the rate of photobleaching through bond breaking. Like the quantum yield of fluorescence, the photobleaching rate is a rather unpredictable property, intrinsic to the dye, that can vary by orders of magnitude from dye to dye. Only a few studies have been conducted on the quantitative photostability of dyes and their photodecomposition products *(2,4)*. Inorganic phosphors, such as those used for television screens, do not photobleach, but all organic dyes do.

3. Chemically Reactive Groups

Many applications require that the fluorescent dye be chemically bonded to a protein, nucleic acid, drug, or other biomolecule which in turn interacts selectively with a receptor. The fluorescent dye provides a means for detection of this interaction. Immunofluorescence is a common application for this type of reactive dye. However the expanding development of fluorescent probes for enzymes, DNA hybridization and fluorescent analogs of peptides, lipids, hormones, growth factors, drugs, and other low-mol wt molecules promises to expand the applications of these chemically reactive dyes.

Chemically-reactive fluorescent dyes used for preparing conjugates and studying protein structure have been reviewed elsewhere *(5,6)*. Multiple references on applications of many of these probes are included in the *Handbook of Fluorescent Probes and Research Chemicals (7)*. In general, the fluorophores are modified so as to introduce a chemical functional group that is designed to react selectively with a site in a biomolecule. The structural lability and solubility of most large biomolecules only in water precludes the use of many chemical reactions that would otherwise be available. Problems that result from the use of low-mol-wt bioligands, such as some peptides and drugs, often can be addressed in organic solvents. A functional group with relatively high energy is typically incorporated in the reactive dye that is to be coupled to one of the reactive groups in the biomolecule. Fluorescent dyes that contain many types of functional groups have been developed for modification of biomolecules. Most of these are listed, with their usual sites of reaction, in Table 1. Reactions that are not possible in water may be possible in organic solvents.

Proteins provide the largest variety of reactive sites; however, probably >90% of the conjugations of proteins with reactive fluorescent dyes occur at either the ε-amino group of lysine or the thiol of cysteine *(8)*. These residues are the most nucleophilic sites in proteins, and most of the reactive functional groups that have been utilized for protein modification, such as isothiocyanates and maleimides, are highly electrophilic. Indeed, it is difficult to modify any other sites in proteins in the presence of these nucleophilic residues, except by using special reagents or in certain situations in which the protein creates special environmental factors that enhance the nucleophilicity of specific residues, such as serine, tyrosine, and histidine.

Amines are common targets for modification in peptides, proteins, ligands, synthetic oligonucleotides, and polymers. Many proteins do not have free thiols, but virtually all have lysines and most have a free amino terminus. The most significant factors in the reactivity of an "amine reagent" with an amine are the structure and basicity of the amine. Aromatic amines, such as anilines, are uncommon in biomolecules. They are weak bases and unprotonated at pH 7. Modification of aromatic amines requires a highly reactive reagent such as sulfonyl chloride or acid chloride, but can be done at any pH of about 4 or above. The "amines" found in nucleotides, such as adenosine or guanosine, and the amide group of peptides are virtually nonbasic and unreactive in aqueous solution. Aliphatic amines, such as the ε- amino group of lysine, are highly basic, but are about 99% protonated and unreactive at pH 7. Amines of lysine have high reactivity with all classes of the "amine reagents" listed in Table 1, but react best above pH 8.5, where a limited fraction of amines are, by equilibrium, unprotonated and reactive. The optimal pH required for modification of lysine amines is always a compromise among four fac tors: degree of protonation,

Table 1

Common Functional Groups and Their Reactive Sites

Functional group	Common reactive site
Typically "amine" reagents	
Isocyanates	Amines, alcohols, phenols
Acid halides and anhydrides	Amines, alcohols, phenols
Acyl nitriles	Amines, alcohols, phenols?
Imido esters	Amines
Sulfonyl chlorides	Amines, phenols
Dichlorotriazines	Amines, phenols
Isothiocyanates	Amines
Acyl azides	Amines (alcohols after Cutius rearrangement)
Succininidyl esters and other "active esters"	Amines
Arylating reagents	Amines, thiols
Aldehydes	Amines, amides?
Typically "thiol" reagents	
Malemides	Thiols, amines
Iodoacetamides	Thiols, amines, imidazole (e.g., histidine), phenols
Aziridines	Thiols, thio ethers (e.g., methionine)
Arylating agents	Thiols, amines
Alkyl halides	Thiols, imizadole, phenols?, methionine?
Disulfides	Thiols
Thiosulfates	Thiols
Typically "carboxylic acid" reagents (usually poor reactivity in water)	
Diazoalkanes	
Carbodiimides	
Aliphatic primary or secondary amines	
Hydrazine derivatives	
Alkyl halides	
Alcohols	
Miscellaneous reactive groups	
Hydrazine derivatives	Aldehydes (e.g., oxidized carbohydrates or RNA)
Glyoxals	Amines, nucleic acids (esp. guanosine)
Alipatic amines + bisulfite	Cytidine
Sulfonyl fluorides	"Serine" esterases
Photoactivated groups (esp. azides)	Relatively indiscriminate
Thiols	Maleimides, idoacetamides

stability of the protein at high pH, stability of the reactive group at high pH, and concentration of both the protein and the reactive dye.

Conjugation of aliphatic amines with isothiocyanates and other reagents, therefore, shows reaction rates that are highly pH-dependent. The α-amino function at the amino terminus is of intermediate pK_a and reactivity of this group can sometimes be selectively modified at neutral pH. A base such as sodium carbonate or triethylamine is usually used to abstract any protons from an amine during synthesis of low-mol-wt fluorescent probes in organic solvents.

The other major factor in the rate of modification of amines is the intrinsic chemical reactivity of the amine reagent. The amine reagents in Table 1 are listed in the approximate order of their decreasing reactivity with amines (but increasing stability in water). Several organic functional groups undergo facile reaction with amines. Sulfonyl chlorides, for instance, are highly reactive. They are unstable in water, especially at the higher pH required for reaction with aliphatic amines, but, once conjugated, they form extremely stable sulfonamides that can survive amino acid hydrolysis (e.g., dansyl end-group analysis). Sulfonyl chlorides also react with phenols, such as tyrosine, to yield relatively stable sulfonate esters. Isocyanates ($R—N=C=O$) give ureas on reaction with amines and urethanes on reaction with alcohols or phenols. Isocyanates are very reactive with all nucleophiles, including water and alcohols, but they are unstable and not in common use as fluorescent probes, except for HPLC derivatization. Isothiocyanates ($R—N=C=S$) form thioureas on reaction with amines. Isothiocyanates are of intermediate reactivity and have quite high stability in water. Succinimidyl esters are selective for aliphatic amines, forming carboxamides of stability similar to the peptide bond. This functional group reacts slowly with aromatic amines and apparently does not react with phenols.

3.1. Chemically Reactive Fluorescent Dyes in Common Use

Probably the most common fluorescent derivatization reagent is fluorescein isothiocyanate (FITC). This reagent is used extensively to prepare antibody conjugates for use in immunofluorescence assays. Since synthesis of FITC and similar fluorescein-derived reagents

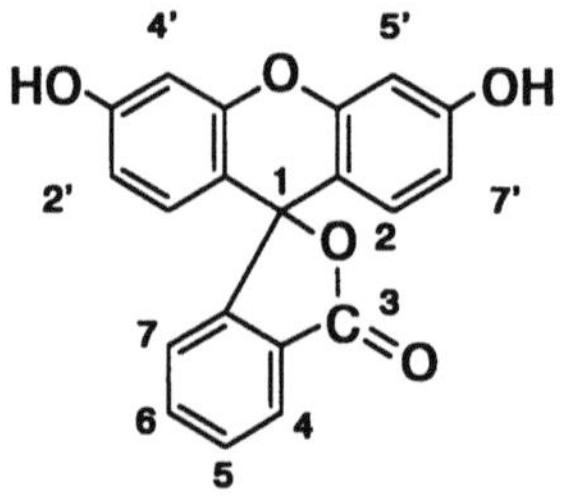

Fig. 1. Numbering system of fluorescein according to Chemical Abstracts.

modified in the "bottom" ring gives two isomers (the 5 and 6 isomers), many fluoresceins and rhodamines are available either as a mixture of two isomers or as separated single isomers. The numbering system for fluorescein and rhodamine (Fig. 1) is based on the lactone form of the molecule. These two isomers are spectrally almost indistinguishable in either wavelength or intensity. The geometry of binding to biomolecules and properties related to elution under electrophoretic or HPLC conditions, however, may require use of the isolated single isomers. The 5 isomer, or "isomer I" of FITC, is the most common, probably because it is the easier of the two isomers to isolate in pure form.

Fluorescein is still one of the best low-mol-wt fluorophores available, with the highest fluorescence/mole in aqueous solution. Advantages of fluorescein include high absorbance, good quantum yield, reasonable chemical stability, water solubility of its reactive derivatives above pH 6, and availability of several reactive forms. In addition to the fluorescein isothiocyanates, amine-reactive dichlorotriazines *(9)*, and succinimidyl esters, such derivatives as thiol-reactive iodoacetamide, maleimide and disulfides, and others have been described. The absorbance of fluorescein closely matches the argon laser line at 488 nm, and its emission quantum yield is quite high, even though it is quenched more than 50% on conjugation with proteins *(10)*. Fluorescein dyes do, however, have some potential disadvantages. The phenolic group in fluorescein has a pK_a of about 6.4. Fluorescein absorbance is significantly quenched on acidification below pH 7 (Fig. 2), which results in the dye having a much lower fluores-

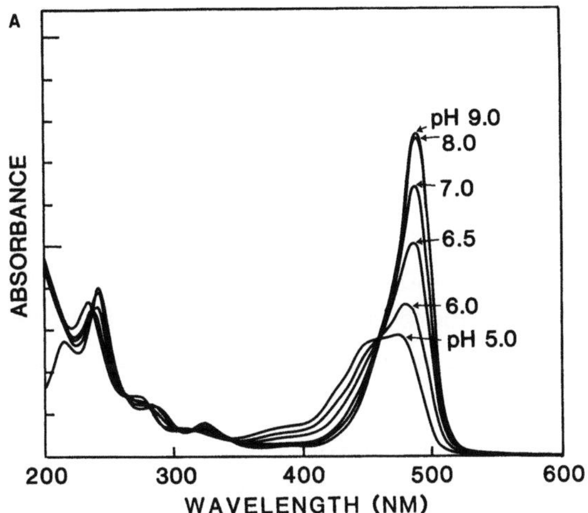

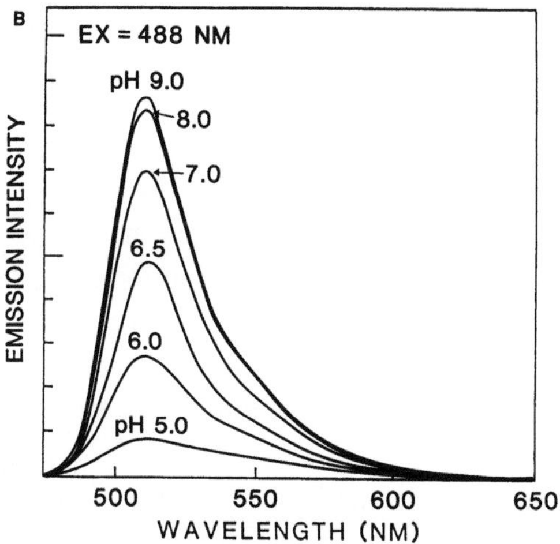

Fig. 2. **(A)** Absorption spectra of fluorescein in 10 m*M* sodium phosphate buffers containing 100 m*M* NaCl at various pH values. The dye concentration is 10 µ*M*. **(B)** Fluorescence emission of fluorescein in 10 m*M* sodium phosphate buffers containing 100 m*M* NaCl at various pH values. The dye concentration is 1 µ*M* and excitation is at 488 nm.

cence intensity in acidic solutions (Fig. 2). Significant photobleaching of fluorescein following illumination has commonly been observed, especially in fluorescence microscopy, in which the intensity of illumination is often very high. Certain protective reagents, including propyl gallate *(11)* and *p*-phenylenediamine *(12)* (both usually in glycerol medium), have been used in fixed-cell preparations to retard bleaching. However, these cannot be used in experiments with living cells.

The absorbance and emission of rhodamines have lower sensitivity to pH than do fluorescein's. However, the fluorescence quantum yield of rhodamine conjugates is usually significantly lower than fluorescein (frequently <0.1). Rhodamines are invariably more stable to illumination than fluoresceins. Like FITC, tetramethylrhodamine isothiocyanate (TRITC) has two possible isomers. The spectra of TRITC conjugates is complex, frequently splitting into two absorption peaks at about 520 and 550 nm (Fig. 3), so the actual degree of substitution of conjugates with this fluorophore is difficult to determine. Excitation into the 520-nm band does not result in much fluorescence, indicating that a nonfluorescent dye complex may be formed, most likely with a second molecule of the dye in the protein conjugates. Since they absorb at longer wavelengths than fluorescein, TRITC conjugates are not well excited by the argon laser, although they are sometimes excited at 514 nm. They are, however, ideally suited for excitation by a 546-nm line in the mercury arc lamp. This very strong excitation source makes tetramethylrhodamine conjugates visibly quite intense in microscopy, despite their relatively low quantum yield. Succinimidyl ester, maleimide, and iodoacetamide derivatives of tetramethylrhodamine are in common use as amine-reactive or thiol-reactive reagents.

Rhodamine B absorbs and emits at somewhat longer wavelengths than does tetramethylrhodamine. This dye differs from tetramethylrhodamine in having ethyl groups rather than methyl groups. This modification shifts the spectra about 10–15 nm toward longer wavelengths, but also reduces the abnormal absorption spectrum from the TRITC conjugates. Sulforhodamine B (also called Lissamine™ rhodamine B, a trademark of ICI, Ltd.) has been commonly used as a sulfonyl chloride. Lissamine™ rhodamine B sulfonyl chloride forms

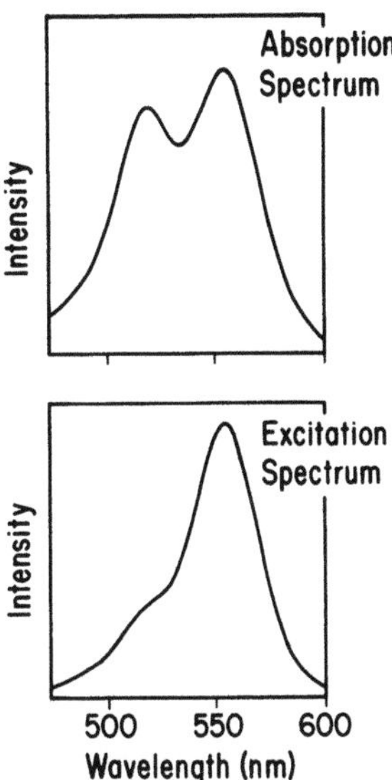

Fig. 3 (Top): Absorption spectrum of tetramethylrhodamine-conjugated bovine serum albumin containing 2.2 dyes/mole of protein, 0.12 mg/mL in 50 mM sodium phosphate buffer, pH 7.4. (Bottom): Fluorescence excitation spectrum of tetramethylrhodamine-conjugated bovine serum albumin, 0.03 mg/mL in 50 mM sodium phosphate buffer, pH 7.4.

sulfonamide bonds that are more stable than those of rhodamine B isothiocyanate (RITC). Removal of the excess reagent by dialysis or gel filtration is also facilitated by the high water solubility of the hydrolysis product, sulforhodamine B.

The reactive dye that has the longest wavelength of those in common use is Texas Red® (registered trademark of Molecular Probes) *(13)*. Texas Red® is derived from sulforhodamine 101. Emission of Texas Red® has a higher quantum yield than that of TRITC or the rhodamine B derivatives, and has the further advantage of having less spectral overlap with that of fluorescein than does tetramethylrhodamine emission (Fig. 4). This property is advantageous for multicolor applications. The enhanced brightness of Texas Red® conjugates is observed only when one uses excitation and emission filters are of longer wavelength than those provided by most equipment makers for "rhodamines." Texas Red® conjugates are very

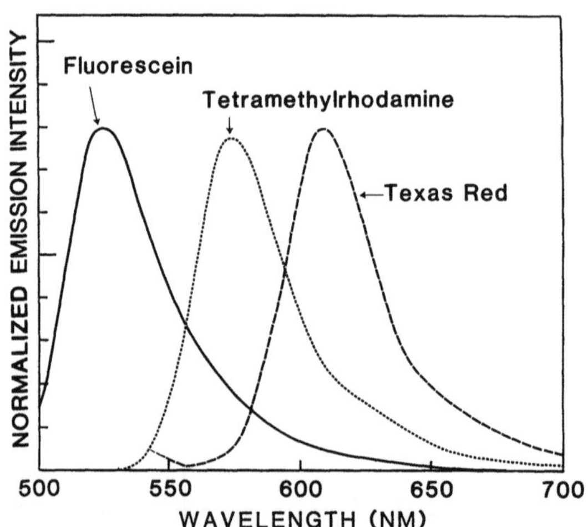

Fig. 4. Emission spectral overlap of fluorescein, tetramethylrhodamine, and Texas Red.® The spectra have been normalized to the same peak intensities. Excitation is at the peak absorption for each dye (495 nm for fluorescein, 550 nm for tetramethylrhodamine, and 596 nm for Texas Red®).

poorly excited by the argon laser at 488 nm (only slightly better at 514 nm); however, they are well excited by dye lasers. The corresponding isothiocyanate derivative, XRITC, has spectra of somewhat shorter wavelength and no particular advantages over Texas Red.® The major disadvantage of Texas Red® is its tendency to cause aggregation and precipitation of proteins to which it is attached and, especially, to precipitate antibodies. This is probably the result of the extended ring structure of Texas Red,® since the more polar TRITC conjugates have less tendency to precipitate. Consequently, Texas Red® is often used as an indirect reagent, such as its avidin, streptavidin protein A, or protein G conjugate.

Many reactive dyes have been described that have absorption in the UV and blue fluorescence ranges. In the visible range, however, relatively few have been used extensively. Recently, the most commonly used has been 7-amino-4-methylcoumarin-3-acetic acid, succinimidyl ester (AMCA) *(14)*. This coumarin dye has bright blue fluorescence when excited in the visible part of the spectrum. In combination with fluorescein and Texas Red,® AMCA has been used for three-color fluorescence for hybridization studies *in situ (15)*.

The remaining group of dyes in common use as fluorescent labeling agents, especially for immunofluorescence, requires more elaborate means to conjugate them to other biomolecules. These are the phycobiliproteins, whose use for preparing fluorescent conjugates was originally described by Oi et al. *(16)*. These intensely fluorescent algal pigments are a group of proteins responsible for light gathering and energy transfer to chlorophyll. In their biologically important function, they have been optimized for fluorescence efficiency. Fluorescence emission in the algal phycobilisome is strongly diminished by energy transfer but, once isolated, the proteins have extinction coefficients up to 2,400,000 cm$^{-1}M^{-1}$ with a quantum yield of 0.98 for B-phycoerythrin *(16)*. On a molar basis, the amount of fluorescence from a phycobiliprotein is about 30–60 times that of fluorescein, although, in practical applications of the phycobiliprotein conjugates, the enhanced sensitivity has commonly been only about fivefold. Of equal importance to the higher sensitivity for application of these dyes is that in the red algae each protein contains more than one type of pigment, and the pigments have different absorption spectra. This results in a broad absorption spectrum (Fig. 5A), particularly in R-phycoerythrin, with a single emission peak, from emission of only the pigment of longer wavelength (Fig. 5B). The enhanced absorption of these proteins at 488 and 514 nm facilitates use of the phycoerythrin conjugates (usually in combination with fluorescein) for multicolor fluorescence applications, such as immunophenotyping, in which it is desired to measure two antigens on a single cell. Because of the low absorbance of TRITC and Texas Red® at the argon laser excitation wavelengths and the low quantum yields of these dyes, it is currently difficult to accomplish two color immunofluorescence by using two low-mol-wt reactive dyes. In addition to being high in mol wt and relatively expensive, the difficulty with use of the phycobiliproteins is that, since they are proteins, the phycobiliproteins do not possess an intrinsic chemical reactivity that permits facile preparation of conjugates. Conjugate formation to another protein commonly requires a three step protocol in which thiols are added to the phycobiliprotein

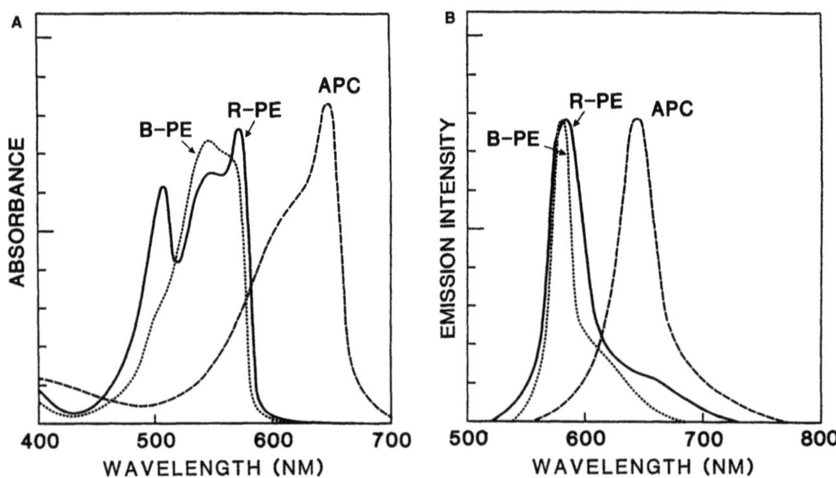

Fig. 5. **(A)** Absorption spectra of phycobiliproteins in 10 m*M* sodium phosphate, 100 mM NaCl, pH 7.4. The absorption peaks have been normalized to the same intensity. B-PE, B-phycoerythrin; R-PE, R-phycoerythrin; APC, allophycocyanin. **(B)** Emission spectra of phycobiliproteins in 10 m*M* sodium phosphate, 100 m*M* NaCl pH 7.4.

with a reagent, such as succinimidyl 3-(2-pyridyldithio)propionate (SPDP) *(17)*, and thiol-reactive maleimides are incorporated in the second protein with a reagent such as succinimidyl trans-4-(*N*-maleimidylmethyl)cyclohexane-1-carboxylate (SMCC) *(18)*. Following the crosslinking reaction, the multiprotein complex is chromatographically purified.

3.2. New Fluorescent Labeling Reagents

There is significant potential for new fluorescent labeling reagents that address some of the problems of current dyes. Although the current fluorescent dyes are quite suitable for many applications in immunofluorescence, there has always been demand for higher sensitivity and specificity, which can be met partially through development of improved dyes. This extra sensitivity can be utilized to search for rare events, to determine low receptor numbers, and to permit use of lower concentrations of difficult-to-isolate materials, such as those required in DNA sequencing. Specifically, the problems with current fluorescent dyes that can potentially be addressed by synthesis of new dyes are the following:

1. Low fluorescence quantum yields and/or low absorbance,
2. Fluorescence quenching on conjugation,
3. Photobleaching and phototoxicity,
4. Spectral overlap,
5. Low Stokes shifts,
6. Lack of suitable long-wavelength reactive dyes, and
7. Insolubility of dyes and precipitation of conjugates.

The principal noninstrumental factors that affect the sensitivity of fluorescence detection for any dye are its absorbance and quantum yield. To achieve the highest net-fluorescence yield, it is necessary to control the factors that result in quenching. As an example, unconjugated fluorescein has a quantum yield of about 0.9 and an extinction coefficient of about 75,000 $cm^{-1}M^{-1}$ in water at pH 8. The loss in fluorescence of fluorescein in an acidic medium is mainly the result of a decrease in the absorbance rather than in the quantum yield (Fig. 2A,B). Photobleaching of fluorescein results in a direct loss of the absorbance, since the dye is converted to products that do not absorb in the visible spectrum. The quenching of fluorescein fluorescence on reaction with proteins, however, is principally the result of a decrease in quantum yield, rather than a change in absorbance. This quenching results from a series of complex interactions of the fluorophore with the proteins, which cannot readily be predicted. For example, we have found that certain proteins, such as avidin, quench dye fluorescence much more than do most antibodies. The quantum yield of fluorescein in aqueous solution is about 0.9. When FITC is conjugated to a protein, the quantum yield for the first dye conjugated is frequently about 0.4 (but about 0.1 on avidin), and decreases further on conjugation with additional dyes *(10)*. When more than about eight to 10 fluorescein molecules are conjugated to an antibody, the total fluorescence usually starts to decrease as additional dyes are conjugated. This quenching results from interaction of the dye with un-

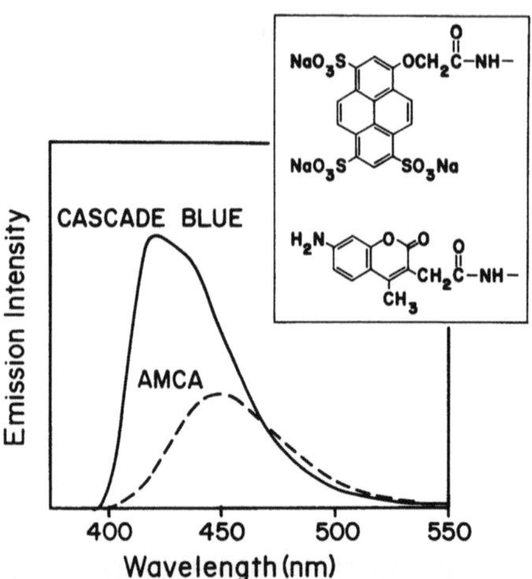

Fig. 6. Emission of Cascade Blue® vs 7-amino-4-methylcoumarin-3-acetic acid (AMCA) at 1 μ*M* concentrations in 10 m*M* sodium phosphate, 100 m*M* NaCl pH 7.4. Excitation is at 365 nm. Inset: structures of Cascade Blue® and AMCA.

known residues in the protein, either in the ground state or in the excited state, or from interactions that occur between fluorophores. This latter type of quenching can occur either over relatively long distances by excited state energy transfer or by the "stacking" of fluorophores to give a nonfluorescent complex. The result is that the quantum yield per fluorophore usually decreases with increasing dye substitution.

There is no general solution to the problem of dye quenching on protein conjugation. We have found one example, however, of a new bright blue fluorescent dye, Cascade Blue® (registered trademark of Molecular Probes). The quantum yield of Cascade Blue® is about the same in the free dye and in its protein conjugates, despite the relatively high overlap of its absorption and emission spectra. For example, it may be compared to the blue-fluorescent tracer AMCA: Cascade Blue® conjugates of rabbit IgG have a fluorescence intensity that increases almost linearly with increasing dye substitution, whereas the fluorescence of AMCA becomes quenched with increasing dye substitution (Fig. 6). The extinction coefficient and quantum yield of Cascade Blue® (about 28,000 cm^{-1}*M*$^{-1}$ and 0.54) is intrinsically higher than that of AMCA (about 16,000 $cm^{-1}M^{-1}$ and 0.3), resulting in the Cascade Blue® conjugates being significantly brighter than AMCA conjugates, particularly when excited at 395 nm.

Many dyes are known that have reasonable fluorescence yields in organic solvents but very low fluorescence yields in aqueous solution, which restricts their utility in preparation of biological conjugates. The quantum yield of most rhodamine dyes in water is less than half of their yield in alcohol. Several potential long-wavelength fluorophores, such as most carbocyanines, oxazines, and styryl dyes, have very high extinction coefficients (often >100,000 $cm^{-1}M^{-1}$) and significant fluorescence when bound to biological membranes, but have quantum yields of <0.01 in water. The choice of suitable fluorophores on which to base new reactive dyes is, therefore, quite limited.

How much more fluorescence intensity can potentially be obtained from better dyes? The practical limit of extinction coefficients for single dyes is about 400,000 $cm^{-1}M^{-1}$ at peak absorbance. Most strongly absorbing dyes in current use have extinction coefficients of <100,000 $cm^{-1}M^{-1}$. Given a quantum yield of 1 as a maximum and a common quantum yield of about 0.2 observed for conjugates, it would seem that the limit for increased sensitivity from new single dyes is about 20-fold, though from experience this is likely to be between five- and 10-fold. To amplify the signal further requires some means of increasing the *number* of dyes that can be conjugated to the biomolecule without resulting in either quenching or interference with the biological activity and specificity of the biomolecule. An example is the covalent attachment of dye-filled liposomes to antibodies, resulting in a many-fold enhancement of the signal vs that of directly dye labeled antibodies *(19)*. For further increases in sensitivity of fluorescence measurements, the improvements will have to come from increases in dye photostability and such instrumental factors as better light collection.

The photostability factor for fluorescent dyes is being increasingly appreciated. The illumination intensity under a microscope is about 1000 times as intense as the sun in Arizona on a summer day. Since light detectors can detect even single photons, the limit for sensitivity is the ability to collect light above any background fluores-

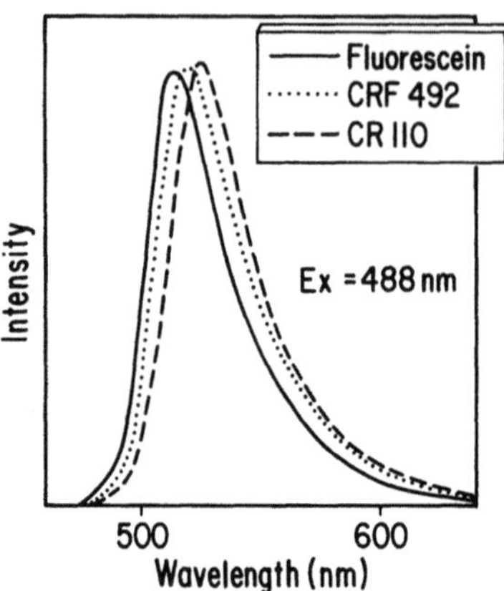

Fig. 7. Fluorescence emission of fluorescein, 5-(and-6)-carboxyrhoda-mine 110 (CR 110), and 5-(and-6-)-carboxyrhodol (CRF 492) at 1 μ*M* concentrations in 10 m*M* sodium phosphate, 100 m*M* NaCl, pH 7.4. Excitation is at 488 nm. The emission peaks have been normalized to the same intensity.

cence signal. The total number of photons that can be collected from a dye is the *integral* of signal intensity and the duration of the signal. Simply stated, the more photostable the dye, the more photons one can collect before the dye photodestructs. Like the quantum yield, each dye has an intrinsic rate of photodestruction. For fluorescein, this results in about 40,000 transitions between the ground and the excited state before the excess energy in the excited state results in molecular catastrophe *(2)*. For rhodamines, this number is closer to 10^6 cycles. Few quantitative measurements have been made for fluorescent dyes; however, photostability is a major factor limiting the detection of single molecules of a dye. Photodegradation can be measured in bulk solutions by the loss of fluorescence, and it is readily observed in microscopy by dimming of intensity of the signal in the field of view. Fluorescein is notorious for its limited photostability. We have recently developed two reactive fluorescein substitutes, CR 110 and CRF 492, that have spectral peaks, extinction coefficients and quantum yields quite similar to fluorescein (Fig. 7), but with significantly higher photostability (Fig. 8). The new Bodipy™ fluorophore (*see below*) and the phycobiliproteins have about the same photostability as fluores-

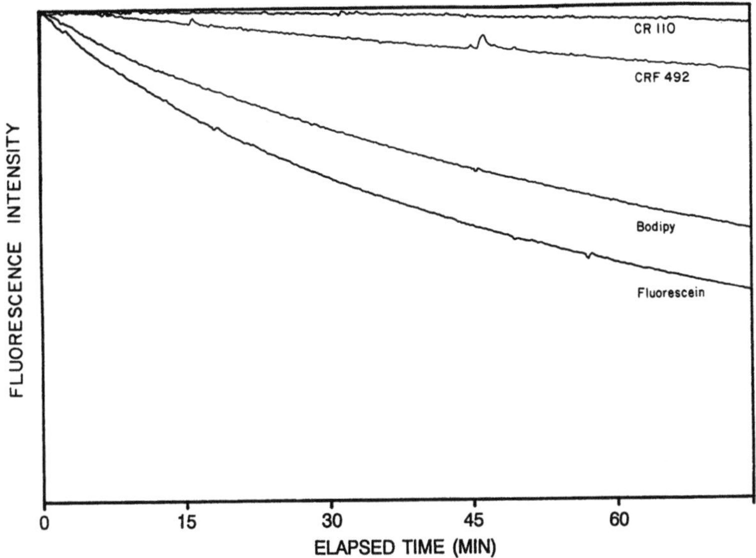

Fig. 8. Relative rates of photobleaching of rabbit IgG conjugates of 5-(and-6)-carboxyrhodamine 110 (CR110) and 5-(and-6-)-carboxyrhodol (CRF492), dimethyl Bodipy™ propionic acid and of fluorescein at 1 µM concentrations in 10 mM sodium phosphate, 100 mM NaCl, pH 7.4. Excitation is at 488 nm. Bleaching is done using the 250-W lamp in an SLM model SPF-500C fluorometer with 20-mm excitation slits and 2.5-nm emission slits and is recorded continuously with stirring.

cein. Since even the new fluorescein substitutes bleach visibly under the microscope, there is clearly a need for new dyes having higher photostability to facilitate quantitative measurements.

Many of the limitations of current dyes are interrelated, and their importance depends on the applications of the dye conjugates. For instance, when working with a single strongly fluorescent dye, such as fluorescein, it is usually possible to optimize the filters and light collection to achieve the sensitivity to detect about 1000 molecules of the dye. It is anticipated, however, that a major direction of research involving fluorescent conjugates will involve simultaneous measurements of multiple dyes, separated either spectrally or spatially. A major consideration in the detection of multiple signals is spectral overlap of emissions. The common means of optically resolving these overlaps is the use of filters with a relatively narrow band pass, which achieve selectivity for detection of a single fluorophore by discarding a major part of the fluorescence intensity. If two spectra overlap in the optical window of the filter, however, it is not possible to determine which dye produced the photon. It is therefore desirable in multicolor applications to have minimally overlapping fluorescence-emission spectra from dyes.

Among the major multicolor applications for fluorescent dyes are flow cytometry, imaging, DNA hybridizations, and DNA sequencing. When considering multicolor applications, there is a clear difference between the types of dyes required for each of these techniques. In flow cytometry, for instance, one has low spatial resolution, necessitating the use of dyes that have very low intrinsic spectral overlap to obtain multiple signals. To perform quantitative measurements of two dyes that have overlapping spectra, it is necessary to compensate for the overlap of the dye emitting at a shorter wavelength with that emitting at a longer wavelength. Fortunately, this is possible if the emission spectrum of both dyes is known. The normalized emission spectra of three common dyes for immunofluorescence, fluorescein, tetramethylrhodamine, and Texas Red® are shown in Fig. 4. At about 525 nm, the emission is 100% from fluorescein, whereas at about 560 nm the emission is primarily from tetramethylrhodamine but with a large contribution from fluorescein (and a slight contribution from Texas Red®). In two-color measurements using fluorescein and tetramethylrhodamine as labels, the quantitative contribution of fluorescein emission at 560 nm to the total signal at 560 is about 30% of the fluorescein peak at 525 nm. After recording the total emission spectrum, this value is subtracted from the total signal at 560 nm to give the emission from only the tetramethylrhodamine dye. Knowing the contribution of tetramethylrhodamine dye and the shape of its emission spectrum, one can then compensate for the overlap of tetramethylrhodamine and fluorescein with the Texas Red® emission at any wavelength. Because the shape of emission spectra may shift with environment, this is an imprecise calculation and would obviously be more precise if dyes could be found that have less spectral overlap.

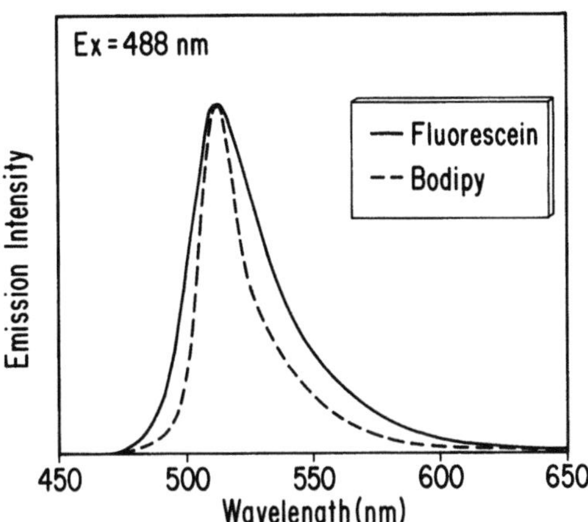

Fig. 9. Fluorescence emission of fluorescein vs dimethyl Bodipy™ propionic acid at 1 µ*M* concentrations in 10 m*M* sodium phosphate, 100 m*M* NaCl, pH 7.4. Excitation is at 488 nm. The emission peaks have been normalized to the same intensity.

It is quite rare to find an entirely new fluorophore that has not previously been used to label proteins. We recently prepared reactive derivatives and conjugates from an unusual fluorophore trademarked Bodipy™ *(20)* that was originally described by Treibs and Krevzer *(21)*. This fluorophore has an extinction coefficient of about 85,000 $cm^{-1}M^{-1}$ and a quantum yield of about 0.9 in aqueous solution. Its spectra are not pH-dependent. Significantly, Bodipy™ fluorescence has an emission bandwidth of only about half that of fluorescein (Fig. 9). Since the quantum yield of a dye is related to the *area* of the emission spectrum, the peak intensity of this dye is actually higher than that of fluorescein. Since fluorescence, particularly in multicolor applications, is often measured through narrow-band optical filters, the intensity within the bandwidth of the filter can be higher than that of fluorescein.

In many multicolor imaging applications of tissues and whole cells (especially in immunofluorescence of cell-surface markers), the dyes are diffusely distributed throughout the cell and, as in flow cytometry, their independent detection and quantitation requires low spectral overlap *(22)*. Some cellular structures, however, such as the nucleus and some other organelles and structural proteins have distinct domains; and because of their size and relatively high spatial separation, these problems can sometimes be resolved using dye combinations that possess a greater overlap, particularly if optical sectioning with a laser scanning microscope is used for the measurement. Consequently, dyes that have higher spectral overlap are potentially suitable. Multicolor fluorescence hybridizations *in situ* usually have high spatial resolution of the signals because of the punctate location of the sites on the chromosomes. DNA sequencing is commonly done by an electrophoretic method in which the dye conjugates are temporally separated by being eluted past the excitation source *(23)*. In this case, it has been reported that dyes having emission spectra separated by only 7 nm can be clearly resolved if the question is only which *one* of the dyes is present in the band *(24)*. Clearly there will always be a trade-off between sensitivity and resolution of signals in multicolor fluorescence measurements.

4. Conclusions

This chapter has reviewed the current status of dyes that can be used to react with biological molecules to form fluorescent conjugates. The intent has been to describe the features of the current dyes and some of their limitations, as well as the direction that is being taken in fluorescent dye research. The development of new reactive fluorescent dyes is being driven by the development of new instrumentation, particularly in imaging and flow cytometry, and by requirements of these applications for greater sensitivity and specificity. The greatest contribution of new dye development will likely be brighter dyes with narrower emission bandwidths which can be readily excited by relatively inexpensive light sources. Some of the newer dyes may also find applications in biosensor developments.

References

1.   Johnson, P. A., Barber, T. E., Smith, B. W., and Winefordner, J. D. (1989) Ultralow detection limits for an organic dye determined by fluorescence spectroscopy with laser diode excitation. *Anal. Chem.* **61,** 861–863.

2. Mathies, R. A., and Stryer, L. (1986) Single molecule fluorescence detection: A feasibility study using phycoerythrin, in *Applications of Fluorescence in the Biomedical Sciences* (Taylor, D. L., Waggoner, A. S., Lanni, F., Murphy, R. F., and Birge, R., eds.), Alan R. Liss, New York pp. 129–140.

3. Weber, G., and Farris, F. J. (1979) Synthesis and spectral properties of a hydrophobic fluorescent probe: 6-Propionyl-2-dimethylaminonaphthalene. *Biochemistry* **18,** 3075–3078.

4. White, J. C. and Stryer, L. (1987) Photostability studies of phycobiliprotein fluorescent labels. *Anal. Biochem.* **161,** 442–452.

5. Haugland, R. P. (1982) Covalent fluorescent probes, in *Fluorescence of Proteins and Nucleic Acids*, 2nd Ed. (Steiner, R. F., ed.), Plenum, NY, pp. 29–58.

6. Kanaoka, Y. (1977) Organic fluorescence reagents in the study of enzymes and proteins. *Angew Chem. Int. Ed. Engl.* **16,** 137–147.

7. Haugland, R. P. (1989) *Handbook of Fluorescent Probes and Research Chemicals.* Molecular Probes, Inc., Eugene, OR.

8. Means, G. E. and Feeney, R. E. (1990) Chemical modification of proteins. *Biocon. J. Chem.* **1,** 2.

9. Blakeslee, D. and Baines, M.G. (1976) Immunofluorescence using dichlorotriazinylaminofluorescein (DTAF). Preparation and fractionation of labeled IgG. *J. Immunol. Meth.* **13,** 305–320.

10. Der-Balian, G. P., Kameda, N., and Rowley, G. L. (1988) Fluorescein labeling of Fab' while preserving single thiol. *Anal. Biochem.* **173,** 59–63.

11. Giloh, H. and Sedat, J. W. (1982) Fluorescence microscopy: Reduced photobleaching of rhodamine and fluorescein protein conjugates by n-propyl gallate. *Science* **217,** 1252–1255.

12. Johnson, G. D., Davidson, R. S., McNamee, K. C., Russel, G., Goodwin, D., and Holborow, E. J. (1982) Fading of immunofluorescence during microscopy: A study of the phenomena and its remedy. *J. Immunol. Methods* **55,** 231–242.

13. Titus, J. A., Haugland, R., Sharrow, S. O., and Segal, D. M. (1982) Texas Red,® a hydrophilic, red-emitting fluorophore for use with fluorescein in dual parameter flow microfluorometric and fluorescence microscopic studies. *J. Immunol. Methods* **50,** 193–204.

14. Khalfan, H., Abuknesha, R., Rand-Weaver, M., Price, R. G., and Robinson, D. (1986) Aminomethyl coumarin acetic acid: A new fluorescent labeling agent for proteins. *Histochem. J.* **18,** 497–499.

15. Nederlof, P. M., Robinson, D. Abuknesha, R., Hopman, A. H. N., Tanke, H. J., and Raap, A. K. (1989) Three-color fluorescence in situ hybridization for the simultaneous detection of muitiple nucleic acid sequences. *Cytometry* **10,** 20–27.

16. Oi, V., Glazer, A. N., and Stryer, L. (1982) Fluorescent phycobiliprotein conjugates for analysis of cells and molecules. *J. Cell Biol.* **93,** 981–986.

17. Carlsson, J., Drevin, H., and Axen, R. (1978) Protein thiolation and reversible protein–protein conjugation. *N*-Succinimidyl 3-(2-pyridyldithio)-propionate, a new heterobifunctional reagent. *Biochem. J.* **173,** 723–737.

18. Yoshitaki, S., Yamada, Y., Ishikawa, E., and Masseyeff, R. (1979) Conjugation of glucose oxidase from Aspergillus Niger and rabbit antibodies using *N*-hydroxysuccinimide ester of *N*-(4-carboxycyclohexylmethyl)-Maleimide. *Eur. J. Biochem.* **101,** 395–399.

19. Truneh, A. and Machy, P. (1987) Detection of very low receptor numbers on cells by flow cytometry using a sensitive staining method. *Cytometry* **8,** 562–567.

20. Haugland, R. P. and Kang, H. C. (1988) Chemically reactive dipyrrom-etheneboron difluoride dyes. US Patent 4,774,339.

21. Treibs, A., and Kreuzer, F. H. (1968) Difluorboryl-komplexe von diund tripyrrylmethenen. *Liebigs Annalen Chem.* **718,** 203–223.

22. DeBiasio, R., Bright, G. R., Ernst, L. A., Waggoner, A. S. and Tayla, D. L. (1987) Five-parameter fluorescence imaging: Wound healing of living Swiss 3T3 cells. *J. Cell Biol.* **105,** 1613–1623.

23. Smith, L. M., Sanders, J. Z., Kaiser, R. J., Hughes, P., Dodd, C., Connell, C. R., Heiner, C., Kent, S. B. H., and Hood, L. E. (1986) Fluorescence detection in automated DNA sequence analysis. *Nature* **321,** 674–679.

24. Prober, J. M., Trainer, G. L., Dam, R. J., Hobbs, F. W., Robertson, C. W., Zagursky, R. J., Cocuzza, A. J., Jensen, M. A., and Baumeister, K. (1987) A system for the rapid DNA sequencing with fluorescent chain-terminating dideoxynucleotides. *Science* **238,** 336–341.

Chemistry and Technology of Evanescent Wave Biosensors

Richard B. Thompson
and Frances S. Ligler

1. Introduction

The purpose of this chapter is to describe fiberoptic biosensors, particularly in relation to the evanescent wave. The basic optical properties of fibers, the configurations of fiberoptic biosensors, the means of attaching biomolecules to optical surfaces, advances in fiberoptics, and related fields of interest are described. More detailed discussions of some of the topics are found in other chapters in this volume.

2. Why Fiberoptics?

The desirable and unique properties of fiberoptics for sensing analytes have been recognized for at least 25 years, since Polanyi and Hehir described an implantable blood-oxygen sensor *(1)* and Hirschfeld demonstrated the potential virtues of fiberoptics in chemical sensing *(2)*. Among these potential virtues are small size, potential for remote and continuous sensing (hence Hirschfeld's term "optrode"), freedom from electrical interference, and relative biocompatibility for in vivo use. However, there are a number of other, subtler advantages, which

Biosensors with Fiberoptics Eds.: Wise and Wingard ©1991 The Humana Press Inc.

are mentioned less often, for incorporating fiberoptics into chemical sensors.

- First is the enormous technology base in optical methods for chemical analysis. There is probably no class of chemical analyte which has not been determined optically by UV/VIS/IR absorption spectrophotometry, emission spectrometry, polarimetry, or photoacoustic spectroscopy.

- Second, the explosive growth in fiberoptic technology, which has made a great variety of fibers, couplers, connectors, sources, and detectors available and, in turn, made advances in sensors possible. Fibers as optical components can have unique properties that are difficult or impossible to obtain in conventional lens/mirror systems. Perhaps chief among these is the ability to carry optical signals long distances through tortuous or oscillating passages independent of shock and vibration. This makes such sensors well-suited for use in manufacturing facilities, in clinics, on moving vehicles, and possibly in the hands of untrained personnel. The high bandwidth arising from low temporal dispersion of single-mode fibers (*see below*) used in communications suggests their use in analytical techniques requiring time resolution, particularly fluorescence methods.

- Third are the recent developments in optical computation and integrated optics, which are well-suited to system integration with optical sensors. Thus, the capability to perform an optical Fourier transform directly on the output of a fiber has application, for instance in infrared spectroscopy *(3)*. Fiberoptics are also well-suited to array detection schemes, and thus to the powerful algorithms used in image analysis and pattern recognition *(4,5)*.

- Fourth, proper adjustment of the refractive indices of the waveguide and surrounding medium permit surface-specific spectroscopies to be performed *(6,7)*, which are otherwise difficult in condensed media.

- Fifth, fibers have a "multiplex" advantage, since they can carry light of different colors simultaneously and in different directions. It is thus possible, at least in principle, to perform different analyses at several wavelengths, all using the same fiber.

For these reasons, the use of fiberoptic sensors is expected to continue to grow.

3. Fiberoptic Fundamentals

Optical fibers are a subset of a class of devices called optical waveguides. Waveguides, as the name implies, direct electromagnetic energy by confining it within a pipe or enclosure; the best-known examples of waveguides are used for directing microwave radiation. Waveguides for use in the optical regime make use of the principle of total internal reflection.

Total internal reflection is described by Snell's law and depicted in Fig. 1. A ray of light striking nearly perpendicular to the interface between two media of differing refractive indices (n_1 and n_2) is refracted in a manner described by Snell's law. At an angle closer to the horizontal, the incident beam is no longer refracted in the second medium, but is totally internally reflected at the interface. Thus, a rod of glass has a higher refractive index than the surroundings and tends to confine light within itself when the rod is in a lower refractive index medium, such as air or water. The light, if introduced into the rod at an angle nearly parallel to the axis of the rod, is totally internally reflected at the glass/air interface; thus the rod acts as a waveguide, in a fashion analogous to a microwave waveguide. The critical angle of incidence (arcsin $[n_1/n_2]$), which defines the onset of total internal reflection, is thus determined by the refractive indices of the media. Thus, if a rod of glass having a higher refractive index is substituted in the above example, the incident light need not be in as narrow a cone around the rod axis to remain confined within the rod.

However, this simple Snell's law description fails to account for several phenomena important in fiberoptic biosensors. Perhaps most important of these is the so-called evanescent wave. In Fig. 1, when the incident light is totally internally reflected, the intensity does not abruptly fall to zero at the interface. Rather, the electromagnetic field intensity decays exponentially with distance starting at the interface and extending into the medium of lower refractive index. The electromagnetic field component close to the interface is called the evanescent wave and is characterized by its penetration depth, the distance from the interface at which it decays to 1/e of its value at the interface. The penetration depth (d_p) is determined by the wavelength of the light

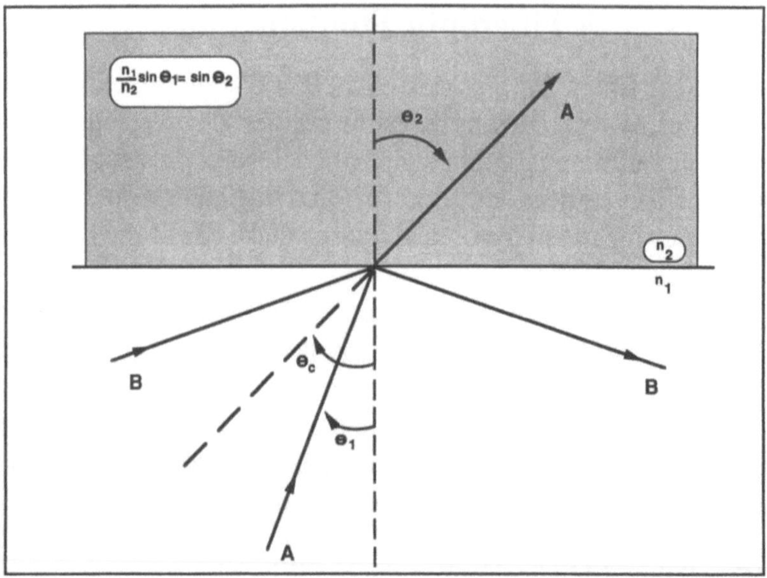

Fig 1. Total internal reflection. Ray A is incident on the interface at angle θ_1 relative to a line perpendicular to the interface between phases of refractive indices n_2 and n_1, where $n_1 > n_2$. Angle θ_1 is less than θ_c, the critical angle, and thus is refracted at angle θ_2. Ray B is incident at an angle greater than the critical angle, and is totally internally reflected at the interface.

(λ), the ratio of the refractive indices (n_1 and n_2), and the angle (θ) of the incident light according to Eq. (1):

$$d_p = \lambda/[4\pi\,(n_1^2\sin^2\theta - n_2^2)^{1/2}] \tag{1}$$

Thus for a glass rod ($n_1 = 1.50$) in air ($n_2 = 1.00$), the penetration depth is 85 nm for light incident at 46.8.° The angle of incidence strongly affects the penetration depth of the evanescent wave. In this example, if the light is incident at an angle 3° closer to the critical angle, the penetration depth increases to 135 nm. This shallow penetration has been used to confer surface specificity on several optical spectroscopic techniques *(6,7)*, since only the small fraction of molecules close to the interface is accessible to the evanescent wave. In one class of fiberoptic biosensors, binding of the analyte to the surface

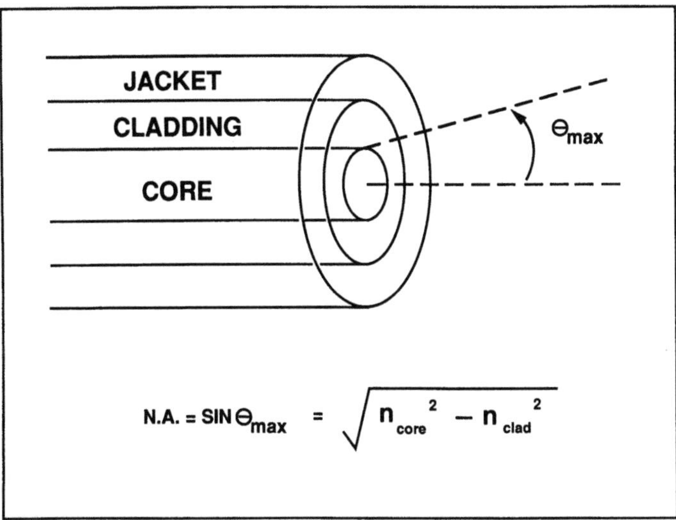

Fig. 2. Generic fiberoptic. The light is largely confined to the fiber core, the refractive index of which is greater than that of the cladding; the whole is protected by a nonoptical plastic jacket. The numerical aperture (NA) defines how narrow the cone angle of acceptance must be; it is a simple function of the refractive indices of the core (n_{core}) and the cladding (n_{clad}). Cores range in diameter from 5–500 μm, and cladding ranges from 100–400 μm in thickness.

of the fiber is detected by a change in fluorescence excited by the evanescent wave near the surface (*see below*).

In the foregoing discussion, little is said about the optical fibers themselves, except that they are a subclass of optical waveguides. The structure of a generic optical fiber is shown in Fig. 2. The light is largely confined to the core, which may be surrounded by a cladding of lower refractive index; the whole is surrounded by a protective, nonoptical jacket made of plastic. Fibers may be constructed of a great variety of transparent materials, including various glasses, fused silica, plastic, and sapphire. The choice depends on the intended application. For instance, in long-distance communications, the glass used in the fiber must be extremely pure to minimize signal attenuation. Most commercially available fibers are used for communication and have glass cores and cladding, whereas others have plastic cladding and silica cores. In the latter, the core may easily be accessed by removing the plastic cladding with a solvent. To remove glass cladding requires etching with hydrofluoric acid. The ratio of refractive indices of the core and cladding determines the Snell's law critical angle, and thus the cone angle within which incident light will become trapped in (launched into) the fiber. The cone angle (*see* Fig. 2) is the arccosine of the numerical aperture (or NA) of the fiber. Some communications fibers have NAs as low as 0.11, and light sources must be well collimated to launch into them, whereas other fibers have NAs as high as 0.55.

For many fibers, especially those used in communications, the geometric optics of Snell's law provide an inaccurate description. It is useful, rather, to consider the fiber as a waveguide or cavity having characteristic modes; i.e., intensity distributions of the electromagnetic field with respect to time and the fiber's physical axes. These modes are exactly analogous to the oscillating modes present in microwave waveguides or laser cavities, and their properties can often be calculated exactly. In waveguides having a large core/cladding ratio of refractive indices ($n_1/n_2 > 1.1$), or a core larger than perhaps 5 μm, many modes will propagate down the waveguide, and such a fiber is termed multimode. However, if the core is very small and the ratio of indices close to one, only a single mode can propagate in the fiber, and it is termed single-mode. Single-mode fibers are important in communications because of their very low temporal dispersion; e.g., a high-frequency (GHz) signal will not become as demodulated passing down a single-mode fiber as it would in a multimode fiber, because of the larger path length variations in the latter. Moreover, single-mode fibers are less susceptible to intensity variations that result from bending, and they can be constructed to conduct selectively light of a particular polarization. However, because of their small core size and low NA, single-mode fibers are more difficult to work with. Note also that in single-mode fibers, the evanescent wave penetrates substantially into the cladding; i.e., a significant proportion of the light in single-mode fibers propagates in the cladding, as well as in the core. It is beyond the scope of this chapter to discuss fiberoptic properties in more detail. An inexpensive, practical text on fiberoptic fundamentals is available *(8)*, as are more thorough treatments *(9)*.

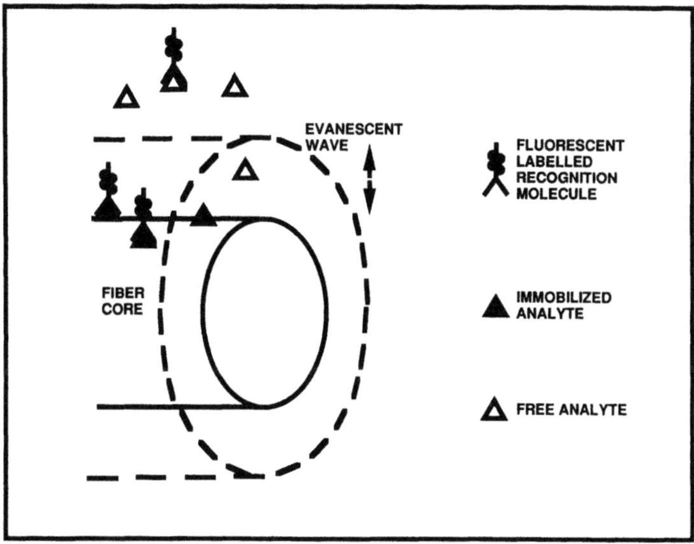

Fig. 3. Waveguide-binding configuration. The binding of fluorescent detector molecules to immobilized analyte generates a fluorescent signal upon excitation by the light in the evanescent wave. Some of the emitted fluorescent light reenters the fiber for subsequent measurement. In the presence of free analyte, some of the labeled recognition molecules are displaced from the surface, and thus the zone of excitation of the evanescent wave; they are no longer excited, and the measured fluorescence decreases.

4. Configurations for Fiberoptic Biosensors

The selectivity and avidity of enzymes and antibodies have long been exploited in chemical analysis and medicine, and several fiberoptic biosensors employing biomolecules have been described. Most employ either fluorescence or absorbance spectroscopic techniques to quantitate the analyte, with fluorescence more prevalent because of its higher intrinsic sensitivity *(10–12)*. The sensors employ one of two fundamentally different optical configurations at the sensing (distal) end:

1. Waveguide-binding configuration, in which molecular recognition occurs at the surface of the fiber core and produces binding or release of a chromophore or fluorophore (Fig. 3), or
2. Distal-cuvet configuration, in which a chromophore or fluorophore is released into the illuminated space at the end of the fiber upon analyte recognition (Fig. 4).

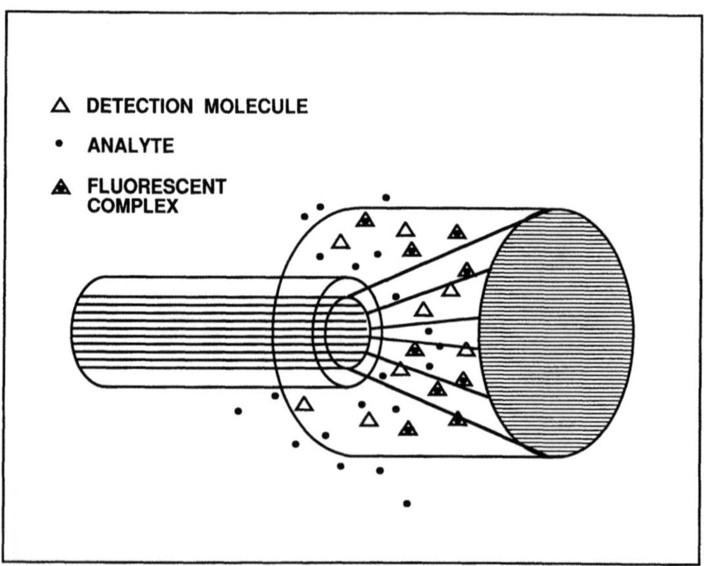

Fig. 4. Distal cuvet configuration. Detector molecules are immobilized at the distal end of the optical fiber in a porous matrix. Upon analyte recognition, an optical signal is generated in the presence of exciting light from the fiber. Emitted light is coupled back into the fiber and collected.

The evanescent wave concept is most important in biosensors because it enables a recognition (binding) event to be transduced as a change in fluorescence intensity. Hirschfeld *(2)* and Harrick and Loeb *(13)* showed that fluorophores near an interface could be selectively excited by the evanescent wave propagating at the surface of a waveguide. This principle of total internal reflection of fluorescence was employed by Kronick and Little in a homogenous immunoassay *(14,15)*. They adsorbed an antigen to the surface of a quartz waveguide and then added a fluorescein-labeled antibody to the solution external to the waveguide *(see* Fig. 3). The labeled antibody bound to the antigen, which is within the zone of the evanescent wave, and thus underwent fluorescence activation. Antigen in the external solution competes for binding sites on the labeled antibody, permitting some of the antibody to diffuse away from the surface and thus away from the evanescent wave. The fluorescent label on these molecules is no longer excited, and the observed intensity of fluorescence decreases.

Antibodies in solution can be similarly quantitated, or antibody can be attached to the waveguide and the analyte measured in a "sandwich" assay *(16)*. This principle was adapted for use in optical fibers instead of planar waveguides by Andrade, Hirschfeld, and others *(17–19)*.

An intuitively simpler arrangement is employed when the sensing apparatus is attached to the distal end of the fiber, which thus acts solely as a light pipe (Fig. 4). Such a configuration requires that the analyte and the biomolecule doing the sensing be brought together at the end of the fiber. Generally, this requires that the biomolecule be immobilized on or enclosed within some porous matrix at the distal end, which permits the analyte to diffuse through to it. Thus the sensors of Zhujun and Seitz *(20)* and Arnold *(21)* employ hemoglobin adsorbed on cation exchange resin and alkaline phosphatase crosslinked to nylon mesh with glutaraldehyde, respectively. There is sometimes considerable skill required to construct such sensors on thin (submillimeter) fibers *(22)*. Since the light beam from the end of the fiber has a relatively small cross-section, the number of molecules accessed by the beam is much increased if the immobilization matrix is transparent. These sensors represent a common configuration for biosensors that employ detection molecules other than antibodies or that are used in vivo *(23,24)*. Sensors of this type are discussed in detail in the chapter by Canassero et al. in this volume.

Both types of fiberoptic biosensors necessitate a photometric apparatus for measurement. An apparatus for fluorometric detection is depicted schematically in Fig. 5. The remainder of this chapter considers only fluorescence, since it is one of the most sensitive and widely used approaches. Although the design in Fig. 5 differs from a standard fluorometer, the criteria for sensitivity do not. In particular, sensitive detection of fluorescence demands efficient excitation of the fluorophores, effective discrimination of excitation from emission, and thorough collection of the emission. The fiber attenuation decreases both excitation and emission intensity, and the fiber itself can be a source of background photoluminescence *(25)*. In both the evanescent wave and distal cuvet geometries, relatively few molecules are excited. In the case of the evanescent wave, the shallow penetration that confers the surface specificity of the technique also ensures that very

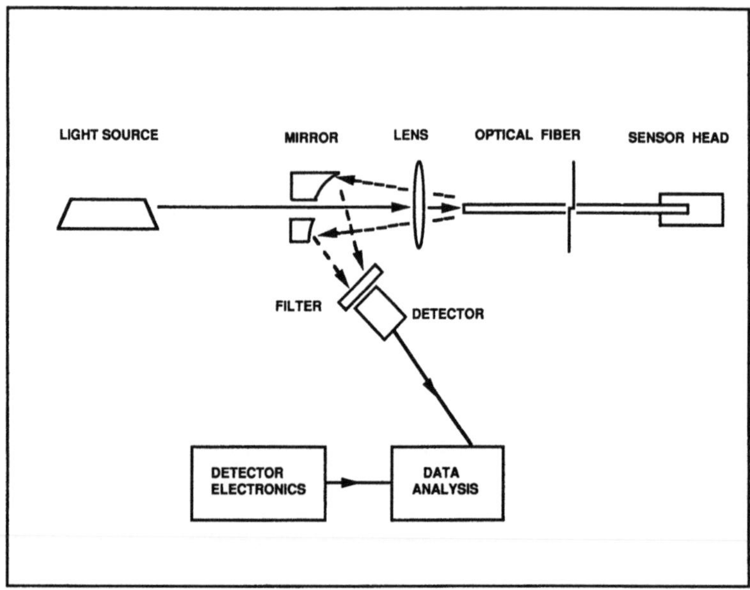

Fig. 5. Generic fiberoptic chemical sensor. Light from the source passes through the perforated off-axis parabolic mirror, and is coupled into the fiber by the objective lens. The light passes down the fiber into the sensor head, which is either a distal-cuvet (Fig. 4) or a waveguide-binding type (Fig. 3). The longer-wavelength fluorescent light from the sensor head is coupled back through the fiber, and reflected off the mirror and into the detector, the output of which is quantitated.

few molecules can absorb light and that the resulting signal is relatively weak. Moreover, for the relatively large (500 µm in diameter) multimode fibers commonly used, only a small fraction (<0.01%) of the power launched into the waveguide ends up in the evanescent field *(26)*. Hence, for milliwatts of light launched into the fiber, the molecules near the surface see microwatts at most. Finally, the light-collection efficiency (e.g., the proportion of fluorescence coupled back into the fiber, which is detected at the near fiber terminus) for fluorophores in the evanescent wave is poorer than for a typical fluorometer with right-angle geometry, although it is better than might be expected based on the geometric optics of the system *(27,28)*.

Clearly the use of the evanescent field geometry requires substantial effort on the part of the designer to assure adequate sensitivity *(29)*.

Two additional configurations for an evanescent-wave device are currently being explored by the authors:

1. A continuous monitoring system in which fluorophores are displaced from antibody-coated beads by the analyte and flow past a fiber (Fig. 6), and
2. Detection molecules immobilized in a separate chamber or precolumn in such a way that analyte recognition results in fluorophore release and subsequent adherence of the fluorophore to the fiber (Fig. 7).

Separating the biological recognition elements from the fiber itself allows easy substitution of a variety of assays into a single biosensor. Also, it permits the biological elements to be stored separately from the fiber; consequently, the disposable portion of the sensor may be less expensive and/or more stable. The need for extensive modification or frequent replacement of the delicate fiber is also eliminated.

Other configurations of fibers and biological detection systems are clearly possible. For example, multiple fibers may be employed for simultaneous detection of different analytes or pattern-recognition approaches to differentiating similar species. Also, several different fluorescent labels corresponding to particular analytes might be used with a single-fiber sensor, in a manner analogous to multicolor fluorescence-activated cell sorting. Finally, optical fibers and biomolecular sensing species may ultimately be integrated into photonic biochips, wherein sensing and computation are combined on the same substrate.

5. Biochemical Architecture of Biosensors

A more detailed examination of the biological "front end" of these fiberoptic-based detection systems indicates that, for any single biosensor configuration, four features of the biochemistry can be adapted for each particular application:

1. The biomolecule used for recognition,
2. The method of signal amplification,
3. The label used to generate the optical signal, and
4. The biochemical method for attaching labels to biomolecules and/or biomolecules to solid supports.

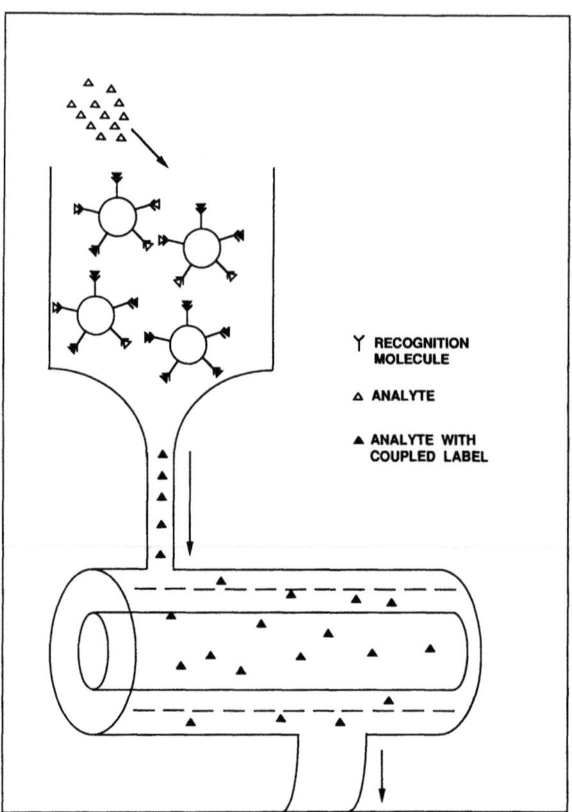

Fig. 6. Continuous flow waveguide-binding configuration. In a small column upstream from the fiber, the recognition molecule is coupled to a solid support and its binding sites are saturated with a labeled form of the analyte. In the presence of test sample containing analyte, the labeled analyte is displaced from the recognition molecule. The displaced labeled analyte is then injected into the capillary surrounding an optical fiber. As the stream flows past the fiber, some of the labeled analyte enters the evanescent-wave region, generating a fluorescent signal.

5.1. The Biomolecule Used for Detection

The types of detector biomolecules available for use in biosensors are discussed in the chapter by Wingard et al. in this volume. The criteria should include not only the ability to recognize a specific analyte, but also the degree of specificity, affinity, reversibility of

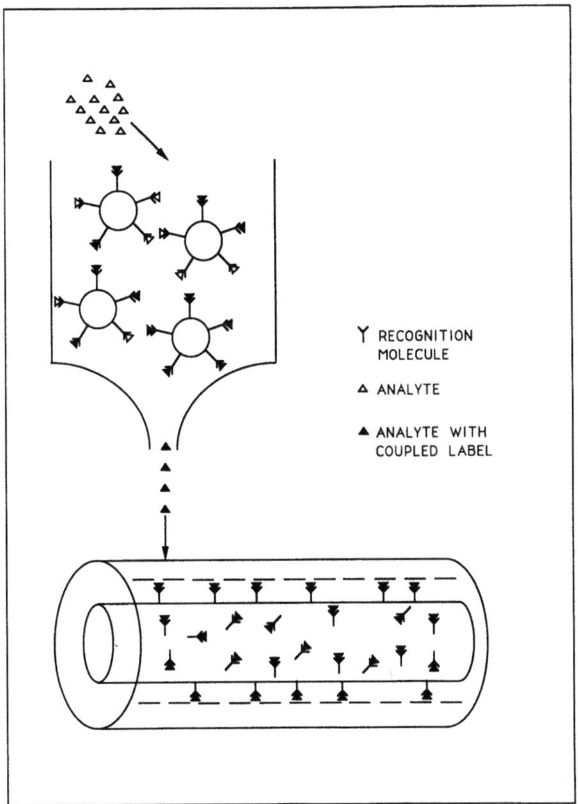

Fig. 7. Waveguide-binding configuration with precolumn. Recognition molecules are immobilized in a precolumn as in Fig. 6. Fluid from the test sample passes through the column and, with any released labeled analyte, enters the capillary surrounding the fiber, which has been coated with the same or another recognition molecule specific for the analyte, the label, or a third component of the analyte–label complex. The labeled analyte thus migrates and remains within the evanescent wave, where a fluorescent signal is generated.

binding, stability, availability, cost, and compatibility with the signal-amplification systems and labels to be used. Antibodies are often the molecule of choice because of their high degree of specificity and reasonable affinity. However, it is difficult to prepare antibodies against very small molecules that are not easily bound to carriers for immunization, are not recognized as foreign entities by the immune system, or are highly toxic.

The affinity with which the biomolecule binds the analyte deserves particular consideration in view of the direct relationship between the kinetics of binding and the speed of response of the biosensor *(19)*. The equilibrium constant for the binding of an analyte to a detector molecule (such as an antigen to an antibody) can be expressed as the ratio of the association rate constant to the dissociation rate constant. As Weber has pointed out *(30)*, the association rate constant of most protein–ligand binding reactions varies little from that to be expected from a purely diffusion-controlled reaction, about $10^7 M^{-1}$ sec^{-1}. Thus the variation observed among binding constants of this sort ($K_a = 10^{15} - 10^3 / M^{-1}$) largely reflects differences in the dissociation rate constants. The tightly bound complex of avidin to biotin dissociates at the rate of 10^{-6}/s, about 1/mo *(31)*, whereas a typical antibody–antigen complex would dissociate in a few minutes. This rate evidently puts a lower limit on the response time for continuous monitoring immunosensors, in that one has to wait for the antibody to release its bound antigen to sense another change in antigen concentration. Moreover, the use of antibodies with higher affinity to improve sensitivity merely compounds this problem. This is not a problem when such sensors are used in a nonequilibrium, two-state, "alarm" mode; the kinetics simply dictate how long it will be before the alarm can be reset.

5.2. The Method of Signal Amplification

Three types of biochemical amplification strategies are useful in signal enhancement in fiberoptic biosensors: linear, multiplicative, and cascade strategies. The most commonly used strategy in fluorescence immunoassays is linear signal amplification in which one recognition event produces a signal from a discrete number of fluorophores. For example, recognition and binding of a large antigen to the surface of the optical fiber might be followed by the binding of an antibody carrying multiple fluorescent dye molecules (Fig. 8A). The number of optical labels (and thus the signal) is linearly proportional to the number of recognition events and does not significantly change with time. Most signal amplification systems using enzyme-triggered colorimet-

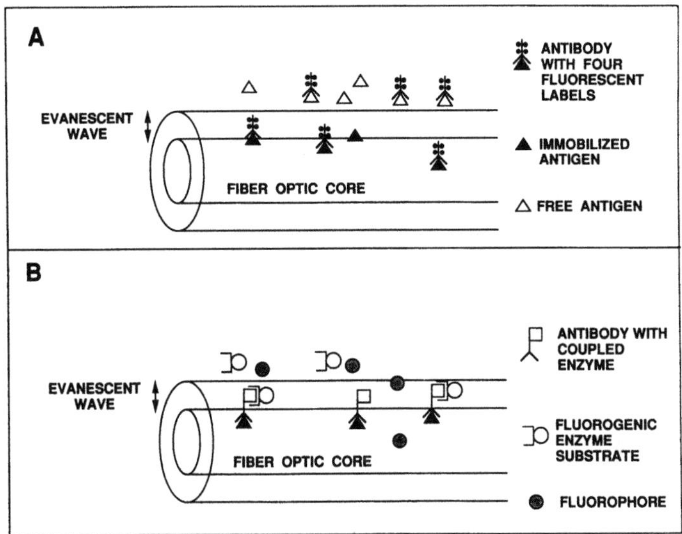

Fig. 8. Amplification strategies. Linear and multiplicative amplification strategies are depicted using a waveguide binding configuration and antibodies as detection molecules. **(A)** Linear amplification is achieved by the binding of one fluorescent-labeled antibody to each antigen bound to the surface of the optical fiber. If each antibody carries four dye molecules, there will be four fluorophores per antigen and little change in signal with time after the initial binding has occurred. **(B)** Multiplicative amplification is achieved by using an enzyme to label the antibody. The enzyme-labeled antibody binds to analyte at the surface of the optical fiber. After washing away excess antibody, chromogenic substrate is introduced. The enzyme cleaves substrate molecules to produce chromophores in a continuous process that lasts as long as sufficient substrate is available. Chromophores within the evanescent-wave region produce a signal. Thus, the signal increases in proportion to both analyte concentration and time.

ric assays follow multiplicative strategies. If this approach is used, the antibody binding to the antigen at the surface of the optical fiber carries an enzyme, such as urease or alkaline phosphatase (Fig. 8B) *(32)*. After the fiber is washed, a fluoro- or chromogenic substrate is added, and fluorophores or chromophores are produced continuously as long as substrate is available. The signal is thus multiplicative, being directly proportional to both the antigen concentration and time. In this case, the label is a catalyst for generating the signal. The disadvantage of this system is that the noise from nonspecific cleavage of substrate also increases with time. Cascade amplification strategies are potentially the most effective, but are the most difficult to implement. An example of a cascade strategy would be a modification of the multiplicative system just described, in which the substrate of the enzyme attached to the antibody is a second enzyme. In this case, the label is a catalyst that creates more catalysts, which in turn produce the signal. After activation by the first enzyme, the second enzyme cleaves a chromogenic substrate to produce an optical signal. In the presence of excess reagents, the signal is proportional to the antigen concentration and the time squared, i.e., there is a logarithmic relationship between antigen concentration and signal. The primary disadvantage of this strategy is the complexity of maintaining excess second enzyme and substrate in addition to coordinating the working range of the assay with the working range of the instrument.

5.3. *The Label Used to Generate the Optical Signal*

The choice of optical label clearly has a large impact on both the molecular architecture of the sensor and the associated hardware. Most fiberoptic chemical sensors utilize visible absorbance or fluorescence changes. In general, luminescence methods are more sensitive than absorbance. In a luminescence experiment, one is comparing light vs dark; moreover, the difference can be increased by increasing the intensity of the excitation *(10–12)*. Thus, whereas the lower limit of detection using absorbance is perhaps $10^{-8}M$, commercially available fluorescence instruments can determine fluorophores at $10^{-12}M$ and, with special instrumentation and fluorophores, at $10^{-15}M$ *(33)*.

In practice, the sensitivity of all these techniques are limited by environmental interferences, background, and instrumental factors. It is often useful to choose a fluorescent label not so much for its high inherent sensitivity (i.e., high quantum yield and extinction coefficient), but rather for its freedom from interference, its photostability, and the difference between the wavelength maxima of excitation and emission (Stokes' shift). This is especially true for analyses performed in complex media, such as sea water, serum, or culture media; the immobilizing materials used at the distal end of sensors like that in Fig. 4 are also a source of interference. Fluorescein is a popular label because of its extinction coefficient, quantum yield, low cost, and relative insensitivity to environmental effects. However, bilirubin interferes with fluorescein emission in serum. Also, the modest Stokes' shift of fluorescein requires careful filtration to minimize interference by scattered light, and fluorescein has well-known photolability. Moreover, the chemistry of the labeling itself, especially its selectivity, is an important problem, which has been recently reviewed *(34)*.

Lasers' intensity, monochromaticity, and easy coupling into fibers strongly suggests that lasers are the preferred light source for fiberoptic sensors and that a label should be chosen more with the excitation in mind than anything else. The ease with which milliwatts of optical power can be launched into a fiber from a laser contrasts sharply with the difficulty of using an incoherent source. Thus, even an intrinsically poor fluorophore might be a better choice as a label if it is suitable for use with a laser. It is also important to consider the attenuation of the fiber at the wavelengths of interest. Typical UV-transmitting silica fibers have approx 100 dB/km attenuation at 300 nm; e.g., a 1-km length of fiber would have an apparent optical density of 10.0. This makes remote sensing using the intrinsic fluorescence of tryptophan *(35,36)* or the very successful lanthanide chelate fluorescent labels *(37,38)* technically difficult; by comparison, attenuation at 500 nm is <20 dB, or 100,000,000-fold better. Therefore, newly developed fluorescent labels, such as phycobiliproteins *(39)* may be very well-suited for use with fiberoptics. Finally, which modes a fiber will guide and their polarization properties are also wavelength-dependent and thus affect the choice of label; for time resolution, a single-mode fiber may be necessary, but such fibers are hard to obtain for wavelengths shorter than 488 nm. Clearly, optical, instrumental, and environmental factors must all be taken into consideration when developing the molecular architecture of the biosensor.

5.4. Labeling and Immobilization of Biomolecules

Most previously described biosensors simply crosslink the protein to silylamine-derivatized fibers using glutaraldehyde. Glutaraldehyde is relatively nonspecific regarding the portion of the protein with which it reacts, derivatizing the side chains of lysine, cysteine, tyrosine, histidine, tryptophan, and arginine, and tending to form intermolecular linkages between the amino groups on the glass or between soluble proteins as well as forming glass–protein links. Thus the reproducibility and yield with which fully functional proteins can be linked to the fiber are often poor. More specific homobifunctional crosslinkers offer some improvement through better control over the group bound in the protein and greater probability for maintaining protein function. The authors have had little success with carbonyldiimidazole, but moderate success with commercially available *bis*-hydroxysuccinimides. Such reagents, however, will still crosslink the molecules within the silyl film in such a manner that they are not available for attachment of proteins.

Various methods for modifying silane coatings have been developed for affinity chromatography purposes. Weetall and Filbert *(41)* have reviewed chemistries to prepare glass beads for the attachment of proteins, and such chemistries are appropriate for modification of silica fibers. Plastic fibers offer added advantages for coating with proteins because of potential modification of the plastic itself for attachment of crosslinkers. Other affinity chromatography methods can be adapted to such systems *(42,43)*.

Better success may be achieved in attaching a protein to a plastic or silyl-coated glass fiber with a heterobifunctional crosslinker. Such a crosslinker may be selected so that it binds by one end to the silyl coating and the other end to the protein. Such crosslinkers are commercially available and utilize amino, sulfhydryl, guanidino, indole and carboxyl groups on the solid phase and in the molecules to be immobilized *(44)*. Silyl compounds are available with a variety of termini, including amino groups, sulfhydryls, and hydroxyls, suggesting the use of such compounds as *N*-succinimidyl-3-(2-pyridyldithio)propionate or succinimidyl-4(*N*-maleimidomethyl)-cyclohexane-1-carboxylate, which are widely used for linking amines to sulfhydryl groups *(44)*.

We have found heterobifunctional crosslinkers and thiolterminal silanes to work particularly well for immobilizing both antibodies and enzymes. Antibodies can be reproducibly immobilized at high densities (1 ng/mm^2) while maintaining the antigen-binding function (0.4–0.5 mol antigen bound/mol antibody) *(45)*.

6. Approaches for Improved Fiberoptic Biosensors

The first generation of fiberoptic biosensors has demonstrated the validity of the concept, but much remains to be done in creating biosensors for the more demanding applications. Perhaps the most important task is to improve sensitivity, since this is a limiting factor for many proposed applications. There are several means of improving the sensitivity of fiberoptic biosensors that are, at present, underexploited; some recently developed hardware and methodological means of doing this are discussed below, particularly with regard to remote sensing. Speed of response is also a priority, but methods of improving this remain unclear. Finally, we suggest that there are a number of other optical techniques beyond visible absorbance and fluorescence that may be well-suited to creating novel fiberoptic sensors with very desirable characteristics.

A variety of sophisticated hardware that may be useful in fiberoptic sensors has appeared recently. We have mentioned the importance of intense, collimated (i.e., laser) sources for maximizing sensitivity; also, it should be noted that the narrow linewidth of most lasers minimizes difficulty with scattered stray light that can interfere with fluorescence assays, especially polarization assays *(46)*. The authors have used a newly introduced helium–neon laser with output at 543 nm in a fiberoptic biosensor with the waveguide-binding configuration. Unlike red helium–neon lasers, the green output of this laser is well-suited to excitation of phycobiliprotein *(37)* and rhodamine labels, often used in immunoassays. Apart from its color, the green helium–neon laser is comparable—in cost, intensity, size, and reliability—to red (633 nm) helium–neon lasers. The ruggedness and portability of such helium–neon lasers makes them good candidates for use in field instruments. Other lasers that likely will have an impact on fiberoptic sensors are the all-solid-state diode-pumped YAG lasers, which when doubled into the green wavelengths (532 nm) are very promising sources. Also, the laser diodes commonly used for fiberoptic communications are cheap, powerful, very small, and rugged. Unfortunately, they emit only in the far red range, and until suitable fluorescent dyes can be developed, their utility is limited. A laser employing titanium-doped sapphire *(47)* offers the prospect of a tunable UV/visible output without the use of a cumbersome dye laser. If successful, it should prove a useful source for sensors requiring several wavelengths for immunoassays with multiple fluorescent labels.

Detector and fiber technologies have also made advances that will enhance the sensitivity of fiberoptic sensors. In particular, the improvements in sensitivity, cost, automatability, and variety of solid-state detectors, such as PIN photodiodes, microchannel plate electron multipliers, photodiode arrays, and charge-coupled devices, argue that solid-state detectors may supplant photomultiplier tubes in many fiberoptic sensors. Fiber technology has been underutilized in the fiberoptic biosensors described in the literature; in this sense, little progress has been made since the developments of Harrick and Loeb *(13)*. Recently, tapered fibers have been shown to improve the sensitivity of absorbance sensors *(48)*, and have displayed encouraging results with waveguide-binding sensors *(29)*. Also, fibers with high numerical aperture should permit construction of optrodes, luminometers, and fluorometers with enhanced light-collection efficiency. Thus, a newly developed sapphire fiber (Saphikon, Inc., Milford, NJ) in water should have approx twice the acceptance angle of a bare fused-silica fiber (Fig. 2). The use of high-numerical-aperture collection optics in the total internal reflection fluorescence experiments of Axelrod and his colleagues *(6)* permitted light collection efficiencies of up to 50%, as compared to approx 5% for ordinary fluorometers.

A number of new fluorescence techniques may be applied to increasing the sensitivity of fiberoptic sensors. We have mentioned the sensitivity enhancements attainable by the use of lanthanide chelates and time-resolved fluorescence techniques *(37,38)*. This technique is based on the slow intrinsic emissive rate of the lanthanides in comparison to ordinary organic fluorophores, which are the main source of background fluorescence. The fluorescence of the sample is excited by a brief pulse of light, but there is a delay before the emitted fluorescence is measured. If the delay is of the order of 500 ns, the background fluorescence decays to about one ten-thousandth its original value, whereas the label fluorescence decays only a few percent. Thus the background is reduced, sharply increasing the signal-to-noise ratio. Other workers have employed time-resolved fluorescence methods for temperature or pressure measurements *(49,50)*, but no fiberoptic biosensor employing such methods has yet been described. As stated above, lanthanide chelate labels are poorly suited for use in remote sensing because of their requirement for UV excitation, which is strongly attenuated. Recognizing this, we *(57)* and others *(52)* have developed the use of ruthenium *tris*(bipryidyl) complexes as fluorescent labels for time-resolved immunoassay methods. These labels overcome many of the drawbacks of the lanthanides, since they can be excited in the visible range, where good laser sources exist and fiber attenuation is lower; have simpler labeling chemistry; and are well-suited to homogenous immunoassay, since they require no time-consuming enhancer step.

Reduction of background is a well-known means of improving sensitivity. Andrade and others have noted the presence of background photoluminescence and Raman scattering from optical fibers and plastic cladding *(53–55)*. This background light is a serious interferent for fluorescence sensors that are operated remotely since, as the fiber lengthens, the signal is attenuated while the background increases. We note also that the lifetime of the photoluminescence (likely phosphorescence) in fibers is comparable to the lanthanide chelate fluorescence lifetime, thus making the use of time-resolved methods (and such dyes) problematical. This background photoluminescence (and Raman intensity) drops with increasing wavelength; taken together with the decreased attenuation at longer wavelengths (up to about 900 nm), the use of redder fluorescent dyes in the 500–700 nm range seems strongly indicated. An additional concern in using time-resolved analytical methods is the temporal dispersion (pulse broadening) that occurs as the light passes through the fiber. For single-mode, gradient index, or short multimode fibers, temporal dispersion (expressed as the frequency–length product) should be unimportant, but for truly remote sensing (>1 km) and/or ns time discrimination, multimode fibers with band widths >100 MHz/km will be found most useful.

Another method of enhancing sensitivity was initially developed by Hirschfeld *(56)*, based on absorption measurements carried out with the sensitivity of fluorescence measurements. Hirschfeld called the underlying principle "photon quantum tunnelling," but the process has been well-known in photosynthesis biophysics as exciton migration. In Hirschfeld's optrode (or a chloroplast), densely packed fluorescent molecules with overlapping absorption and emission spectra tend to rapidly pass energy from absorbed photons among themselves by a nonradiative route. Thus, the delocalized excitation (exciton) stays in the fluorescent-molecule array until the energy is emitted as a photon or quenched by absorption by a nonemitting molecule (a trap). Since colorimetric assays are based on the change in absorbance of some reagent molecule as a result of the presence of the analyte (for instance, binding ferrous ion causes an increase in visible absorbance of phenanthroline as a result of formation of a complex), colorimetric assays ordinarily are limited in sensitivity by the extinction coefficient of the reagent molecule. In this optrode, such a reagent molecule would operate as a nascent trap, effectively quenching the fluorescence of many dye molecules only in the presence of the analyte. Thus, a change in absorbance is detected as a change in fluorescence of many dye molecules, with concomitant increase in sensitivity. This tactic is obviously suited for use in solution as well as in fiberoptic sensors, and is destined for wide application.

We have mentioned that fiberoptic sensors are well-suited to array detection schemes, wherein the output signals from several sensors are analyzed together, rather than treated autonomously. The reason for doing this is primarily mathematical. By analyzing the output of several sensors together using least-squares or pattern-recognition algorithms, remarkable multicomponent analyses can be obtained, even though the individual sensors have mediocre selectivity *(4,5,57)*. Giuliani has described a multielement fiberoptic sensor that anticipates this trend *(58)*. The fiberoptic biosensors described in the literature generally utilize visible absorption or fluorescence. However, there are other optical sensing techniques that may prove adaptable, if not well-suited, to fiberoptic biosensors, especially infrared absorption spectroscopy. A few reports have appeared describing fiberoptic absorbance sensors at particular wavelengths *(59)*, and an all-fiber Fourier transform infrared spectrometer has been built *(60)*. Available silica or chalcogenide glasses unfortunately have substantial attenuation in certain regions of the infrared, limiting their utility *(61)*, and a coherent broad-band source remains a problem. Recently developed sapphire- or fluoride-glass *(62)* fibers may prove more useful in this respect. Raman spectroscopy has been little used in fiber sensors *(55)* because of its inherent insensitivity and fiber background *(53)*. Nevertheless, more sensitive Raman techniques, such as surface-enhanced and resonance Raman spectroscopy, have proven useful in chemical analysis and may be adaptable to a biosensor. Similarly, the commercial availability of polarization-preserving optical fibers suggests their use in such methods as circular dichroism, polarimetry, and especially fluorescence-polarization immunoassay. Changes in turbidity (light scattering) have also been made the basis of commercial immunoassays and would seem to be adaptable to use with optical fibers *(63)*. Unfortunately, interesting new spectroscopic methods employing nonlinear optical effects *(64)* or two-photon transitions *(65)* require very high optical powers that may be incompatible with fibers and many biomolecules. However, damage-resistant fibers developed for laser machining and surgery may prove useful in this regard. We know of only one chemical sensor that uses the extremely sensitive interferometric techniques *(66)*. This is not surprising, since it is necessary to transduce the chemical recognition event into a mechanical (and thus optical) distortion of a significant length of the fiber. This requires enough analyte to interact with that length of fiber, compromising sensitivity. Finally, little has appeared on the use of fibers with chemiluminescence or phosphorescence detection *(67)*. The former is particularly interesting, since it has very high inherent sensitivity (attomolar in favorable cases) (this volume). The number and diversity of the techniques mentioned above clearly suggest possibilities for expansion in the field of fiberoptic biosensors in the future.

7. Summary

We have discussed some of the virtues and problems of fiberoptic biosensors and suggestions for their improvement. It remains to be seen whether workers in the field, including ourselves, can develop fiberoptic biosensors that will live up to this potential, but the effort appears to be worth making.

Acknowledgments

The authors would like to thank the Office of Naval Technology, the Federal Aviation Administration, and the US Army Medical Research Institute of Infectious Disease for their support; Susan McBee and Lynne Kondracki for help with the figures and references; and Carl Villarruel for helpful discussions.

References

1. Polanyi, M. L., and Hehir, R. M. (1962) In vivo oximeter with fast dynamic response. *Rev. Sci. Instrum.* **33,** 1050–1054.
2. Hirschfeld, T. (1965) Total reflection fluorescence. *Can. J. Spectr.* **10,** 128.
3. Lee, S. H. (ed.) (1981) *Optical Information Processing* (Springer-Verlag, NY).
4. Stetter, J. R., Jurs, P. C., and Rose, S. L. (1986) Detection of hazardous gases and vapors: Pattern recognition analysis of data from an electrochemical sensor. *Anal. Chem.* **58,** 860–866.
5. Carey, W. P., Beebe, K. R., and Kowalski, B. R. (1987) Multicompartment analysis using an array of piezoelectric crystal sensors. *Anal. Chem.* **59,** 1529–1534.
6. Axelrod, D., Burghardt, T. P., and Thompson, N. L. (1984) Total internal reflection fluorescence. *Ann. Rev. Biophys. Bioeng.* **13,** 247–268.
7. Harrick, N. J. (1979) *Internal Reflection Spectroscopy* (Harrick Scientific, Ossinning, NY).
8. Anonymous. (1986) *Projects in Fiberoptics* (Newport Research, Mountain View, CA).
9. Snyder, A. W. and Love, J. D. (1983) *Optical Waveguide Theory* (Chapman and Hall, London).
10. Schulman, S. G. (ed.) (1985) *Molecular Luminescence Spectroscopy* (Part I) (Wiley, NY).
11. Udenfriend, S. (1962,1969) *Fluorescence Assay in Biology and Medicine* (2 vols.) (Academic, NY).

12. Hercules, D. M. (ed.) (1965) *Fluorescence and Phosphorescence Analysis* (Wiley-Interscience, NY).
13. Harrick, N. J. and Loeb, G. I. (1973) Multiple internal reflection fluorescence spectrometry. *Anal. Chem.* **45,** 687–691.
14. Kronick, M. N. and Little, W. A. (1975) A new immunoassay based on fluorescence excitation by internal reflection spectroscopy. *J. Immun. Methods* **8,** 235–240.
15. Kronick, M. N. and Little, W. A. (1976) Fluorescent immunoassay employing total reflection for activation. US Patent 3,939,350.
16. Woodhead, J. S., Addison, G. M., and Hales, C. N. (1974) The immunoradiometric assay and related techniques. *Br. Med. Bull.* **29,** 44–49.
17. Hirschfeld, T. E. (1984) Fluorescent immunoassay employing optical fiber in capillary tube. US Patent 4,447,546.
18. Andrade, J. D. and Van Wagenen, R. A. (1983) Process for conducting fluorescence immunoassays without added labels and employing attenuated internal reflection. US Patent 4,368,047.
19. Andrade, J. D., Van Wagenen, R. A., Gregonis, D. E., Newby, K., and Lin, J. N. (1985) Remote fiber-optic biosensors based on evanescent-excited fluoroimmunoassay: Concept and progress. *IEEE Trans. Elec. Dev.* **32,** 1175–1179.
20. Zhujun, Z. and Seitz, W. R. (1986) Optical sensor for oxygen based on immobilized hemoglobin. *Anal. Chem.* **58,** 220–222.
21. Arnold, M. A. (1985) Enzyme-based fiberoptic sensor. *Anal. Chem.* **57,** 565,566.
22. Peterson, J. I. and Vurek, G. G. (1984) Fiber-optic sensors for biomedical applications. *Science* **224,** 123–127.
23. Angel, S. M. (1987) Optrodes: Chemically selective fiber-optic sensors. *Spectroscopy* **2 (4),** 38.
24. Seitz, W. R. (1984) Chemical sensors based on fiberoptics. *Anal. Chem.* **56,** 16A–34A.
25. Thompson, R. B., Levine, M., and Kondracki, L. (1990) Component selection for fiberoptic fluorometry. *Appl. Spectrosc.* **44,** 117–122.
26. Thompson, R. B. Fluorescence-based fiberoptic sensors, in *Fluorescence Spectroscopy, vol. II: Biochemical Applications* (Lakowicz, J. R.,ed.), Plenum, New York, in press.
27. Lee, E.-H., Benner, R. E., Fenn, J. B., and Chang, R. K (1979) Singular distribution of fluorescence from liquids and monodispersed spheres by evanescent wave excitation. *Appl. Opt.* **18,** 862–868.
28. Thompson, R. B. and Kondracki, L. (1989) Waveguide parameter for waveguide-binding fiberoptic biosensors, in *Proceedings of the Eleventh International Conference of the IEEE Engineering in Medicine and Biology Society,* Kim, Y. and Spelman, F. A., eds., (Institute of Electrical and Electronic Engineers, NY), p. 1102,1103.

29. Marcuse, D. (1988) Launching light into fiber cores from sources in the cladding. *IEEE J. Lightwave Technol.* **LT-6,** 1273–1279.

30. Weber, G. (1975) Energetics of ligand binding to proteins. *Adv. Prot. Chem.* **29,** 1–83.

31. Green, N. M. (1963) Avidin: 1. The use of ^{14}C-biotin for kinetic studies and for assay. *Biochem. J.* **89,** 585–591.

32. Voller, A., and Bidwell, D. E. (1985) Enzyme immunoassays, in *Alternative Immunoassays* (Collins, W. P., ed.), Wiley, NY, chap. 6.

33. Bador, R., Dechaud, H., Claustrat, F., Desuzinges, C. (1987) Eu and Sm as labels in time-resolved immunofluorimetric assay of follitropin. *Clin. Chem.* **33,** 48–51.

34. Haugland, R. P. (1983) Covalent fluorescent probes, in *Excited States of Biopolymers* (Steiner, R. F., ed.), Plenum, NY, pp. 29–58.

35. Hlady, V., Reinecke, D. R., and Andrade, J.D. (1986) Fluorescence of absorbed protein layers. *J. Coll. Interface Sci.* **111,** 555–569.

36. Newby, K., Andrade, J. D., Benner, R. E., and Reichert, W. M. (1986) Remote sensing of protein adsorption using a single optical fiber. *J. Coll. Interface Sci.* **111,** 280–282.

37. Soini, E., and Hemmila, I. (1979) Fluoroimmunoassay: Present status and key problems. *Clin. Chem.* **25,** 353–361.

38. Lovgren, T., Hemmila, I., Pettersson, K., and Halonen, P. (1985) Time-resolved fluorometry in immunoassay, in *Alternative Immunoassays,* (Collins, W. P., ed.), Wiley, NY, chap. 12.

39. Oi, V. T., Glazer, A. N., and Stryer, L. (1982) Fluorescent phycobiliprotein conjugates for analyses of cells and molecules. *J. Cell Biol.* **93,** 981–986.

40. Tromberg, B. J., Sepaniak, M. J., Vo-Dinh, T., and Griffin, G. D. (1987) Fiber-optic chemical sensors for competitive binding fluoroimmunoassay. *Anal. Chem.* **59,** 1226–1230.

41. Weetall, H. H. and Filbert, A. M. (1974) Porous glass for affinity chromatography applications. *Methods Enzymol.* **34,** 59–72.

42. Wilchek, M., Miron, T., Kohn, J. (1984) Affinity chromatography. *Methods Enzymol.* **104,** 3–55.

43. Scouten, W. H. (1983) *Solid Phase Biochemistry: Analytical and Synthetic Aspects* (Wiley, NY).

44. Ji, T. H. (1983) Bifunctional reagents. *Methods Enzymol.* **91,** 580–.

45. Bhatia, S., Shriver-Lake, L., Prior, KJ., Georger, J. H., Calvert, J. M., Bredehorst, R., and Ligler, F. S. (1989) Use of thiol-terminal silanes and heterobifunctional crosslinkers for immobilization of antibodies on silica surfaces. *Anal. Biochem.* **178,** 408–413.

46. Lakowicz, J. R. (1983) *Principles of Fluorescence Spectroscopy* (Plenum, NY), chap. 5.

47. Lacovara, P., Esterowitz, L., and Allen, R. (1985) Flash-lamp pumped Ti:Al$_2$O$_3$ laser using fluorescent conversion. *Opt. Lett.* **10,** 273–275.

48. Villarruel, C. A., Dominguez, D. D., and Dandridge, A. (1987) Evanescent wave fiberoptic chemical sensor. *Proc. SPIE* **798,** 225–229.

49. Bosselmann, T., Reule, A., and Schroder, J. (1984) Fiber-optic temperature sensor using fluorescence decay time, in *Proceedings of the Second International Conference on Optical Fiber Sensors, Stuttgart* (VDE-Verlag, Berlin), pp. 151–154.

50. Bertil, H. and Jonsson, L. (1984) Pressure sensor with fluorescence decay as information carrier, in *Proceedings of the Second International Conference on Optical Fiber Sensors, Stuttgart* (VDE-Verlag, Berlin), pp. 391–394.

51. Thompson, R. B. and Vallarino, L. (1988) Novel fluorescent label for time-resolved immunoassay. *Proc. SPIE* **909,** 426–433.

52. Muller, F. and Schmidt, D. (1986) Ruthenium complexes useful as carriers for immunologically active materials. US Patent 4,745,076.

53. Newby, K., Reichert, W. M., Andrade, J. D., and Benner, R. E. (1984) Remote spectroscopic sensing of chemical adsorption using a single multimode optical fiber. *Appl. Opt.* **23,** 1812–1815.

54. Dakin, J. P. and King, A. J. (1983) Limitations of a single optical fiber fluorimeter system due to background fluorescence, in *Proceedings of the First International Conference on Optical Fibre Sensors* (Institute of Electrical Engineers, London), p. 195.

55. Schwab, S. D., McCreery, R. L., and Gamble, F. T. (1986) Normal and resonance Raman spectroelectrochemistry with fiberoptic light collection. *Anal. Chem.* **58,** 2486–2492.

56. Jordan, D. M., Walt, D. R., and Milanovich, F. P. (1987) Physiological pH fiber-optic chemical sensors based on energy transfer. *Anal. Chem.* **59,** 437–439.

57. Neal, S. L., Patonay, G., Thomas, M. P., and Warner, I. M. (1986) Data analysis in multidimensional luminescence spectroscopy. *Spectroscopy* **1**(3), 22.

58. Giuliani, J. F. and Bey, P. P. (1987) Multielement optical waveguide sensor, in *Digest of Papers of the Fourth International Conference on Solid-State Sensors and Actuators* (Institute of Electrical Engineers of Japan, Tokyo), p. 195.

59. Chan, K., Ito, H., and Inaba, H. (1984) An optical-fiber-based gas sensor for remote absorption measurement of low-level CH_4 gas in the near-infrared region. *IEEE J. Lightwave Technol.* **2** (3), 234–237.

60. Kersey, A. D., Dandridge, A., Tsveten, A. B., and Giallorenzi, T. G. (1985) Single-mode fiber Fourier transform spectrometer. *Electronics Lett.* **21,** 463,464.

61. Musikant, S. (1985) *Optical Materials* (Marcel Dekker, New York), chaps. 3 and 4.

62. Tran, D. C., Sigel, G. H., and Bendow, B. (1984) Heavy metal fluoride glasses and fibers: A review. *IEEE J. Lightwave Technol.* **2**(5), 566–586.

63. van Hell, H., Leuvering, J. H. W., and Gribnau, T. C. J. (1985) Particle im-

munoassays, in *Alternative Immunoassays*, (Collins, W. P., ed.), Wiley, New York, chap. 4.

64. Williams, D. J. (ed.) (1983) *Nonlinear Optical Properties of Organic and Polymeric Materials*, American Chemical Society Symposium Series No. 233, Washington, DC.

65. Anderson, B. E., Jones, R. D., Rehms, A. A., Ilich, P., and Callis, P. R. (1986) Polarized two-photon fluorescence excitation spectra of indole and benzimidazole. *Chem. Phys. Lett.* **125,** 106–112.

66. Butler, M. A. (1984) Optical fiber hydrogen sensor. *Appl. Phys. Lett.* **45,** 1007–1009.

67. Freeman, T. M. and Seitz, W. R. (1978) Chemiluminescence fiberoptic probe for hydrogen peroxide based on the luminol reaction. *Anal. Chem.* **50,** 1242–1246.

Optical Characteristics of Fiberoptic Evanescent Wave Sensors

Theory and Experiment

Walter F. Love, Leslie J. Button, and Rudolf E. Slovacek

1. Introduction

The availability of low-loss optical fibers with the associated benefits of freedom from electromagnetic interference, small size, light weight, and wide bandwidth has led to the development of sensors, both intrinsic and extrinsic. Commercial offerings of the extrinsic type, in which the sensing element is attached to the end of the fiber, are becoming readily available for the measurement of many physical and chemical parameters, e.g., temperature, pressure, and pH. The intrinsic sensors, in which the optical fiber itself functions as the sensor, potentially offer more sensitivity and versatility. In some cases, distance or location information may be obtained with intrinsic sensors by means of optical-pulse techniques. The intrinsic sensors are usually more complex in their instrumentation or environmental requirements, and their entry into the commercial market has proceeded more slowly.

The evanescent wave sensor is an intrinsic sensor, since the core of the optical fiber interacts directly with the material being analyzed. A dielectric cylindrical fiber immersed in a medium of smaller refrac-

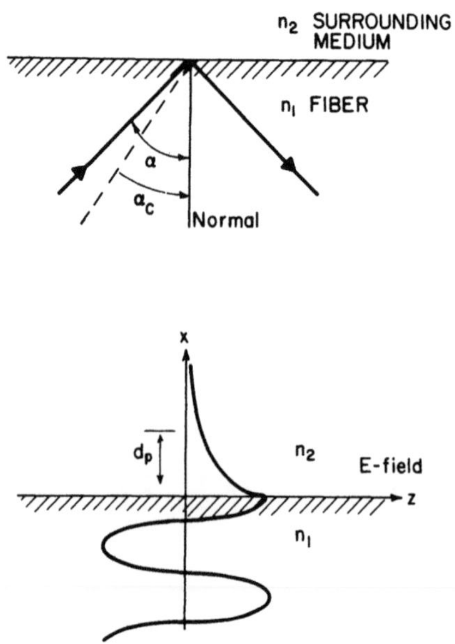

Fig. 1. Light ray incident on a dielectric interface at angle α—incident and evanescent fields. The refractive index of the waveguide is n_1, the surrounding medium is n_2, d_p is the penetration depth, and α_c is the critical angle.

tive index forms an optical waveguide. If the surrounding medium is fluorescent, then the fiber can be used as a sensor, which both excites and collects fluorescent radiation. This is illustrated in Fig. 1, in which a planar interface is shown for simplicity. A light ray is incident on the interface between two dielectric media, having refractive indices n_1 and n_2 with $n_1 > n_2$. When the ray angle, α, exceeds the critical angle, α_c, relative to the normal to the interface, the incident light ray is totally internally reflected, with no refracted light wave entering the surrounding medium. This critical angle, α_c, is given by

$$\alpha_c = \sin^{-1}(n_2/n_1) \tag{1}$$

In the medium, with refractive index n_2, there is a nonpropagating or evanescent wave that is nonzero for distances close to the interface. In particular, this wave decays exponentially to zero with distance, δ, from the interface (Fig. 1) as

$$E = E_s \exp[-\delta/d_p] \tag{2}$$

where E_s is the electric field amplitude at the surface, viz., $\delta = 0$, and d_p is the penetration depth, or distance for the electric field amplitude to fall to $1/e$ (0.37) of its value at the interface. The penetration depth, d_p, is given in Eq. (3) according to ref. *(1)*

$$d_p = \lambda/[2\pi n_2 (n_{rel}^2 \sin^2 \alpha - 1)^{1/2}] \tag{3}$$

where $n_{rel} = n_1/n_2$. The penetration depth becomes large when α approaches α_c, indicating the transition from an evanescent to a refracting wave in the surrounding medium. Typically, $d_p < \lambda$, the wavelength of light in vacuum.

The evanescent electromagnetic field can interact with molecules in the surrounding medium (i.e., with refractive index n_2) near the interface, thereby producing absorption and fluorescence with incident light of the appropriate wavelength. Hirschfeld *(2)* demonstrated this effect for dilute solutions of fluorescein dye using a planar geometry. Highly absorbing or turbid (scattering) solutions can be more easily measured using evanescent wave excited fluorescence, since the light beam does not pass directly through the solution, where reabsorption and scattering would otherwise take place. Because of these effects, the linearity of the detected fluorescence with chemical concentration is further extended to the higher concentration ranges. Alternatively, at the lower concentration ranges, the sensitivity can be increased by multiple internal reflections. An optical fiber has a definite advantage over the planar geometry in this regard, since the internal reflections are confined to two of the three dimensions rather than one. Spreading of light into the unconfined transverse dimension is thus eliminated, and low-noise, small-area photodetectors can be used. The total number of reflections, N, in an optical fiber of length, L, and radius, a, is given by

$$N = (L/2a) \cot \alpha \tag{4}$$

where α is the angle shown in Fig. 1, which is referred to the normal to the core-surrounding medium interface. Although Eq. (4) applies rigorously only for meridional rays, it is instructive to calculate N for the experimental conditions used in Section 4. Thus, for diameter $2a =$

500 µm, $L \approx 60$ mm, and $\alpha = 65° \approx \alpha_c$, $N \approx 60$ reflections. It will be shown in the section on theory that both the absorption and emission of fluorescing molecules in the surrounding medium are maximized near the critical angle. Furthermore, the fluorescent light can be collected and guided by the optical fiber, and imaged through appropriate optics onto a low-noise photodetector. The evanescent wave technique is thus ideally suited to the identification and quantification of chemical species that are bound to the surface of the fiber or retained within the evanescent wave region characterized by penetration depth d_p. Since d_p is typically <1 µm, the fluorescence occurring near the fiber surface is preferentially collected over fluorescence occurring in bulk solution, which significantly reduces background fluorescence. This bulk fluorescence constitutes a noise source when surface fluorescence is required and can limit the sensitivity of detection.

A complete and quantitative optical theory specifically for fiberoptic evanescent wave fluorescence sensors is not currently available, although some details and principles have been established. Sensitivity with an eighth-power dependence on launch numerical aperture (i.e., $\sin^8\theta$) was shown *(3)*, and the importance of maintaining high-angle light propagation for maximum fluorescence signal was demonstrated both experimentally and theoretically *(4)*. A recent article *(5)* outlines a theory for the effect of launch conditions on both absorption and emission processes, but does not establish quantitative details for the complete system.

In this chapter, a quantitative optical theory of fiberoptic evanescent wave sensors is developed from first principles. The theoretical predictions are discussed in Section 3. and are compared with experimental results in Section 4. In Section 5., the key results and conclusions are summarized.

2. Theory of Evanescent Wave Excitation and Detection of Fluorescence

2.1. Overview

When a waveguide has a diameter that is large compared to the wavelength of light, the individual reflections of the guided light rays from the core–cladding interface can be described to a good approximation as reflections from a planar interface. The cylindrical sensor described in Section 1. can thus be characterized by adding together contributions from individual rays, with each ray analyzed as though it interacts with a plane. The interactions of individual light rays with a dielectric interface have been carefully analyzed by Lee et al. *(6)*. A similar treatment is presented here, adapted for use with a continuous set of ray angles. The physical configuration to be examined first is the planar one, in which fluorescent molecules are situated on one side of a planar dielectric interface, in the region of lower refractive index, n_2. Light impinges on the interface from the opposite side at an angle, α, with respect to the normal, with $\alpha > \alpha_c$ (Fig. 1). The light undergoes total internal reflection, without loss of power, in the absence of absorbent molecules. However, power may be removed from the incident beam as a result of the interaction between the evanescent electric field and the electric dipole moments of molecules located in the medium of lower index. A semiclassical analysis of the electric field/molecular dipole interaction accurately describes the absorption process *(7)*.

Likewise, there is a finite probability that radiation emitted by a dipole located in the medium of lower index will propagate as a plane wave above the critical angle within the denser medium. In quantum-mechanical terms, there is a precise reciprocity between the absorption and emission of radiation because of the symmetry of the Hamiltonian operator. In the classical picture presented here, this reciprocity is exhibited by the exact correspondence of the angle dependences for absorption and emission.

The emission probability into any range of solid angle is obtained from the classical radiation pattern for a dipole near a dielectric interface. This can be calculated in detail *(8–10)*. Combining the results for absorption and emission gives the fluorescent signal per unit area of interface caused by an incident plane wave.

Using the planar geometry results as a building block, the case of the cylindrical waveguide immersed in a fluorescent medium can be studied. Each ray ("local plane wave") launched into the waveguide is considered an independent producer of fluorescence; the net signal is obtained by adding the signals generated by each ray. The only approximation required is that the waveguide diameter be very large compared to the wavelength of the light, so that the results from the planar case may be applied. The cylindrical geometry of the waveguide is accounted for in the sense that it defines a particular set of rays that excite fluorescence and that can be collected and detected. The waveguide also provides multiple reflections for each ray. Details of the complete calculation are provided in Sections 2.2.–2.5. The following is an overview of the steps involved:

- First, the power absorbed from a local plane wave or ray at each layer of the fluorescent medium is calculated, using the magnitude of the electric field obtained from the Fresnel equations and the optical absorption coefficient of the fluorescent medium. This "local absorption function" is denoted $P_{abs}(\delta;\alpha)$, where α is the angle of the ray to the normal (as in Fig. 1) and δ is the distance of fluorescent material from the interface in the medium of lower index. The result is presented as Eq. (9) and derived in detail in Section 2.6.
- Then the probability that a dipole will radiate into the cone of rays that are within the detection optics of the optical system is determined as a function of distance δ from the waveguide surface. This function is denoted $p_{emit}(\delta;\theta_0^{max})$, where θ_0^{max} is a "maximum detection angle" defined by the detection optics (*see* Section 2.3.). For the optical system and experiments discussed in Section 4., the maximum launch and detection angles are the same.
- Assuming that all the power absorbed by the dipoles is reradiated, the fluorescent signal per unit area of interface for the ray is given by Eq. (5) (*see* Section 2.4.).

$$I_{fluor}(\alpha) = \int_0^\infty P_{abs}(\delta;\alpha)p_{emit}(\delta;\theta_0^{max})d\delta \qquad (5)$$

In Eq. (5), a distribution function for the density of fluorescent molecules may be included inside the integral; this allows a treatment of both the bulk-fluorescence case, in which the molecules are uniformly distributed in solution, and the thin-film case, in which all of the absorption takes place in a film layer near the interface. The equation can also be made more realistic by including a multiplicative factor for the quantum efficiency for conversion of absorbed radiation to fluorescent radiation.

- Thus, the total signal generated by a single ray is

$$dS_{ray}(\alpha) = NI_{fluor}(\alpha)\, da\, d\Omega \tag{6}$$

where N is the total number of reflections (interactions with the interface) that the ray experiences, da is the projection of the ray's (infinitesimal) cross-section on the fiber's surface, and $d\Omega$ is the element of solid angle in which the ray direction is contained (*see* Section 2.4., Eq. [21]).

- Finally, the total fluorescent signal emanating from one end of the waveguide is obtained by summing the previous results over all the rays launched by the input optics. This amounts to an integral over solid angle for the ray directions, and over the end face of the fiber. Various source characteristics can be dealt with easily. *See* Section 2.5., where Eq. (28) describes the calculation for a Lambertian source.

Section 2.6. is a mathematical appendix containing explicit formulae for refracted fields and for dipole radiation patterns in the presence of an interface.

2.2. Excitation of Fluorescence

The excitation of fluorescence is described in this section as simple absorption or dissipation of power from an incident light ray. Each ray is assumed to be randomly polarized as it approaches the interface; the net result for the power absorbed from the ray is then a simple average over the two independent polarizations. Note that Gaussian units for electromagnetic quantities are used in this and the following sections.

In an absorbing medium, the phenomenon of power dissipation or energy absorption means that the optical irradiance (J, power per unit area) obeys Lambert's law, $J(x) = J_0\exp(-\gamma x)$, where γ is the absorption constant. In a small layer of thickness Δx, the irradiance lost is therefore $\Delta J = \gamma J \Delta x$. The power flux is given by $J = cn|E|^2/8\pi$, where E is the electric field, n is the refractive index, and c is the speed of light, so the power absorbed (lost from the beam) per unit volume can be written

$$P_{abs} = \Delta J/\Delta x = (\gamma cn/8\pi)|E|^2 \tag{7}$$

The calculation pertains only to the case of weak absorption, $\gamma\lambda \ll 1$, where λ is the wavelength in vacuum. Although Eq. (7) was implicitly derived for propagating plane waves, it is also correct for evanescent waves *(1)*.

The geometry of Fig. 1 represents the physical situation, in which a plane wave is incident on a dielectric interface with the wave vector at an angle, α, with respect to the normal. The upper region contains the fluorescent (absorbing) medium. Fields for the reflected and transmitted waves may be calculated in terms of the incident field using the Fresnel relations (*see* Section 2.6.). Then the absorption in the fluorescent region (the upper half-space $x > 0$) is obtained from the transmitted fields and Eq. (7). The major qualitative feature of the transmitted fields is that, if α is greater than the critical angle α_c, the fields decay exponentially into the upper half space, with a decay length equal to d_p, defined by Eq. (3).

The absorbed power for a molecule at a distance, δ (on the x axis), for each incident polarization is obtained from Eq. (7) using the fields calculated in Section 2.6. Integrating over δ provides the total power absorbed per reflection. If this result is divided by the incident power flux normal to the interface, given by Eq. (8)

$$I_{ray} = (cn_1/8\pi)E_0^2 \cos \alpha \tag{8}$$

the familiar "effective thickness" formulas from the theory of internal reflection spectroscopy *(1)* may be obtained. The important result here is that the local power absorption at a given distance, δ, from the interface is given, for randomly polarized light, by Eq. (9):

$$P_{abs}(\delta;\alpha) = 2\gamma I_{ray} \cos \alpha\{[n_{rel}\exp(-2\delta/d_p)]/(n_{rel}^2 - 1)\}$$

$$\times \{1 + (2n_{rel}^2\sin^2 \alpha - 1)/[(n_{rel}^2 + 1) \sin^2 \alpha - 1]\} \tag{9}$$

The second line of Eq. (9) arises from averaging over the two possible incident polarizations. As described by Glass et al. *(5)*, this factor (which is denoted in the referenced paper as $1 + g[\alpha]$) varies by only about 2% over the range of α values of interest, and it can be considered a constant equal to its average value for all practical purposes.

Equation (9) allows the incident flux of the ray and the absorption constant to be factored out, and defines a dimensionless "geometrical absorption factor" F_{abs} by

$$F_{abs}(\delta;\alpha) = P_{abs}(\delta;\alpha)/\gamma I_{ray}(\alpha)$$
$$= 2\cos\alpha\{[n_{rel}\exp(-2\delta/d_p)]/(n_{rel}^2 - 1)\}$$
$$\times \{1 + (2n_{rel}^2\sin^2\alpha - 1)/[(n_{rel}^2 + 1)\sin^2\alpha - 1]\} \tag{10}$$

The significant feature of F_{abs} is its exponential dependence on δ/d_p. The absorption factor is much larger for angles near the critical angle, where $d_p \to \infty$.

The specific case in which $n_1 = 1.46$ and $n_2 = 1.33$, corresponding to fused silica and water, respectively, is illustrated in Figs. 2 and 3. Figure 2 is a graph of calculated F_{abs} vs the incident angle, α, for several values of δ/λ. Figure 3 shows the exponential dependence of F_{abs} on δ for several values of α; the curve with $\alpha = 65.7°$ shows the strong enhancement of absorption for light near the critical angle ($65.6°$ in this case).

The calculation can be generalized to a case in which the fluorescent molecules have a preferred orientation. The absorption is maximized when the electric field vector is parallel to the molecular dipole moment and, in general, depends linearly on the cosine of the angle between the field vector and the dipole moment. In Fig. 4, an ellipsoidal molecule with its dipole moment along its long axis is indicated. It can be seen that incident fields polarized parallel to the interface interact best with those dipoles lying in the plane of the interface. In Fig. 4, neither polarization of the incident waves shown will interact effectively with the molecule in the displayed orientation.

For the case of antibody/antigen molecules that attach closely to the surface of an optical fiber, we assume random molecular orientation. Hence, averaging over orientations will yield a constant numerical factor. The reader should be alerted that this may not be the result in highly symmetrical configurations. For example, it is shown in ref. *(7)* that elongated fatty acid molecules can be arranged to lie in the plane of the interface and, arranged in this manner, will not interact with the polarization perpendicular to the interface.

2.3. The Emission Probability

This section is concerned with the construction of the function p_{emit}, which describes the probability that an oscillating electric dipole

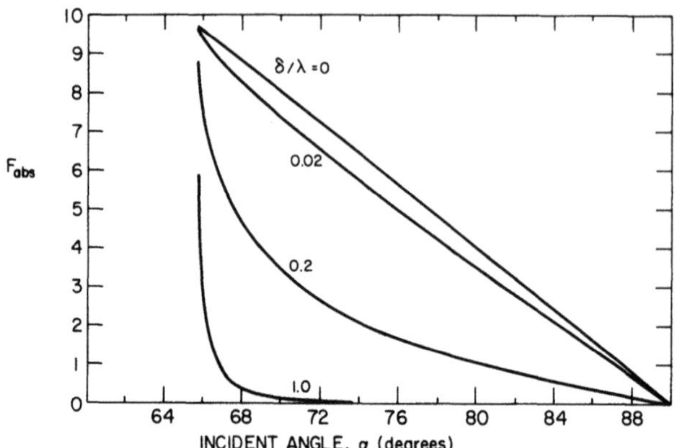

Fig. 2. Calculated geometrical absorption factor, F_{abs}, as a function of angle of incidence for several distances δ/λ and for $n_1/n_2 = 1.1$.

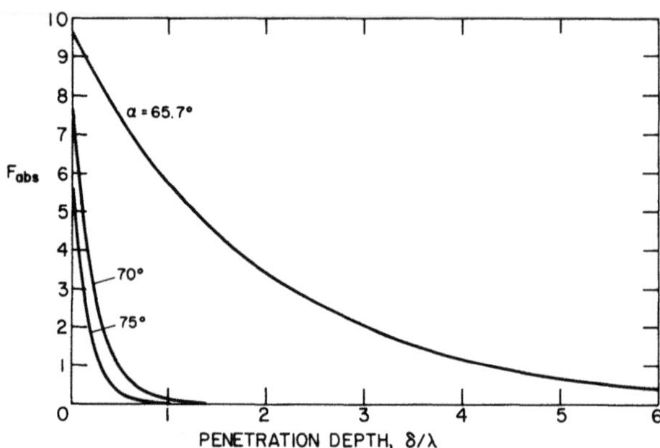

Fig. 3. Variation of calculated geometrical absorption factor, F_{abs}, with distance from the interface for several values of incidence angle α and for $n_{rel} = 1.1$.

at a distance, δ, from a dielectric interface will emit radiation into a certain cone of solid angles. This cone is defined by the experimental system and can be varied systematically.

To calculate p_{emit}, the radiation from a classical dipole is needed. Consider the following geometry, as shown in Fig. 5: A dipole of strength D_0, represented by the dark oval, is located at a distance, δ, from

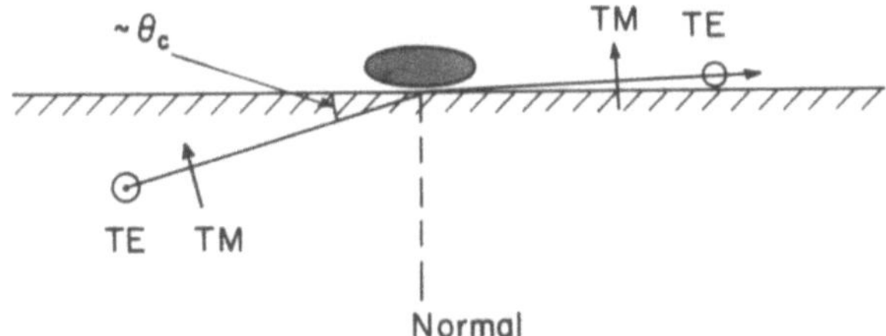

Fig. 4. Polarization dependence of dipole/field interaction near the critical angle. Fields in the plane of incidence (TM) interact primarily with dipoles oriented at a normal to the interface; fields perpendicular to the plane of incidence (TE) interact with dipoles parallel to the interface.

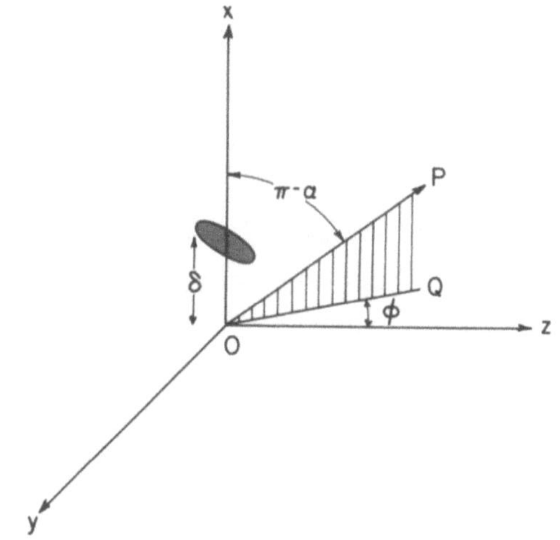

Fig. 5. Geometry for dipole radiation calculation. The black oval symbolizes a molecular dipole. OP corresponds to the outgoing ray direction, and OQ is the projection of OP, which lies in the *yz* plane. The angle ϕ is the azimuthal angle between OQ and the *z* axis.

a dielectric interface. The half-space in which the dipole is contained, i.e., the region "above" the interface, contains material with index of refraction n_2; the other half-space has index n_1. The normal pointing from the interface toward the dipole defines the *x* axis. As defined previously, α is the angle measured from the normal pointing into the medium of higher index (the negative *x* axis).

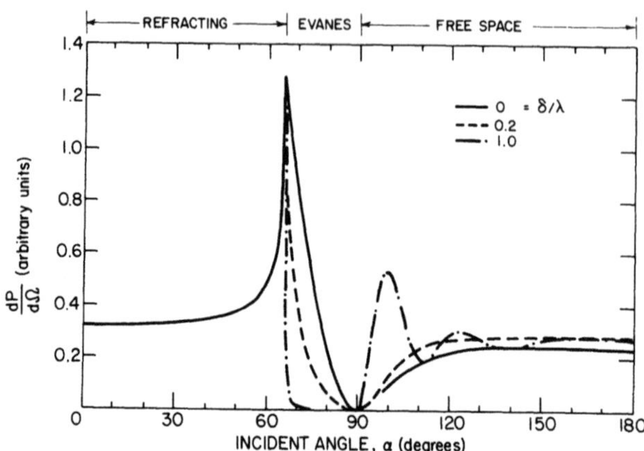

Fig. 6. Radiation pattern of dipoles located at the interface and at two distances δ/λ from the interface. The radiation pattern is averaged over an isotropically random distribution of dipole orientations.

The total power radiated per unit solid angle for a dipole of arbitrary orientation may be calculated explicitly *(8–10)*. The mathematical technique consists of expanding the dipole radiation field in a set of homogeneous and inhomogeneous (evanescent) plane waves. The reflection and refraction of each of these plane and evanescent waves incident on the interface are then considered separately, in such a way that the boundary value problem is solved. From the electromagnetic fields thus obtained in the two half—spaces, the angular distribution of the optical power emitted by the dipole is calculated. The final results for arbitrary angle α and arbitrary dipole orientation are summarized in Section 2.6., namely Eqs. (33–39). The quantity $dP/d\Omega$ thus obtained is the optical power emitted in a given direction per unit solid angle. By averaging over the orientations of the dipole, the power per unit solid angle for an isotropically random collection of dipoles has been obtained. Figure 6 is a graph of this result for the cases $\delta/\lambda = 0, 0.2$, and 1.0. For values of δ that are small compared to a wavelength, appreciable power is emitted into the region $\alpha_c \leq \alpha \leq \pi/2$.

In Section 2.6., Eq. (40), the specific case of angles within the region $\alpha_c \leq \alpha \leq \pi/2$ has the form

$$dP/d\Omega = (cn_2 k^4 D_0^2/6\pi)[n_{rel}^3 \cos^2 \alpha/(n_{rel}^2 - 1)] \exp (-2\delta/d_p)$$
$$\times \{1 + (2n_{rel}^2\sin^2\alpha - 1)/[(n_{rel}^2 + 1)\sin^2\alpha - 1]\} \tag{11}$$

This region of angles corresponds to emitted plane waves that are evanescent in the upper half-space. The structure of this formula is remarkably similar to the power absorption result in Eq. (9). This is an explicit case in which the reciprocity between absorption and emission of radiation is clearly visible. Indeed, if these results for absorption and emission rates are reformulated in terms of transition probabilities, exact equality is obtained *(11)*.

The results for the power radiated by a classical dipole are not immediately applicable to a quantum-mechanical dipole transition, since the total energy radiated by the molecule in a quantum transition is a well-defined quantity, whereas radiation of power does not continue once the quantum system emits a photon. However, by Bohr's correspondence principle, the total classical radiation rate is inversely related to the quantum lifetime. Similarly, the classical radiation pattern may be used to determine an angular probability distribution for photon emission: The radiation pattern is proportional to the probability density for emission into a given element of solid angle, as follows: The total radiated power is

$$P_{tot} = 2\pi\int_0^\pi (dP/d\Omega) \sin \alpha \, d\alpha \tag{12}$$

so the probability density for emission into the solid angle element $d\Omega$ is $P_{tot}^{-1} (dP/d\Omega)$.

To obtain the probability of emission into the set of "detected" rays, the set of rays within the detection optics must be determined. The sensor systems considered here have the property that both launch and detection of light from the end face of the fiber are carried out with the same optics so that the "cone of detection" in which fluorescent radiation can be detected is equal to the "cone of excitation" into which probe light can be launched.

The rays that strike the end face of the fiber are restricted to within a cone of half-angle θ_0^{max}, where $\sin \theta_0^{max}$ is the numerical aperture (NA) of the launch lens, which is experimentally controlled. Angles with a subscript of "0" are defined to be angles exterior to the fiber, such as

                                                    Love, Button, and Slovacek

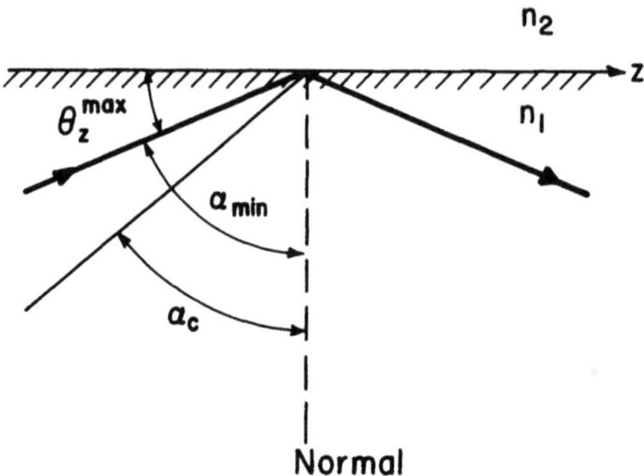

Fig. 7. Light ray shown inside the fiber with maximum detectable angle θ_z^{max}. The angles θ_z^{max} and α_{min} are complementary angles. The interior angle θ_z^{max} and the exterior launch angle θ_0^{max} are related by Eq. 24.

launch angles. The external angle θ_0^{max} corresponds by Snell's law to a certain internal angle, α_{min}, at which rays meet the core-surrounding-medium interface; that is, launched and detected rays meet the interface with an angle to the normal in the range $\alpha_{min} \leq \alpha \leq \pi/2$ (*see* Fig.7). The angle α_{min} is determined explicitly from θ_0^{max} in Section 2.5. The condition that α_{min} should be greater than the critical angle α_c is shown there to be equivalent to the condition that only light within the material NA, i.e. $NA = [n_1^2 - n_2^2]^{1/2}$, of the waveguide is launched.

At this point, the cylindrical nature of the waveguide plays an important role. Not all the rays that are incident on the core-surrounding-medium interface in the range of angles $\alpha_{min} \leq \alpha \leq \pi/2$ are in fact detected by the system optics. Indeed, some of those rays are not properly guided by the waveguide, even though the rays nominally appear to be totally internally reflected. This phenomenon is a crucial consequence of the curvature of the waveguide's core-surrounding-medium interface. This curvature causes only those rays within a cone oriented along the waveguide axis to be in fact guided *(12)*; the rest form the so-called tunneling rays. Figure 8 contains a schematic representation of the half-cone of solid angle in which rays are bound and "detectable." The lower half of the figure defines the various angles used in the calculation.

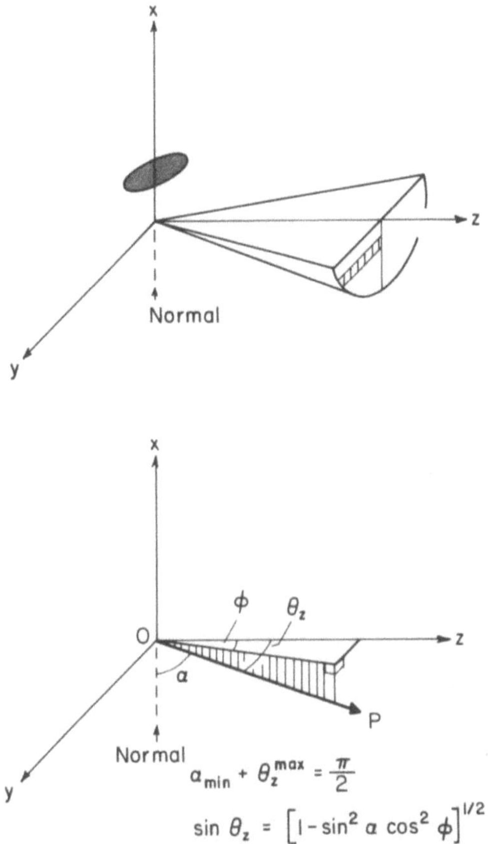

$$\alpha_{min} + \theta_z^{max} = \frac{\pi}{2}$$

$$\sin \theta_z = \left[1 - \sin^2 \alpha \cos^2 \phi \right]^{1/2}$$

Fig. 8. The half-cone of rays emitted by the dipole (dark oval) that are bound within the waveguide and are detectable. This half-cone forms the region over which the dipole radiation intensity is integrated to obtain p_{emit} (Eq. 13).

To be explicit, assume that $\varphi = 0$ and $\varphi = \pi$ are the directions that point along the waveguide axis. According to Figs. 7 and 8, the condition that light be contained within the cone corresponding to $\sin\theta_z^{max}$ can be written as $\sin\theta_z = (1 - \sin^2\alpha\cos^2\phi)^{1/2} \leq \sin\theta_z^{max} = \cos\alpha_{min}$. Hence the launch/detection cone limits for $\alpha_{min} \leq \alpha \leq \pi/2$ are $\varphi = \pm\cos^{-1}$ (sin α_{min}/sin α) and $\varphi = \pi \pm\cos^{-1}$(sin α_{min}/sin α). The two ranges for φ correspond to forward- and backward-directed rays, respectively.

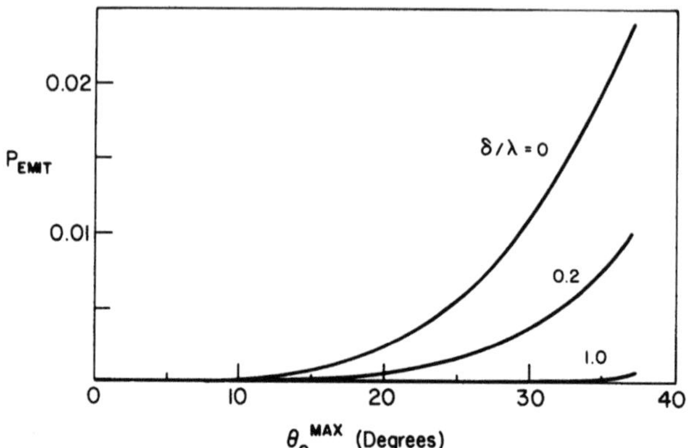

Fig. 9. Probability p_{emit} of emission into bound rays as a function of maximum launch angle for several distances δ/λ of dipole from waveguide and for $n_{rel} = 1.1$.

To define the probability of launching a detectable ray, the probability density for the radiation direction, defined by $P_{tot}^{-1}(dP/d\Omega)$, is integrated over the ray angles that are within the launch/detection cone, as defined by the above limits on α and φ. The result is the total probability that a photon emitted by the dipole will propagate as a detectable forward ray of the waveguide. (Only one of the two ranges of solid angle is used, since detection takes place only at a single end of the optical fiber.)

Denote the probability obtained $p_{emit}(\delta; \theta_0^{max})$:

$$p_{emit}(\delta; \theta_0^{max}) = (2/P_{tot})\int_{\alpha_{min}}^{\pi/2} \cos^{-1}(\sin \alpha_{min}/\sin \alpha)(dP/d\Omega) \sin \alpha \, d\alpha \qquad (13)$$

where the relation between α_{min} and θ_0^{max} is determined in Section 2.5., Eq. (25).

Equation (13) cannot be integrated analytically, but a numerical integration allows it to be calculated conveniently. Figure 9 is a graph of p_{emit} for the specific case of $n_1 = 1.46$ and $n_2 = 1.33$, with $\delta/\lambda = 0, 0.2$, or 1. Note that p_{emit} becomes large for a collection cone with a maximum angle approaching the critical angle. This is similar to the response found for the dipolar absorption, as expected from the reciprocity principle *(7)*. For optimum conditions and $\delta/\lambda = 0$, approx 2% of the emitted light is captured and guided by the fiber.

Figure 10 shows the dependence of p_{emit} on δ/λ for $n_1 = 1.46$ and $n_2 = 1.33$, with θ_0^{max} equal to 15°, 30°, or 37°. (The maximum detection angle possible is $\sin^{-1}[n_1^2 - n_2^2]^{1/2} = 37.03°$.) Particularly notable is the decreased slope in decay of p_{emit} with δ when $\sin\theta_0^{max}$ approaches the fiber's NA, i.e., when light near the critical angle is included in the detection cone. This calculation can be generalized to cases other than the complete isotropic randomness considered above by averaging the general results of Lukosz *(10)* over the appropriate orientation distribution for the dipoles.

2.4. Fluorescent Signal from a Ray

This section develops the calculation of I_{fluor}, the fluorescent signal per unit area of interface from a single incident plane wave, and dS_{ray}, the fluorescent signal from a given ray summed over the length of the waveguide.

Put simply, I_{fluor} is the power absorbed at a depth, δ, in the fluorescent medium times the probability that a photon emitted at δ will propagate as a detectable waveguide ray, integrated over δ

$$I_{fluor}^{(Bulk)}(\alpha) = \int_0^\infty P_{abs}(\delta;\alpha)p_{emit}(\delta;\theta_0^{max})d\delta$$

$$= \gamma I_{ray}(\alpha) \int_0^\infty F_{abs}(\delta;\alpha)p_{emit}(\delta;\theta_0^{max})d\delta \tag{14}$$

In the case in which bulk fluorescence is being detected, it is important to understand this expression—it does *not* allow the fluorescent signal to be decomposed into an expression involving the total absorbed power (at all distances from the interface) times some probability for detection of radiated light. Instead, contributions to the fluorescent signal from thin layers of dipoles are doubly attenuated with distance from the interface. First, the amount of power absorbed in a layer at distance δ decreases exponentially with δ. Second, the probability p_{emit} for detection of radiation emitted from a layer at distance δ also decreases in a nearly exponential manner.

A thin-film fluorescent layer of thickness d at the surface may be treated by simply integrating over the layer (the difference in refractive index between the film and the rest of the surrounding medium is assumed to be small, so the same expression for "local absorption" may be used):

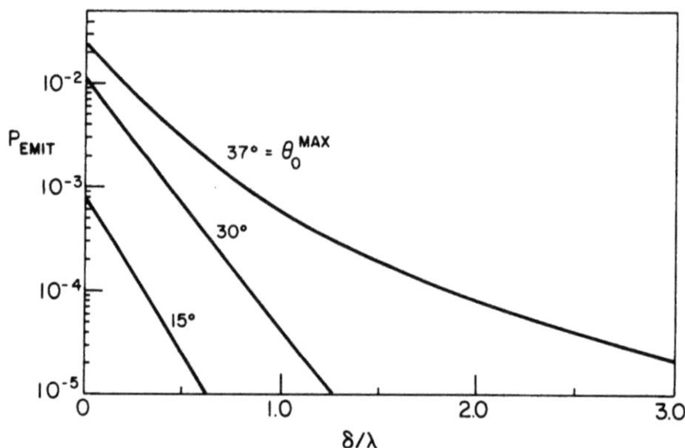

Fig. 10. Probability p_{emit} of emission into bound rays as a function of distance δ/λ for several values of maximum launch angle and for $n_{rel} = 1.1$.

$$I_{fluor}^{(Film)}(\alpha) = \gamma I_{ray}(\alpha)\int_0^h F_{abs}(\delta;\alpha)\, p_{emit}(\delta;\theta_0^{max})\,d\delta \qquad (15)$$

where h is the film thickness.

If h is much smaller than a wavelength, so that P_{abs} and p_{emit} are approximately constant over the layer, we obtain

$$I_{fluor}^{(Film)}(\alpha) = \gamma h \times I_{ray}(\alpha)\, F_{abs}(\delta = 0;\alpha)\, p_{emit}(\delta = 0;\theta_0^{max}) \qquad (16)$$

In this case, the fluorescent signal *does* decouple into a simple product of the absorbed power and the emission probability. In practice, this approximation is correct only for films that are on the order of 0.01 wavelengths thick. Even for a film thickness of 0.1 λ, there is significant error in using Eq. (16), leading to an overestimate of $I_{fluor}^{(Film)}$ of approx 50%.

To calculate the total signal from a given ray, the number of reflections, N, taken by the ray is needed, since only the signal from a single reflection has been considered until now. N is most easily described in terms of some parameters from the ray theory for optical waveguides *(12)*.

A bound ray of the cylindrical waveguide is characterized by two angles, as shown in Fig. 11: θ_z, the angle with respect to the

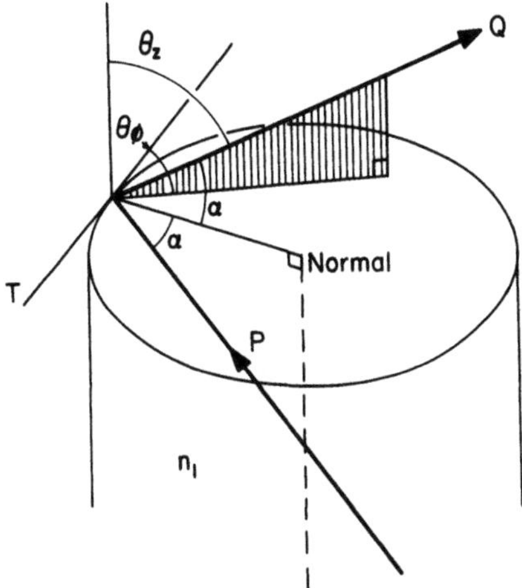

Fig. 11. Ray geometry in a cylindrical dielectric waveguide. Rays P and Q make equal angles α with the tangent T to the cylinder.

waveguide axis, and θ_φ, the azimuthal angle about the normal to the core-surrounding-medium interface.

These angles are invariants, the same at each reflection. The condition that the ray be guided is $\theta_z < \pi/2 - \alpha$, where α_c is the critical angle (given in Eq. [1]).

Let the angle α continue to represent the angle of the ray with respect to the normal to the core-surrounding-medium interface. Its value is determined from

$$\cos \alpha = \sin \theta_z \sin \theta_\varphi \qquad (17)$$

The half-period of the ray, the distance along the waveguide between successive reflections, is

$$z_p = 2a \sin \theta_\varphi / \tan \theta_z \qquad (18)$$

where a is the fiber or waveguide radius. The total number of reflections in a length of fiber, L, is therefore

$$N = L/z_p = (L \tan \theta_z)/(2a \sin \theta_\varphi) \qquad (19)$$

Finally, the element of area $d\bar{A}$ along the core-surrounding-medium interface is needed, since all the results to this point have been expressed per unit area. The area element can be projected back to the area element, dA, on the end face of the fiber (as will be required once launch of power onto the end face is considered). From Fig. 12

$$d\bar{A} = (d\bar{A}/dA)dA = \cot \theta_z \, dA \tag{20}$$

Using these waveguide theory results, the total signal from a ray originating in the area element dA on the end face of the fiber with a direction determined by the angles α and θ_φ is given by

$$dS_{ray}^{(Bulk)} = I_{fluor}(\alpha) \, (L/2a \sin \theta_\varphi) \, dA$$
$$= \gamma I_{ray}(\alpha) \, (L/2a \sin \theta_\varphi) \, dA \int_0^\infty F_{abs}(\delta;\alpha) \, p_{emit}(\delta;\theta_0^{max}) \, d\delta \tag{21}$$

The thin film fluorescent layer is treated by using Eq. (16) instead of (14) for I_{fluor}.

2.5. Summing over the Rays

The rays that are excited into the waveguide are dependent on the illumination of the end face *(12)*. The following assumptions are made about the illumination. The light that strikes the end face of the fiber is assumed to be produced by a diffuse, Lambertian source. The light is focused so that it uniformly illuminates a spot centered on the end face of the fiber. The radius of the spot is r_{max}, where r_{max} is less than or equal to the radius of the fiber.

The optical power flux per unit area within the spot, per unit solid angle $d\Omega$, is

$$dP_{inc} = I_0 \cos \theta_0 \, dA \, d\Omega \tag{22}$$

where θ_0 is the polar angle measured from the fiber axis, dA is the area element on the fiber end face, and I_0 is the optical radiance *(13)*, which is expressed in units of power per unit area per unit solid angle. The launch lens has a specific NA that cuts off the power at some angle θ_0^{max} (which is never allowed to be more than the NA of the optical waveguide).

Let (r, φ) be polar coordinates on the end face of the fiber defining the position of the entering rays, and let (θ_0, ψ) describe the spherical

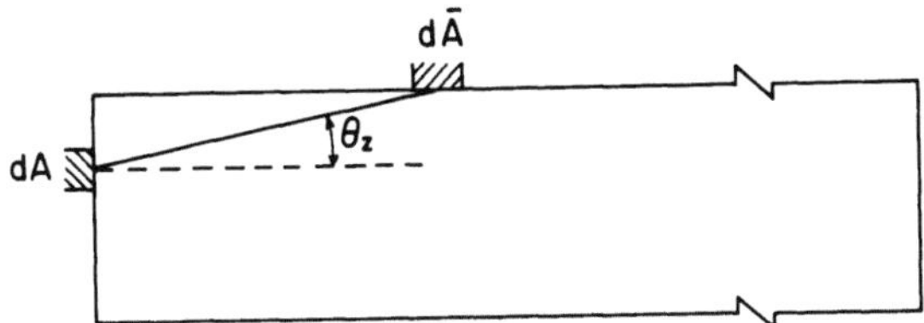

Fig. 12. Projection of end face area, dA, onto lateral waveguide surface, $d\bar{A}$.

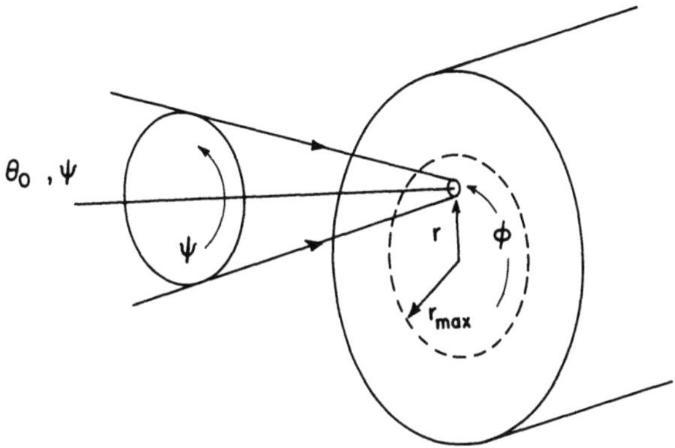

Fig. 13. Definition of launch variables. Incident ray angles within the launch cone of light are described by θ_0, φ, and the fiber end face is described by polar coordinates r, φ. The rays are incident only within a spot of radius r_{max}, with $r_{max} < a$.

polar directions for rays entering at the point (r, ψ) (*see* Fig. 13). The medium from which the end face is illuminated is assumed to be air.

Because of the azimuthal symmetry about the fiber axis, the origin of the ψ coordinate is arbitrary. It is defined here so that the projection of the ray on the end face makes an angle ψ, with respect to the azimuthal (φ) direction. It can then be shown, using the law of sines (*see* Fig. 14), that ψ is related to the second invariant ray angle, θ_φ, by

$$(r/a) \sin \psi = \cos \theta_\varphi \qquad (23)$$

The value of θ_φ determined by Eq. (23) is restricted to the physically allowed range of $0 < \theta_\varphi \le \pi$. Note that, as ψ travels through its range of

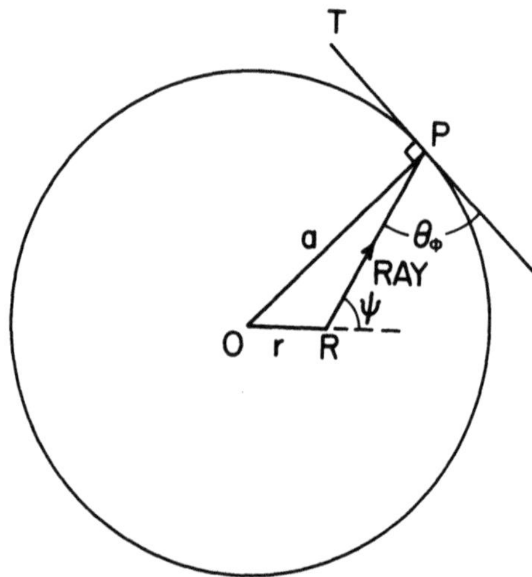

Fig. 14. Projection of ray (RP) onto the end face of the fiber. The ray invariant $\theta\psi$ is determined from the launch point coordinate r, and the launch angle ψ (Eq. 23).

0–2π radians, the range of allowed θ_φ for the launched rays is traversed twice. The $\theta_\varphi = \pi/2$ direction corresponds to *meridional* rays, that is, rays that pass through the center of the waveguide. Other values of θ_φ define *skew* rays: those with $\theta_\varphi < \pi/2$ correspond to a counterclockwise sense of propagation, and those with $\theta_\varphi > \pi/2$ correspond to clockwise propagation.

A ray that is incident with the angles (θ_0, ψ) has, upon transmission, an angle θ_z with respect to the fiber axis given by Snell's law:

$$\sin \theta_z = (1/n_1) \sin \theta_0 \tag{24}$$

From these results, the value of α_{min}, the minimum angle to the normal in Section 2.3., can be determined. Rays that are incident at α_{min} on the core-surrounding-medium interface correspond to rays that are launched (or detected) at the maximum launch angle, θ_0^{max}. The maximum launch angle defines a maximum θ_z value from Snell's law: $\sin \theta_z^{max} = \sin \theta_0^{max}/n_1$. For $\theta_\varphi = \pi/2$ (so the ray is meridional), θ_z and the angle α with respect to the normal are complementary angles (Fig. 11), so α_{min} is determined from

$$\cos \alpha_{min} = (1/n_1) \sin \theta_0^{max} \tag{25}$$

In fact, not all of the power incident on the fiber end face is transmitted to rays propagating within the fiber, since there is a finite amount of reflected power. The magnitude of this power is small in practical cases and is, furthermore, a weak function of angle, so it will be neglected. By ignoring the small retroreflection, the power flux in the ray launched at angles $(\theta_0, \theta_\varphi)$ into the element dA is seen to be $I_0 \cos \theta_0$. Projecting this flux onto the normal to the core-surrounding-medium interface gives the power flux per unit solid angle:

$$I_{ray} = I_0 \cos \theta_0 \times \cos \alpha \, d\Omega \tag{26}$$

In Eq. (26), α is the angle to the normal determined by Eq. (17), and θ_z is determined from θ_0 by Eq. (24).

Substituting these results into Eq. (21), the total signal from the ray (i.e., the total signal per unit fiber end face area per unit solid angle) is

$$dS_{ray}^{(Bulk)} = \gamma I_0 \cos \theta_0 \cos \alpha \, (L/2a \sin \theta_\varphi) \, dA \, d\Omega$$
$$\times \int_0^\infty F_{abs}(\delta;\alpha) \, p_{emit}(\delta;\theta_0^{max}) \, d\delta$$

$$= \gamma I_0 \cos \theta_0 \sin \theta_0 \, (L/2n_1 a) \, dA \, d\Omega$$
$$\times \int_0^\infty F_{abs}(\delta;\alpha[\theta_0,\psi, r/a]) \, p_{emit}(\delta;\theta_0^{max}) \, d\delta \tag{27}$$

where $\alpha[\theta_0,\psi, r/a] = \cos^{-1}[\sin \theta_0 \sin \theta_\varphi/n_1]$, and θ_φ is determined from r/a and ψ in Eq. (23). The thin-film case is similar; the integral over δ is replaced by the expression appearing in Eq. (16).

The total fluorescent signal is the sum of dS_{ray} over the rays launched within the illuminated spot A on the fiber end face and within the cone of angles $\theta_0 \leq \theta_0^{max}$:

$$S_{tot} = (\pi\gamma I_0 L/an_1)\int_0^{r_{max}} r \, dr \int_0^{2\pi} d\psi \int_0^{\theta_0^{max}} \sin^2\theta_0 \cos \theta_0 \, d\theta_0$$
$$\times \begin{cases} \int_0^\infty F_{abs}(\delta; \alpha[\theta_0,\psi, r/a]) \, p_{emit}(\delta;\theta_0^{max}) \, d\delta & \text{(bulk case)} \\ h \times F_{abs}(\delta=0; \alpha[\theta_0,\psi, r/a]) \, p_{emit}(\delta=0;\theta_0^{max}) & \text{(thin-film case)} \end{cases} \tag{28}$$

The units for S_{tot} are determined by I_0, L, a, and r_{max}. Thus S_{tot} is expressed in units of power, e.g., watts.

2.6. Mathematical Details

2.6.1. Evanescent Fields and Absorption

The power absorbed from the evanescent region of a plane wave incident on a dielectric interface can be obtained, correct to the first order in the absorption constant, by using the fields obtained from the nonabsorbing case and the power absorption given by Eq. (7). According to Born and Wolf *(14)*, the transmitted electric fields, **E**, for the two orthogonal polarizations are

$$\mathbf{E}^{(TE)} = E_0\,\{(2n_{rel}\cos\alpha)\,/\,[n_{rel}\cos\alpha + i(n_{rel}^2\sin^2\alpha - 1)^{1/2}]\}$$
$$\times \exp\{k[izn_1\sin\alpha - xn_2(n_{rel}^2\sin^2\alpha - 1)^{1/2}]\} \times \hat{\mathbf{e}}_y \qquad (29)$$

$$\mathbf{E}^{(TM)} = E_0\,\{(2n_{rel}\cos\alpha)/[\cos\alpha + in_{rel}(n_{rel}^2\sin^2\alpha - 1)^{1/2}]\}$$
$$\times \exp\{k[izn_1\sin\alpha - xn_2(n_{rel}^2\sin^2\alpha - 1)^{1/2}]\}$$
$$\times [i(n_{rel}^2\sin^2\alpha - 1)^{1/2}\,\hat{\mathbf{e}}_z - (n_{rel}\sin\alpha\,\hat{\mathbf{e}}_x)] \qquad (30)$$

where $n_{rel} = n_1/n_2 > 1$ is the relative index, $k = 2\pi/\lambda$ is the free-space wavenumber, E_0 is the amplitude of the incident wave, α is greater than the critical angle, and the geometry is given in Fig. 1. The x axis is taken along the normal, pointing into the lower-index (n_2) medium, and the z axis is taken to be along the plane of the interface and in the plane of incidence (the plane defined by the incident and reflected wavevectors). The TE polarization has its electric field perpendicular to the plane of incidence; the TM polarization has its field in the plane of incidence.

By taking the magnitude $|E|^2$ of these fields and substituting into Eq. (7), the power absorbed per unit volume is obtained:

$$P_{abs}^{(TE)} = (\gamma c n_2 E_0^2/8\pi)[4n_{rel}^2\cos^2\alpha/(n_{rel}^2 - 1)] \times \exp[-n_2kx(n_{rel}^2\sin^2\alpha - 1)^{1/2}] \qquad (31)$$

$$P_{abs}^{(TM)} = (\gamma c n_2 E_0^2/8\pi)\{4n_{rel}^2\cos^2\alpha(2n_{rel}^2\sin^2\alpha - 1)\,/$$
$$(n_{rel}^2 - 1)\,[(n_{rel}^2 + 1)\sin^2\alpha - 1]\} \times \exp[-n_2kx(n_{rel}^2\sin^2\alpha - 1)^{1/2}] \qquad (32)$$

assuming unpolarized incident light is equivalent to averaging the absorbed power for the two polarizations; the average of the above results combined with the definition of I_{ray} in Eq. (8) produces Eq. (9).

2.6.2. Dipole Radiation near a Dielectric Interface

The radiation fields from a dipole of arbitrary orientation can be obtained explicitly *(10)*. In the conventions of Fig. 5, α and φ are spherical polar coordinates describing the direction of the radiation flux. The angle α is measured from the *negative x* axis to maintain consistency with the notation for the absorption geometry; that is, $0 \leq \alpha \leq \pi/2$ describes the lower half-space, or the waveguide interior.

Let the dipole's orientation in space be described by the spherical angles χ and ψ. Then the power flux is given by:

- for $0 \leq \alpha \leq \alpha_c$,

$$dP/d\Omega = (cn_2 k^4 D_0^2/8\pi)\{|t_\perp|^2 \sin^2 \chi \sin^2 (\psi - \varphi) + |t_\parallel|^2$$
$$\times [n_{rel} \cos \chi \sin \alpha + (1 - n_{rel}^2 \sin^2 \alpha)^{1/2} \sin \chi \cos(\psi - \varphi)]^2\} \quad (33)$$

- for $\alpha_c \leq \alpha \leq \pi/2$,

$$dP/d\Omega = (cn_2 k^4 D_0^2/8\pi) \exp[-2n_2 k\delta(n_{rel}^2 \sin^2 \alpha - 1)^{1/2}]$$
$$\times \{|t_\perp|^2 \sin^2 \chi \sin^2 (\psi - \varphi) + |t_\parallel|^2 [n_{rel}^2 \cos^2 \chi \sin^2 \alpha$$
$$+ (n_{rel}^2 \sin^2 \alpha - 1) \sin^2 \chi \cos^2(\psi - \varphi)]\} \quad (34)$$

- and for $\pi/2 \leq \alpha \leq \pi$,

$$dP/d\Omega = (cn_2 k^4 D_0^2/8\pi)\{\sin^2 \chi \sin^2 (\psi - \varphi)[1 + r_\perp^2 + 2r_\perp \cos (2n_2 k \delta \cos \alpha)]$$
$$+ \sin^2 \chi \cos^2 \alpha \cos^2 (\psi - \varphi) [1 + r_\parallel^2 - 2r_\parallel \cos (2n_2 k \delta \cos \alpha)]$$
$$+ \cos^2 \chi \sin^2 \alpha \ [1 + r_\parallel^2 + 2r_\parallel \cos (2n_2 k \delta \cos \alpha)]$$
$$+ (1/2) \sin 2\chi \sin 2\alpha \cos (\psi - \varphi) [1 - r_\parallel^2]\} \quad (35)$$

where the Fresnel coefficients are

$$r_\perp = [\cos \alpha + (n_{rel}^2 - \sin^2 \alpha)^{1/2}] / [\cos \alpha - (n_{rel}^2 - \sin^2 \alpha)^{1/2}] \quad (36)$$

$$r_\parallel = [n_{rel}^2 \cos \alpha + (n_{rel}^2 - \sin^2 \alpha)^{1/2}] / [n_{rel}^2 \cos \alpha - (n_{rel}^2 - \sin^2 \alpha)^{1/2}] \quad (37)$$

$$|t_\perp|^2 = \begin{cases} 4n_{rel}^3 \cos^2 \alpha / [(1 - n_{rel}^2 \sin^2 \alpha)^{1/2} + n_{rel} \cos \alpha]^2 & \text{for } 0 \leq \alpha \leq \alpha_c \\ 4n_{rel}^3 \cos^2 \alpha / (n_{rel}^2 - 1) & \text{for } \alpha_c \leq \alpha \leq \pi/2 \end{cases} \quad (38)$$

$$|t_\parallel|^2 = \begin{cases} 4n_{rel}^3 \cos^2 \alpha / \{n_{rel}(1 - n_{rel}^2 \sin^2 \alpha)^{1/2} + \cos \alpha\}^2 & \text{for } 0 \leq \alpha \leq \alpha_c \\ 4n_{rel}^3 \cos^2 \alpha / \{(n_{rel}^2 - 1)[(n_{rel}^2 + 1) \sin^2 \alpha - 1]\} & \text{for } \alpha_c \leq \alpha \leq \pi/2 \end{cases} \quad (39)$$

The reflection coefficients in Eqs. (36) and (37) have an unfamiliar appearance compared to those in Born and Wolf *(14)* because the coefficients are used here with the angle α in the quadrant $\pi/2 \leq \alpha \leq \pi$, where $\cos \alpha < 0$, instead of the more usual quadrant, $0 \leq \alpha \leq \pi/2$. This result can be averaged over the dipole orientations (i.e., over the angles χ, ψ) to obtain the power per unit solid angle radiated by an isotropically random distribution of dipoles:

$$\frac{dP}{d\Omega} = \frac{cn_2 k^4 D_0^2}{24\pi} \begin{cases} \{|t_\parallel|^2 + |t_\perp|^2\} & \text{for } 0 \leq \alpha \leq \alpha_c \\[2mm] \exp\{-2n_2 k\delta(n_{rel}^2 \sin^2 \alpha - 1)^{1/2}\} \\ \quad \times \{|t_\parallel|^2(2n_{rel}^2 \sin^2 \alpha - 1) + |t_\perp|^2\} & \text{for } \alpha_c \leq \alpha \leq \pi/2 \\[2mm] 2 + r_\perp^2 + r_\parallel^2 + 2(r_\perp - r_\parallel \cos 2\alpha) \\ \quad \times \cos(2n_2 k\delta \cos \alpha) & \text{for } \pi/2 \leq \alpha \leq \pi \end{cases} \quad (40)$$

Note that, in these equations, $\pi/2 \leq \alpha \leq \pi$ corresponds to radiation into the upper half-space, the lower-index (n_2) medium. The range $0 \leq \alpha \leq \pi/2$ corresponds to radiation into the higher-index (n_1) medium.

3. Results Calculated from Theory

There are a number of conclusions that can be predicted for the total fluorescent signal (Eq. [28]):

1. The fluorescent signal is directly proportional to the fiber length L, the absorption constant γ, and the input radiance I_0. In the weak absorption limit considered here, these are physically intuitive results.
2. The signal is approximately proportional to r^2_{max}, or the area of the illuminated spot on the fiber's end face. This proportionality to r^2_{max} is to be expected, since the total input power for a uniform source radiance is proportional to r^2_{max}.
3. The signal depends inversely on the fiber radius a, because the number of reflections per unit length decreases with larger radius.
4. The signal should have a strong dependence on θ_0^{max}, since both F_{abs} and p_{emit} are strongly enhanced in the large angle region near the critical angle.

These and other conclusions are shown using calculations based on a fiberoptic waveguide of core index $n_1 = 1.46$ (fused silica) and radius 250 μm. Numerical integration is used to calculate the fluores-

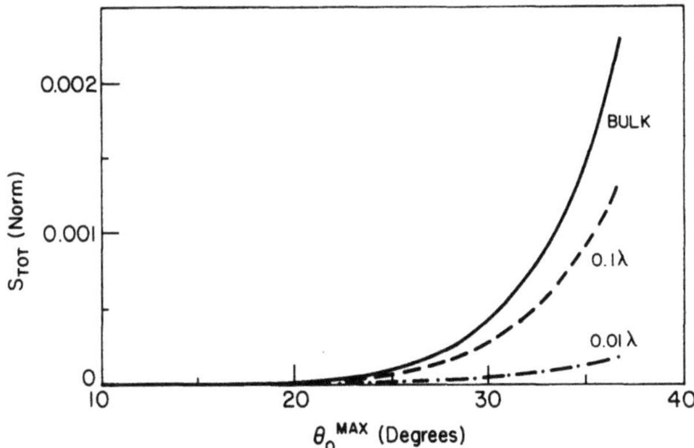

Fig. 15. Variation of total fluorescent signal, S_{tot}, with maximum launch angle. The fluorescent material is located either in films of thickness 0.01 or 0.1 λ, or in the bulk surrounding medium.

cent signal S_{tot} under various conditions. The results presented here have been normalized by dividing S_{tot} by $\gamma I_0 a \lambda L$.

The fluorescent signal does in fact depend strongly on the maximum launch angle, as already noted by Love and Slovacek *(4)* and Glass et al. *(5)*. Figures 15 and 16 show this dependence in the case in which the index of the surrounding medium corresponds to water or to dilute aqueous solution, $n_2 = 1.33$. The value of r_{max}/a for launched light is 0.5 and in all cases is 1.0 for the collected fluorescent light. The maximum launch angle possible in this case (to stay within the cone of bound rays) is $\theta_0^{max} = 37.03°$. There are three curves, which differ depending on the location of the fluorescent material. These locations include films of thickness 0.01 or 0.1 wavelength and the bulk surrounding medium. The logarithmic graph (Fig. 16) of the signal vs $\sin\theta_0^{max}$ shows very little curvature, except for the bulk curve near the maximum launch angle. The slope is approx 8, so the signal depends on the angle via $S_{tot} \propto \sin^8(\theta_0^{max})$. The slope of the curve for the bulk case increases somewhat beyond 8 as $\sin \theta_0^{max}$ approaches the NA of the fiber.

Figure 17 shows the dependence of signal on the size of the illuminated spot. Once again, the results are presented for fluorescent material located in films of thickness 0.01 or 0.1 wavelength or in the

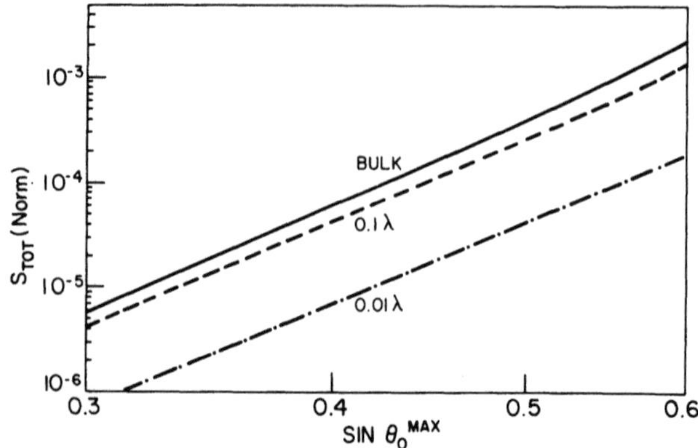

Fig. 16. Variation of total fluorescent signal, S_{tot}, with maximum launch angle, displayed on log–log axes (cf Fig. 15). The fluorescent material is located either in films of thickness 0.01 or 0.1 λ, or in the bulk surrounding medium.

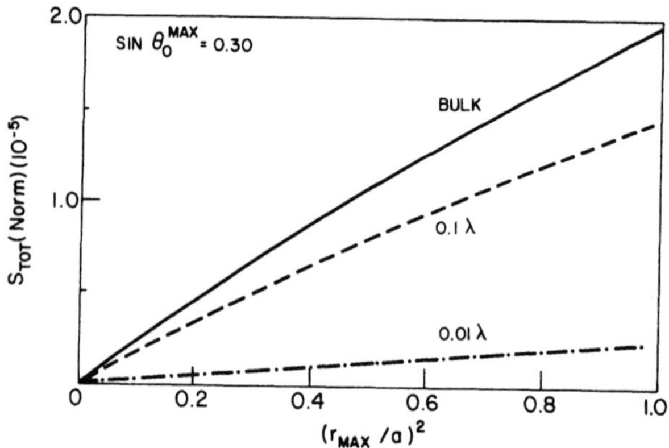

Fig. 17. Variation of total fluorescent signal, S_{tot}, with size of illuminated spot, r_{max}.

bulk surrounding medium. The refractive indices were the same as given above, and sin θ_0^{max} was 0.3. The lines show a slight curvature, but are approximately linear with respect to $(r_{max}/a)^2$. The curvature results from the term multiplying $r\,dr$ in the integral in Eq. (28), which depends weakly on r.

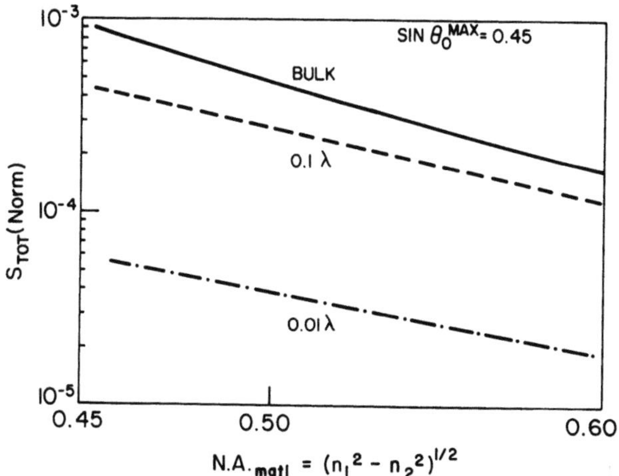

Fig. 18. Variation of total fluorescent signal, S_{tot}, with material NA of the waveguide at a fixed maximum launch angle of 26.7°. The fluorescent material is located either in films of thickness 0.01 or 0.1 λ, or in the bulk surrounding medium.

Figure 18 illustrates the dependence of the fluorescent signal on the refractive index of the surrounding medium or the material NA. In the calculation, a fixed launch angle of 26.7° (corresponding to a launch NA of 0.45) was used for the input launch cone, and a value of 0.5 for r_{max}/a was used. The core index was held fixed at n_1= 1.46, and the index of the surrounding medium was increased from 1.33 to 1.38. This is equivalent to narrowing the cone of bound rays of the system downward from 37° to the launch angle 27°. Once again, the results are shown for the fluorescent material located in films of thickness 0.01 or 0.1 wavelength or in the bulk surrounding medium. The material NA is used as the independent variable. The slopes of the lines in Fig. 18 for the thin-film case are very nearly constant and equal to -4. This can be attributed to the $1/(n_{rel}^2 - 1) \propto NA^{-2}$ terms in the Fresnel transmission coefficients, which appear in both F_{abs} and in p_{emit}. Since the total signal depends on the product of F_{abs} and p_{emit}, S_{tot} should be proportional to NA^{-4}. There is some curvature for the bulk surrounding medium case, and the average slope over the region considered is approx -5.9. This enhanced dependence on the material NA is based on the increased penetration depth, enabling a larger number of dipoles to participate in the absorption and emission processes. Thus the increase of signal for the bulk case with decreasing material NA is enhanced over and above the effect of the Fresnel transmission coefficients.

4. Results Based on Experiments

This section describes the optical apparatus, sample chamber, and results from the experimental characterization of the evanescent wave fluorescence technique. The optical launch conditions, viz., launch NA (sin θ_0^{max}) and spot size, were varied to determine how the total fluorescent signal depends on these parameters. The material NA was also varied by changing the refractive index of the surrounding medium. These results are compared with the theoretical predictions of Section 3.

4.1. Optical Research Apparatus

A block diagram of the optical characterization system is shown in Fig. 19. Light from a tungsten–halogen lamp is spectrally filtered by a visible light, single grating monochromator with a 10 cm focal length. The collimated beam is further filtered by a 10 nm wide interference filter centered at 488 nm, which is close to the peak of the fluorescein isothiocyanate (FITC) absorption band. This was required to reduce the spectral crosstalk between excitation and fluorescent wavelengths. The absorption and emission bands of FITC, a common biological labeling dye used in this study, are shown in Fig. 20. It can be seen that the absorption and emission bands measured by using a standard cuvet fluorimeter and the present characterization system are in good agreement. Returning to Fig. 19, the light is then focused on a pinhole (typically 1 mm in diameter), collimated, and relayed to the objective lens by the beam splitter. This 30×, 0.60 NA objective lens focuses the excitation light down onto the end of the optical fiber and fills the fiber NA. The NA $= [(1.46)^2 - (1.33)^2]^{1/2} = 0.60$ when aqueous solution ($n_2 = 1.33$) serves as the surrounding medium and fused silica serves as the waveguide fiber ($n_1 = 1.46$). The light-spot diameter (r_{max})

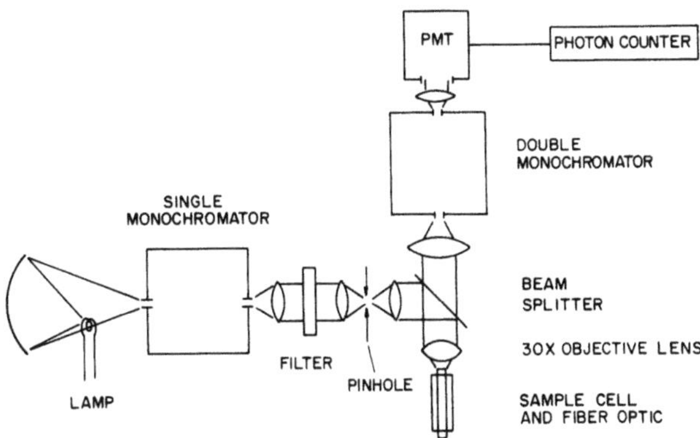

Fig. 19. Schematic diagram of the optical research apparatus (*see* text for description).

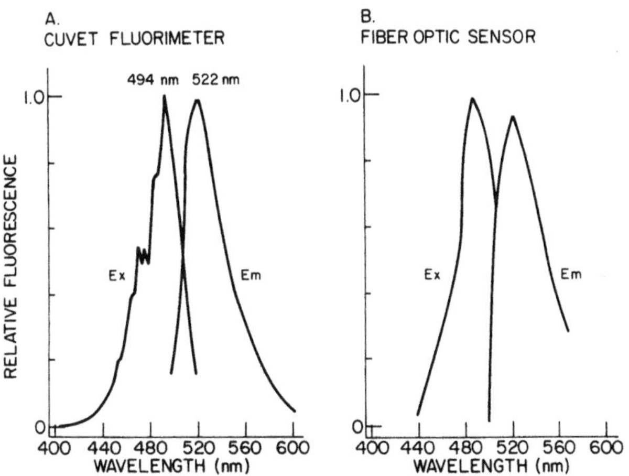

Fig. 20. FITC absorption and emission bands obtained by (A) conventional fluorimetry and (B) fiberoptic evanescent wave sensing.

incident on the nominally 500 μm diameter optical fiber is given by 1 mm times the ratio of the focal lengths of the objective and pinhole collimating lenses, and is approx 250 μm. The fluorescent light obtained by evanescent coupling to FITC placed in the aqueous solution surrounding the fiber is also collected in retroreflection by this same

30× objective lens and passed through the beam splitter to the 10 cm, focal length visible light, double-grating monochromator. The double monochromator has excellent stray light rejection (10^{-9}) and is utilized to obtain low crosstalk between excitation and emission wavelengths. A cooled photomultiplier (PMT) having a wide spectral response (200–900 nm) is used with photon counting detection. Because of the low PMT dark count, ~3 counts per second (cps), light intensities of ~10^{-18} watts are detectable with integration times of several seconds. This makes the optical system extremely sensitive and easily wavelength-selectable. These were indeed the design objectives for this research apparatus.

The excitation and emission light that propagates along the fused-silica fiber is easily perturbed and stripped out of the waveguide, depending on the seals or supports that touch the waveguide core. This is especially true for light rays that propagate at higher angles relative to the fiber axis and that sample the lateral surface at a frequent rate. It is these light rays that have larger values for evanescent transmission and penetration and, thus, are important for sensitivity of evanescent wave excitation and detection. A measurement of the far-field angular distribution of light emerging from the distal end of the optical fiber would thus serve as a diagnostic tool for evaluating these effects. Such a measurement method has been set up using a silicon photodetector with an aperture in front of its photosensitive area. The NA_{90} is obtained by moving the photodetector unit away from the distal end and along the fiber axis until 90% of the total optical power is detected. The angle θ_{max} subtended by the aperture can readily be computed from the distance to the fiber and the aperture diameter; thus, $NA_{90} = \sin \theta_{max}$ is obtained. Measurements of NA_{90} are more precise than NA_{100} determinations, which suffer from small refraction changes from edge chips or imperfections on the fiber face. The definition of $\sin \theta_{max}$ in terms of NA_{90} is used in all the subsequent figures. Since perturbations to the fiber have been thoroughly minimized, this experimental external angle θ_{max} corresponds closely to the theoretical angle θ_0^{max}.

The optical fiber sensor and flow-through cell are shown in Fig. 21. A peristaltic pump is used to fill and remove liquids in contact with the fiber. Attention to detail has resulted in an $NA_{90} = 0.54$, which

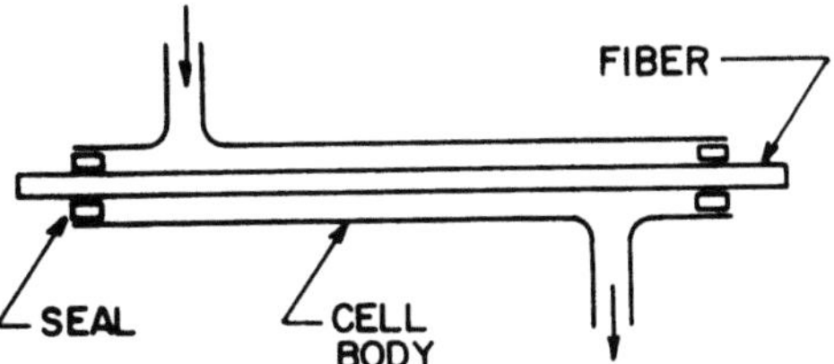

Fig. 21. Sensor flow-through cell. FITC solutions are pumped into the jacketed region that surrounds the fiber.

is considerably greater than that obtained by using a partially plastic-clad silica fiber for which $NA_{90} \approx 0.30$. The dye FITC was dissolved in phosphate-buffered aqueous saline (PBS) solution to maintain the pH constant and to serve as a model system applicable to fluorescent immunoassay (*see* chapter by Bluestein et al., this volume, for details). The fiber is carefully cleaned in Nochromix™ and sulfuric acid, and then in a dilute (10%) nitric acid rinse and distilled water. Subsequent handling of the fiber is done with surgical gloves or finger cots. This has resulted in very low background fluorescence contributed by the fiber. The total background fluorescence, B, for an uncoated silica fiber in distilled water is $B \approx 35$ counts per second (cps), which includes the dark count, stray light, and parasitic fluorescence from the launch objective lens. The emission-detection double monochromator is set for a detection wavelength of 526 nm, which is close to the FITC emission peak (Fig. 20). The monochromator resolution is 4 nm.   With a silanized fiber and PBS solution, typically $B \approx 100$ cps. In con-trast, the Fresnel reflection from the fiber proximal end, measured at 488 nm, is $\sim 2 \times 10^8$ cps. This gives a working contrast ratio of 2×10^6. In summary, the optical system has high sensitivity and precision   (2%) with no significant spectral crosstalk and very low background fluorescence.

4.2. Launch Conditions and Sensitivity

The FITC sensitivity curves for various maximum launch angles are shown in Fig. 22. The signal minus background at 526 nm in cps is plotted vs the molar concentration of FITC in PBS buffer solution. The FITC fluorescence varies approximately linearly with concentra-

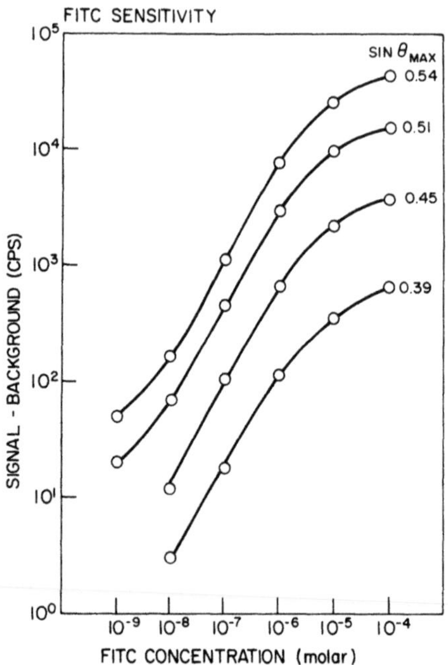

Fig. 22. Experimentally measured variation of fluorescent signal with FITC concentration for different maximum launch angles and for material NA of 0.60.

tion in the middle range, and shows curvature at high and low concentrations. The high-concentration effect is well known and is a consequence of fluorescence quenching. The low-concentration effect is less well known, and may result from fluorescent impurities in the PBS solution. It can be seen that excellent sensitivity is obtained down to $10^{-9}M$.

Figure 23 displays the approximate eighth-power dependence of the normalized signal on the sine of the maximum launch angle. The data in the figure were normalized by dividing the signal minus background by the Fresnel reflection of the incident light (at 488 nm) from the fiber proximal end face. The value of $\sin\theta_{max}$ is controlled by an aperture in the back plane of the launch objective lens, but is measured in the fiber far-field as the throughput NA_{90}. These angular measurements were made only for the background case (PBS buffer), since

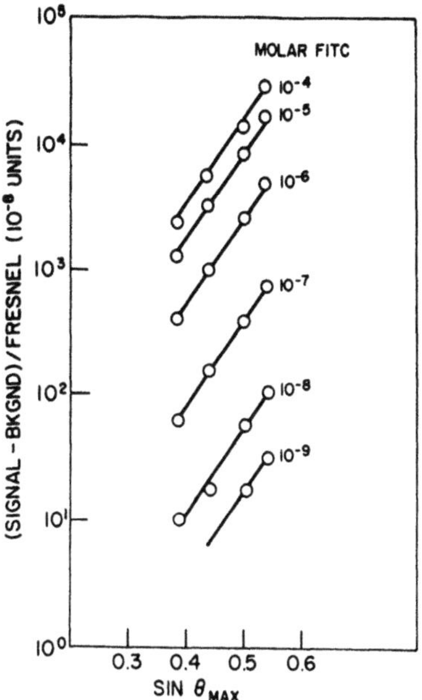

Fig. 23. Experimentally measured variation of normalized (by Fresnel reflection) fluorescent signal with maximum launch NA (sin θ_{max}) for different FITC concentrations.

there is no measurable reduction in the angle of the output light from FITC absorption at concentrations $< 10^{-4}M$. It can be seen from Fig. 23 that the dependence on sin θ_{max} is very strong; the normalized signal varies approximately as $\sin^8 \theta_{max}$ nears 0.50 and is independent of the concentration value. This implies that the launch optics and the solution chemistry are statistically independent variables, as would be expected intuitively. Referring to Fig. 16, the theoretical graph of S_{tot} vs sin θ_0^{max}, it can be seen that the signal S_{tot} varies as the eighth power of sin θ_0^{max} for the fluorescent material in the bulk surrounding medium, except for a slight increase in slope at angles very close to the critical angle. Since the launched spot of light is not a perfectly truncated Lambertian source (in angle) because of some vigneting of the incident beam, some rounding or smoothing is expected to take place. Thus, the variation of the measured signal compares favorably to the theoretically predicted variation within experimental uncertainties.

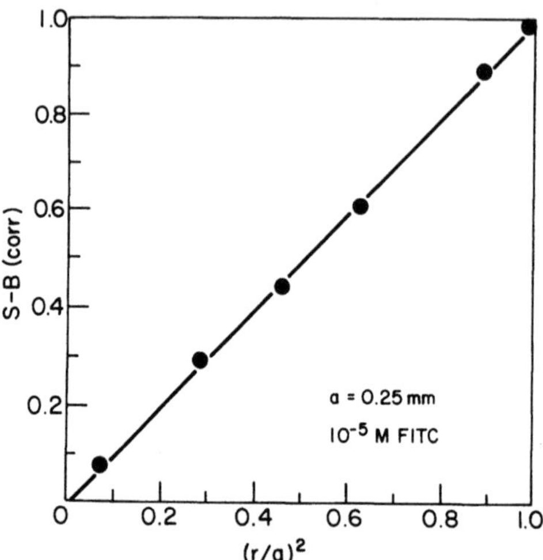

Fig. 24. Variation of experimentally measured fluorescent signal with area of illuminated spot. The launch NA is fixed at 0.30.

The experimental results of varying the input or launch spot size at a fixed launch NA on the fluorescent signal is shown in Fig. 24. A concentration of $10^{-5}M$ FITC was used with a lower power (20×, 0.3 NA) launch objective lens in order to be able to fill the fiber diameter completely with light. The input-spot size is varied by changing the size of the pinhole (Fig. 19) in the excitation-beam optics. The output fluorescent light is totally collected from the fiber end face for launch angles up to 0.30 NA. Thus, intuitively, the fluorescent signal should vary linearly with incident optical power, viz., $(r_{max}/a)^2$. This is approximately the case according to Fig. 24. The measured signal minus background, $S - B$, is corrected according to

$$S - B(\text{corr}) = S - B(\text{meas}) \, [I_{\text{throughput}}(\text{corr})/I_{\text{throughput}}(\text{meas})] \qquad (41)$$

where $I_{\text{throughput}}(\text{meas})$ is the measured throughput optical power from the distal end of the fiber and $I_{\text{throughput}}(\text{corr})$ is proportional to $(r_{max}/a)^2$ for a uniformly illuminated spot. This is required to obtain the signal minus background for a Lambertian-spot launch, which is modeled in the theory section and displayed in Fig. 17. Typically this correction factor is close to unity (within 5%), indicating that the experimental launch spot is reasonably uniform. The agreement between Fig. 24 and

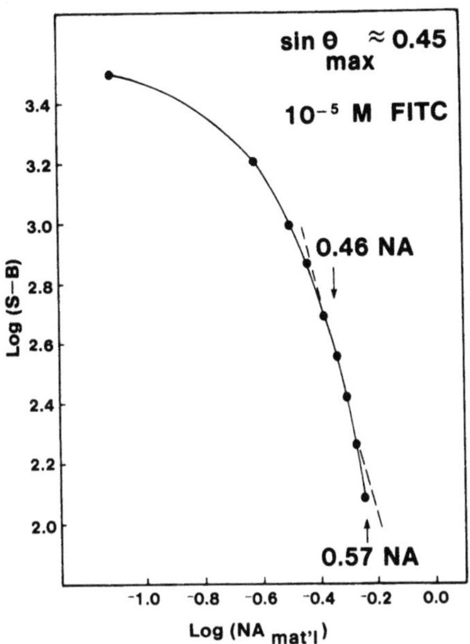

Fig. 25. Variation of experimentally measured fluorescent signal with material NA, $(n_1^2 - n_2^2)^{1/2}$, at a fixed maximum launch angle (sin θ_{max} = 0.45).

the theoretical predictions from Fig. 17 is reasonably good, considering the experimental precision of several percent.

4.3. Material Effects

Besides varying the optical launch conditions, the material properties of the sensor can be changed. This is most easily done by varying the refractive index of the surrounding medium solution that contains the dissolved dye. Sucrose solutions of various concentrations were prepared, and the refractive index was measured directly using an Abbe refractometer. The material NA, defined by $NA_{matl} = (n_1^2 - n_2^2)^{1/2}$, was thus varied from 0.6 to 0.1 with increasing sucrose concentration. The dye concentration was held constant at $1 \times 10^{-5} M$ FITC and the launch condition sin $\theta_{max} \approx 0.45$ was maintained. The results of this experiment are shown in Fig. 25. The log of the signal minus background, $S - B$, is plotted vs the log of the material NA. The background corresponds to the signal detected in the absence of FITC, and is not significantly changed by the sucrose concentration. As the NA_{matl}

decreases from 0.57 to 0.46, the fluorescence increases, since the launch NA, $\sin \theta_{max}$, is being approached and light rays are now becoming more closely incident at the critical angle. With further decrease in NA_{matl}, the incident light more than fills the cone of acceptance for bound light rays, and refraction of light occurs. This explains the curving or approach to saturation for the fluorescence at the smaller values of NA_{matl}, since the incident light is gradually being lost from the core of the sensor. In the region between 0.46 and 0.57 NA (Fig. 25), $S-B$ is proportional to NA_{matl}^{-4}, as indicated by the straight dashed line. Comparison with the theoretically predicted response can be obtained by referring to Fig. 18. The bulk case has a local slope of approx -6.0 at 0.5 NA_{matl} and increases with decreasing NA_{matl}. The surface-film cases, however, show a uniform slope of -4.0 and appear to be in better agreement with the experimental results. It is possible that adsorption of the FITC to the surface was occurring, especially in the presence of sucrose solutions. Similar results, namely NA_{mat}^{-4}, were found in another experiment using a larger waveguide. Glass et al. *(5)* have observed a similar adsorption effect. Further experimentation is needed to verify this, but there is no reason to question the theoretical expectations.

5. Summary and Conclusions

We have developed a quantitative optics model from fundamental optical principles. This theoretical model predicts that approx 2% of the evanescent fluorescent light is coupled back into the optical fiber under optimum conditions. It also predicts the launch NA ($\sin \theta_0^{max}$), spot size, and NA_{matl} dependences of the total fluorescent signal that are in good agreement with the experimental results discussed in Section 4.

Thus, for fluorescent molecules bound to the lateral surface of the fiber, namely, $\delta / \lambda = 0$,

$$S_{tot} \propto I_0 L a \, (r_{max}/a)^2 \sin^8 \theta_0^{max}(NA_{mat}^{-4}) \qquad (42)$$

where S_{tot} is the total fluorescent signal obtained from one end of the fiber, I_0 is the optical source radiance, L is the fiber length, a is the fiber radius, r_{max} is the radius of the launch spot, θ_{max} is the maximum (external) angle of the launched light cone, and NA_{matl} is the material NA, $(n_1^2 - n_2^2)^{1/2}$. This result clearly emphasizes the importance of launching and maintaining high-angle light through the fiber. It also allows evaluation and examination of trade-offs involved in using other materials or geometrical parameters. Finally, the results indicate that evanescent wave sensing with optical fibers can give sensitivity for free FITC solution to approx $10^{-9}M$ with reasonable signal-to-noise ratio. Further improvements in sensitivity may be possible with the use of lasers and other dye materials.

6. Symbols

a: fiber radius

B: background signal, i.e., without dye solution

c: speed of light

D_0: dipole strength

d_p: evanescent field penetration depth ($1/e$ distance, Eq. [3])

dA: element of area on fiber end face

$d\bar{A}$: element of fiber lateral surface area

dP_{inc}: differential element of incident optical power on the fiber end face (Eq. [22])

$dP/d\Omega$: optical power emitted in a given direction/unit solid angle by electric dipole radiator

dS_{ray}: differential element of total fluorescent signal from a single ray, from a given element dA of fiber end surface and a given solid angle element $d\Omega$ of ray angles (Eq. [6])

E: electric field amplitude

E_s: electric field at dielectric surface

$\mathbf{E}$: electric field vector

E_0: amplitude of incident wave

$\hat{e}_x, \hat{e}_y, \hat{e}_z$: unit vectors

$F_{abs}(\delta;\alpha)$: normalized absorption function at distance δ into the evanescent region (surrounding medium), for light rays incident at angle α

h: thickness of fluorescent film on fiber lateral surface

I_0:   launch radiance (units of power/unit area/unit solid angle, Eq. [22])

$I_{fluor}(\alpha)$:   fluorescence/unit area of interface (Eq. [5])

I_{ray}:   incident power flux normal to the interface for a given ray (Eq. [8])

J: irradiance (optical power flux, i.e., energy transmitted/unit area/unit time)

J_0:   reference or initial irradiance

k: free space wavenumber, $2\pi/\lambda$

L: fiber length

N: number of internal reflections of a ray in a length of fiber (Eqs. [4,19])

n_1:   refractive index of fiber

n_2:   refractive index of surrounding medium

n_{rel}: n_1/n_2

NA: numerical aperture

NA_{matl}:   material NA, $(n_1^2 - n_2^2)^{1/2}$

NA_{launch}:   sine of the maximum launch angle, $\sin\theta_0^{max}$

NA_{90}:   experimentally measured throughput NA, $\sin\theta_{max}$ (*see* text for definition); correlates closely with launch NA

$P_{abs}(\delta;\alpha)$:   power absorption/unit volume, at distance δ into the evanescent region (surrounding medium), for rays incident at angle α (Eq. [9])

$p_{emit}(\delta;\theta_0^{max})$:   probability for emission of power into the cone of detectable rays (defined by θ_0^{max}) for a dipole at distance δ into the evanescent region (Eq. [13])

P_{tot}:   total power radiated by a dipole (Eq. [12])

x: coordinate axis perpendicular to the dielectric interface

z: coordinate axis parallel to the fiber longitudinal axis

z_p: ray half-period (distance along the waveguide between successive reflections, Eq. [18])

$r_\perp, r_\parallel, t_\perp, t_\parallel$:   Fresnel coefficients (Eqs. [36–39])

α: angle between light ray and normal to interface

α_c: critical angle (Eq. [1])

γ: absorption coefficient ($1/\gamma$ is $1/e$ decay length for optical power)

δ: distance of dipole from interface

λ: wavelength of light in vacuum

θ_0^{max}: maximum angle with respect to the fiber axis for the cone of external light rays that are launched into or collected from the fiber; cone is defined by the experimental system

θ_z^{max}: maximum angle with respect to the fiber axis for the cone of internal light rays (Eq. [24] defines the relation between the internal and external cones)

θ_0: exterior launch angle (integration variable that labels ray angles)

θ_φ: angle between ray projection onto fiber end face and tangential line to the fiber at the point of the ray reflection (Fig. 14)

φ: polar coordinate for ray directions in calculation of dipole radiation patterns (Figs. 5,8)

ψ: azimuthal angle defining (with θ_0) the direction of rays launched into the fiber (Figs. 13,14)

$d\Omega$: element of solid angle

Acknowledgments

We gratefully acknowledge Ralph Westwig and Don Keck for stimulating discussion, Tom Cook for expert technical assistance, and Dick Falb and John MacPhee for their encouragement. We are particularly grateful to Al Luderer for his assistance in the initial stages of this work.

References

1. Harrick, N. J. (1967) *Internal Reflection Spectroscopy* (Wiley, New York).
2. Hirschfeld, T. (1965) Total reflection fluorescence. *Can. Spectroscopy* **10**, 128.
3. Hirschfeld, T. (1987) Apparatus for improving the numerical aperture at the input of a fiber optics device. US Patent 4,654,532.
4. Love, W. F. and Slovacek, R. E. (1986) Fiber optic evanescent sensor for

fluoroimmunoassay. Paper 6.6 presented at the 4th Int. Conference on Optical Fiber Sensors (OFS '86, Tokyo, Japan).

5. Glass, T. R., Lackie. S., and Hirschfeld, T. (1987) Effect of numerical aperture on signal level in cylindrical waveguide evanescent fluorosensors. *Appl. Opt.* **26**, 2181.

6. Lee, E. H., Benner, R. E., Fenn, J. B., and Chang, R. K. (1979) Angular distribution of fluorescence from liquids and monodispersed spheres by evanescent wave excitation. *Appl. Opt.* **18**, 862.

7. Carniglia, C. K., Mandel, L., and Drexhage, K. H. (1972) Absorption and emission of evanescent photons. *J. Opt. Soc. Am.* **62**, 479.

8. Lukosz, W. and Kunz, R. E. (1978) Light emission by magnetic and electric dipoles close to a plane interface. I. Total radiated power. *J. Opt. Soc. Am.* **67**, 1607.

9. Lukosz, W. and Kunz, R. E. (1978) Light emission by magnetic and electric dipoles close to a plane interface. II. Radiation patterns of perpendicular oriented dipoles. *J. Opt. Soc. Am.* **67**, 1615.

10. Lukosz, W. (1979) Light emission by magnetic and electric dipoles close to a plane interface. III. Radiation patterns of dipoles with arbitrary orientation. *J. Opt. Soc. Am.* **69**, 1495.

11. Carniglia, C. K. and Mandel, L. (1971) Quantization of evanescent electromagnetic waves. *Phys. Rev. D.* **3**, 280.

12. Snyder, A. W. and Love, J. D. (1983) Bound rays of fibers (Chapter 2), pp. 26–50; Fiber illumination and pulse shape (Chapter 4), pp. 63–88; Leaky rays (Chapter 7), pp. 134–153; in *Optical Waveguide Theory* (Chapman and Hall, New York).

13. Boyd, R. W. (1983) *Radiometry and the Detection of Optical Radiation* (Wiley, New York).

14. Born, M. and Wolf, E. (1980) Reflection and refraction of a plane wave, in *Principles of Optics*, 6th Ed. (Pergamon, Elmsford, NY), pp. 36–51.

Evanescent Wave Immunosensors for Clinical Diagnostics

Barry I. Bluestein, Mary Craig,
Rudolf Slovacek, Linda Stundtner,
Cynthia Urciuoli, Irene Walczak,
and Albert Luderer

1. Introduction

1.1. Immunoassay Development

With the recent trend in clinical diagnostics toward cost containment and decentralization, medical personnel and deliverers of health care have looked to new technological advances for simpler, more rapid testing systems. Nowhere is this need more pronounced than in the field of quantitative immunoassay.

The use of antibodies (Ab) directed against specific substance antigens (Ag) offers an analytical technology capable of measuring concentrations to picomolar (10^{-12}) or lower levels. The development of radioactively labeled proteins and haptens for use in immunoassay opened an era of high-sensitivity analyte analysis *(1)*. There have been developed solid-phase systems that simplify the separation of labeled antigen–antibody (Ag–Ab) complexes from unbound tracer species at the termination of the binding reaction. These systems function by immobilization of either the Ag or Ab to a solid support, such as glass particles or plastic beads *(2)*. Following incubation of sample with the solid-phase Ab and a known quantity of radioactive Ag or Ab, the solid phase is washed to remove unbound label and quantitated directly in a gamma or scintillation counter.

The desire to eliminate the use of radioisotopes has led to the evolution of solid-phase enzyme immunoassays wherein the labeled Ab or Ag is composed of an enzyme conjugate instead of a radioactive label. Following completion of the primary Ag–Ab binding reaction, excess unbound conjugate is removed by washing and a second reaction is initiated by addition of enzyme substrate. Changes in optical absorbance resulting from colored-product formation or substrate consumption are then measured with the rate of change proportional to the original analyte concentration.

However, although the costs and hazards associated with isotopes have been removed, these systems require increased time and additional processing steps in the performance of immunoassay procedures. The key drawback of requiring separation of bound and free label prior to final system readout still exists. System steps required to perform a standard enzyme immunoassay are summarized in Table 1.

1.2. Characteristics for an Immunoassay Biosensor

In order for a biosensor to meet requirements for an immunoassay device, it should have the following characteristics. First it should contain either immobilized Ag or Ab as the biospecific component. Second, a change induced by Ag binding to Ab should be immediately convertible to an electrical signal that can be amplified, stored, and displayed. In an optical biosensor, conversion would be detected through changes in light intensity at specific wavelengths, which in turn would be converted to an electrical signal through a photomultiplier tube or solid-state silicon diode. In line with direct signal transduction is the real-time nature of the response. As soon as the primary bioreaction occurs, there is an immediate detectable signal output.

These criteria preclude the use of secondary signal-developing reagents after the primary reaction has occurred and confer upon the

Table 1
Process Steps in Enzyme Immunoassay

Aliquot immunoreagents
Aliquot sample
Incubate to end point
Separate bound from unbound tag
Add enzyme substrate
Incubate to allow color development
Stop color-development reaction
Measure optical density

system an extremely easy-to-use format, since no additional process steps are required by either the user or the instrument once the primary reaction has been initiated. If one can read the signal directly as the Ag and Ab bind to each other, the response times also will be rapid, usually on the order of seconds to minutes.

Although some sensors have traditionally been reusable, with the capability of handling many samples prior to regeneration, for any practical analytical application, Ag–Ab immunosensing needs to be considered a technique of single use and irreversible binding. The reason for this is the high affinity constants of Ag–Ab binding (10^{10}–$10^{12}/M$), which confer upon immunologic systems extremely high levels of sensitivity. The price to be paid is one of essential irreversibility. Most attempts at dissociating Ag–Ab complexes utilizing drastic pH changes or chaotropic agents result in some denaturation of the ability of the Ab to rebind Ag. Thus, the capacity of the Ab to rebind Ag is continuously altered, making calibration of a reusable immunosensor extremely difficult.

Within the above context, current heterogeneous immunoassay protocols (Fig. 1) cannot be considered to fulfill the criteria of direct signal readout by virtue of the required multiple reagent addition steps and user-mediated separation of bound from free tag. Although some homogeneous, nonseparation immunoassay procedures wherein Ag–Ab binding can be measured directly in solution have been developed in recent years, these systems depend on steric-activity differences between appropriately labeled small-molecule tags free in solution

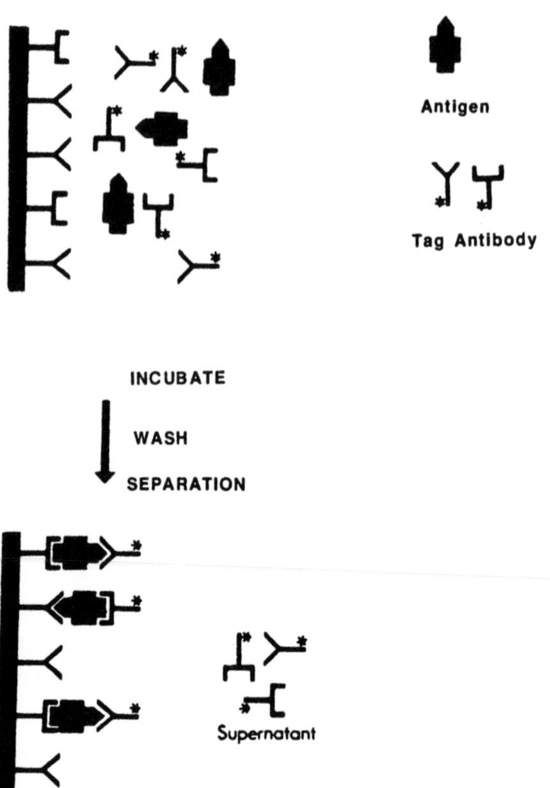

Fig. 1. Schematic diagram depicting the processing steps for a solid-phase Ab–Ag–tagged antibody heterogeneous immunometric assay.

and tags bound to their high-mol-wt antibodies (M_r 160, 000). Such systems utilize fluorescence polarization *(3)*, apoenzyme reconstitution *(4)*, or enzyme activity modulation *(5)*, but are ineffectual in measuring high-mol-wt proteins that confer steric constraints equivalent to Ab binding. For the most part these systems are currently used to measure small-mol-wt substances, such as drugs and steroid hormones.

Requirements for the ideal immunosensor are as follows:

1. Ability to measure high-mol-wt analytes ($M_r > 5000$) present in ultralow concentration in blood (10^{-9}–10^{-12} M) as well as low-mol-wt analytes,

2. No additional steps needed beyond addition of patient sample to the system, and
3. Rapid response times.

Compared to traditional chemical sensors for such electrolytes as K^+ and Na^+, the analyte concentrations to be measured are present at 10^6- to 10^9-fold-lower values. Limitations of present approaches to developing various types of immunosensors are discussed by North *(6)*.

1.3. Fiberoptic Chemical Sensing

Optical sensors based on the use of fiberoptics fall into two distinct categories, distal and evanescent wave sensing. The basic concept of distal sensing (described in detail by Cámara et al. in this volume) is to immobilize light-responding chemicals either directly at the distal end of the fiber or in a miniature cell placed directly under the distal end of the fiber. These chemicals may be either absorbing or fluorescent in nature. In either case, the fiber is used in a classic mode as a light pipe, and the energy of total internally reflected light emerging from the end of the fiber tip is used to probe a chemical reaction external to the fiber. Probe energy changes, absorption, or fluorescence generation are then transmitted back through the fiber to an appropriate optical detection system. These approaches have not met with success in immunoassay development for at least two reasons:

1. The surface area at the tip of the fiber is insufficient to immobilize enough Ab or Ag to result in a discernible response. This is readily understandable, since all solid-phase immunoassay systems are specific concentrators of mass, and signal generation is therefore directly related to the total number of molecules accumulated on the surface. In the case of fluorescent immunoassay, in order to concentrate sufficient fluorescent molecules on the surface to generate discriminatory responses across the appropriate concentration range, there must be sufficient Ab present to bind the fluorescent tag.
2. Nonspecific fluorophores present in patient serum and unbound fluorescent tag also will be excited by the probe radiation. In general this causes background signals too high and variability too great to develop a precise analytical system. Several reviews discuss distal-sensing fiberoptic sensors *(7–10)*.

The second category for optical sensors based on fiberoptics is to use the evanescent wave component of the internally reflected light as a source of probe radiation *(11)*. If a fluorescent molecule is placed in solution alongside the outer surface of a bare fiber that is stripped of its plastic coating and devoid of any cladding of lower refractive index, then this fluorescent material can be excited by the evanescent wave that is generated perpendicular to the reflected light rays within the core of the fiber *(12)*.

A detailed discussion of the physics of the evanescent wave is given elsewhere *(13)* and in the chapters by Love et al. and by Lackie et al. in this volume. The key advantage of using this component of electromagnetic radiation is that, since the amplitude of this wave decays exponentially with distance into the surrounding solution, only fluorescent compounds positioned within a fraction of a wavelength from the fiber–solution interface will be exposed to the full intensity of the radiation. Fluorophores located in solution beyond this evanescent zone will not be exposed to excitation radiation and will not fluoresce. Likewise, endogenous fluorophores present in a patient sample will be excited only if they are located within the evanescent zone of radiation. Since this zone is very thin, the contribution of endogenous-sample bulk fluorescence is minimized. Likewise, the influence of chromophores present in solution, which may quench the fluorescent signal as a result of spectral absorption of emitted fluorescence (inner filter effect) *(14)*, is minimized by virtue of the short path length.

If an antibody is chemically immobilized to the lateral fiber surface, as a fluorescent tagged Ag or Ab–Ag complex binds to the immobilized Ab, it will be excited and, in turn, emit fluorescent light. This forms the basis of a physical-separation (heterogeneous) immunoassay system that requires no user-mediated removal of unbound tag, because phase separation of solid-phase bound and free fluorescently tagged molecules can be detected while the primary reaction is occurring, in consequence of the unique nature of the restricted zone of the excitation beam.

The earliest approaches to the development of immunoassays utilizing this technique used flat, total-internal-reflection quartz plates (rather than cylindrical fibers) as the solid phase to which the antibodies were immobilized *(15)*. In that instance, fluorescent emissions from the solid-phase, Ab-bound fluorescent tag that propagated through the solution were collected by a detector placed externally and at right angles to the internal reflecting element. That approach, although allowing nonseparation immunoassay to be performed (as a result of detector orientation) had sensitivity limitations of approx $10^{-7} M$ analyte detection.

Recent studies with flat-plate internal-reflection elements utilizing detection of fluorescent emissions that have been tunneled back into the planar element and directed by a waveguide back through the element to a detector have been described *(16,17)*. Although sensitivity improvements to approx $2 \times 10^{-8} M$ have been shown, this level is insufficient for measurement of most clinical analytes. Furthermore, in one instance *(16)* this sensitivity required a two-stage reaction in which Ag was first incubated with the solid-phase Ab followed by washing and a second incubation of fluorescein isothiocyanate (FITC) labeled Ab directed against the Ag was required.

Compared to cylindrical fibers, flat-plate waveguides have an inherent sensitivity limitation attributable to geometrical considerations. In a cylindrical waveguide, fluorescent light is guided back into the fiber and collected in multiple planes, whereas in a flat plate, only a single plane is used. In addition, with appropriate optics, nearly the entire surface area of the fiber is used for excitation and fluorescent sensing with all fluorophores bound to the fiber surface receiving similar excitation power *(13)*.

The development of a fiberoptic immunoassay utilizes optical principles developed by Block and Hirschfeld *(18–20)*, including excitation of solid-phase bound fluorophores and capture of tunneled fluorescent emission at the same proximal end of the fiber. This approach has the advantage of reducing the amount of optical filtration required of emerging excitation light in order to detect the lower-intensity, higher-wavelength fluorescence emitted when the detector is placed at the distal end of the fiber. The fiber on which Ab is immobilized is typically a 500-µm diameter fused-quartz rod with optically polished ends. Reagentless systems based on intrinsic fluorescence of protein tyrosine and tryptophan residues *(21)* or on changes in scattered light detected at the surface on Ag–Ab binding *(22)* have been proposed, but the system described in this chapter utilizes a single fluoresceinated (FITC) compound as a tracer species to minimize nonspecific signal and to enhance sensitivity.

Preliminary studies for a two-site immunometric assay for the high-mol-wt protein ferritin and a competitive immunoassay for the cardiac glycoside digoxin ($M_r = 780$) utilizing this approach have been described previously *(23,24)*. Presented here is a detailed description of the two-site immunometric fiberoptic evanescent wave immunosensor (EWS), including characterization of the various reagents used.

2. Materials and Methods

2.1. Basic Strategy

A two-phase strategy was used in the development of an immunoassay system for the quantitative measurement of ferritin. This molecule was chosen because it has a high molecular weight ($M_r = 450,000$), a broad dynamic clinical range (from 5 ng/mL to 1000 ng/mL, or 10^{-11}–$10^{-9}M$), and a natural occurrence in human serum. Quantitation of this molecule is of significance in evaluating disease states, including anemia and hepatoma.

The first level of evaluation was based on isolation of the fiber as a solid support matrix for immobilization of Ab in the absence of any of the variables involved in optical measurement. Such variables include, for example, efficient coupling of high-angle light into the fiber for fluorescent excitation, and variations in optical transmission characteristics from fiber to fiber. In this isolation phase of the study, ^{125}I-labeled antiferritin Ab was used as the tag species, in place of an FITC-labeled Ab. With knowledge of the radiospecific activity of the labeled Ab, precise tests were made of the ability of the Ab-coated sensor to concentrate tagged probe molecules. The absolute coating capacity of the fibers was also investigated utilizing nonspecific ^{125}I-labeled rabbit immunoglobulin (IgG) to differentiate total protein loading of the fiber from retained capability of an Ab to specifically bind Ag. Parameters investigated at this stage are given in Table 2. For optical studies, the radioactive Ab was replaced in the system with an FITC-labeled Ab of identical concentration.

Table 2
Parameters Investigated with Isotopic Probes

Methods of chemically coupling proteins to optical fibers
Capacity of the solid phase to bind Ag
Absolute protein-loading capability vs retained immunologic activity
Fiber-to-fiber reproducibility
Stability of fiber Ab coatings
Reactant concentrations of Ag and labeled Ab
Diffusion-limited processes
Correlation studies of patient samples vs commercial assays

2.2. Optical Fibers

Immunosensors were prepared from precision redrawn fused-quartz rods (Type 214, General Electric, Cleveland, Ohio). Rods were redrawn to a mean diameter of 0.5 mm ± 2% and a length of 65 mm, although only 50-mm lengths were utilized in the immunoassay evaluation. Residual length was required in order to position the fiber in 50-mm chambers containing sample and tagged Ab and to align the sensor in the optical beam.

Both ends of each fiber were polished to an optical mirrored surface as evaluated by monochromatic light and interference banding. Prior to organic silanization, fibers were cleaned in NO-CHROMIX™ (Godax Corp., New York, NY)/concentrated sulfuric acid and rinsed exhaustively in distilled water.

2.3. Antibody Preparation and Immobilization

When covalent coupling procedures were used, fibers were first reacted batchwise with 3-aminopropyltriethoxysilane (10% [v/v] in distilled water) according to procedures described by Weetall *(25)*. Following reaction for 2–3 h at 75°C in a water bath, fibers were rinsed exhaustively in distilled water and heat-polymerized at 100°C overnight. Fibers were stored in a clean, dry vessel prior to protein (Ab) coating.

Partially purified rabbit Ab to human ferritin was utilized for coupling to fiber surfaces in preparation of the immunosensor. Rabbit serum was first treated with 33% (v/v) saturated ammonium sulfate to precipitate immunoglobulins. Following centrifugation, the pellet was reconstituted to the same starting volume of serum and dialyzed against $0.02M$ K_2HPO_4, pH 8.0, to remove residual ammonium salt.

This fraction was then applied to an ion-exchange column of DEAE-AffiGel Blue (Biorad, Richmond, CA). Purified IgG was eluted directly in $0.02M$ K_2HPO_4 buffer, dialyzed against Dulbecco's phosphate-buffered saline (PBS), and diluted in PBS to 50–150 µg/mL final protein concentration for use in coupling to optical fibers. Protein concentrations were assessed by the method of Lowry et al. *(26)* using bovine serum albumin (BSA) as standard. Quantitative titration analysis of purified Ab compared to a purified monoclonal antiferritin Ab reference showed approx 0.5% of the total IgG to be ferritin-specific immunoglobulin.

Antibody was covalently coupled to fibers by a two-step glutaraldehyde procedure. Fibers were placed in a small rectangular polypropylene tray (100 fibers/tray) and covered with 2.5% (v/v) glutaraldehyde (Sigma, St. Louis, M0) in PBS buffer. The solution was gently swirled around the fibers for 1–2 h at room temperature, the fluid removed, and the fibers washed with 4–5 changes of distilled water to remove residual glutaraldehyde. Following this activation, fibers were transferred to a second rectangular container and sufficient Ab solution added to cover the fibers. The approximate volume required was 1–1.5 mL/fiber. Fibers were agitated by gentle swirling for several hours at room temperature followed by overnight incubation at 4°C. Care was taken to avoid excessive agitation, which tended to chip and scratch the optical surfaces of the fibers.

Following incubation, fibers were washed in 2 vol of PBS, 1vol of PBS with $1.5M$ NaCl to remove loosely adsorbed protein, and then another 2 vol of PBS. Fibers were stored in bulk in 50-mL conical tubes containing 0.1% BSA–PBS buffer with 0.2% sodium azide as preservative. Protein in this buffer served to block any residual available reactive sites on the fiber surface.

Studies utilizing radioactive rabbit IgG indicated that the protein-binding capacity of each fiber was 1.2–1.5 µg for a total surface area of 78.5 mm². Since only 1–2% of the total protein offered was bound to the fiber, solutions of Ab were reused at least 4–5 times without any demonstrable reduction in protein loading.

When direct adsorption of protein to the fiber surface was carried out, identical procedures were utilized, with the exception of preactivation of the fiber surface with glutaraldehyde.

2.4. Labeled-Antibody Preparation
2.4.1. Radioiodinated Antibodies

In an immunometric (sandwich) assay architecture, multisite (epitope) antigens are capable of being detected, since they will bind more than one Ab. If one Ab is immobilized to a solid support matrix, a second labeled Ab, also directed against the Ag, may be utilized to detect the presence of Ag on binding to the solid-phase. Thus, when the solid matrix is quantitated by measurement of tagged Ab the concentration of analyte is directly proportional to the bound tagged Ab.

Iodinated mouse monoclonal antiferritin Ab (Hybritech, La Jolla, CA) was prepared by a modification of the method of Greenwood et al. *(27)* utilizing ^{125}I-Na (Amersham, Chicago,IL) and chloramine T as an oxidizing agent to incorporate radioactive iodine into tyrosine and histidine residues of the molecule. Reaction conditions were optimized to provide a final specific activity of 1–2 uCi/μg protein. Counting of fibers in a gamma spectrometer after immunoassay allowed precise quantitation of the mass (moles) of Ab bound under various reaction conditions. This served as a basis for comparison of the binding characteristics of the immunosensor (a) as a solid-phase support and (b) under optical EWS conditions, when fluorescent Ab was used in place of radioactive Ab.

2.4.2. Fluoresceinated Antiferritin Antibody

Fluorescent Ab was prepared by a modification of the procedure of Kawamura *(28)*. Briefly, 1 mg of mouse monoclonal Ab identical to that used for radioiodination (1 mg/mL) was dialyzed overnight at 4°C in 0.05M bicarbonate–saline buffer, pH 9.5. Dialysis tubing containing the Ab was transferred to a beaker containing 0.1 mg/mL FITC (isomer I, Sigma) and allowed to react for 2 d at 4°C. After conjugation, FITC–Ab conjugate was separated from free FITC by passage through a 10-mL Sephadex G-25 column (Pharmacia, Piscataway, NJ). One-milliliter fractions were eluted with PBS. The protein peak was assessed by optical absorbance at 280 nm and was pooled. Comparison of the ratios of optical absorbance at 280 and 495 nm showed a fluorescein: protein (F:P) ratio of 8–10 mol/mol Ab. Conjugates were diluted into 1% BSA–PBS, pricr to storage.

Estimates of reactive immunologic activity of FITC–Ab conjugates were determined by an Ag-specific radio-titration analysis of the labeled conjugate and compared to a titration curve of the parent monoclonal Ab prior to fluoresceination. These studies were performed utilizing a solid-phase radioimmunometric assay system for ferritin (IMMOPHASE™, Ciba-Corning Diagnostics, Medfield, MA), wherein the solid-phase Ab was saturated with ferritin prior to analysis. Comparisons of dilutions of monoclonal Ab (unlabeled or fluoresceinated) that inhibited 50% of the radiolabeled Ab from binding to the solid-phase were made, assuming that native unlabeled Ab represented 100% activity. By this method, it was determined that the FITC–Ab retained 20–30% of its initial immunologic activity. Utilizing this approach, it was possible to quantitate FITC–Ab used in EWS on a concentration basis by referring back to ng/mL protein of the fully immunologically active parent monoclonal Ab rather than an arbitrary dilution value. Likewise, direct comparisons could be made of both fluorescent and radioactive concentrations of antiferritin Ab utilized in the above test systems.

Antibodies prepared in this fashion were stable for at least 1 yr in buffer when stored at 4°C.

2.5. Immunoassay Test Systems

2.5.1. Immunoradiometric Assay (IRMA) of Fibers

Human-serum ferritin standards were prepared by doping pooled, defibrinated plasma (endogeneous-ferritin content, 57 ng/mL) with human-spleen ferritin (Behring-Calbiochem, La Jolla, CA). To prepare low-concentration standards, ferritin was stripped from serum utilizing an antiferritin affinity immunoadsorbent. Standards and patient samples were assessed for accuracy and calibrated utilizing a precise commercial immunoradiometric assay system for ferritin.

An IRMA assay utilizing the fiber as a solid phase was developed under emulation conditions identical to those used subsequently in the EWS assay. In a typical assay, equal volumes of serum standards containing various concentrations of ferritin from 0 ng/mL to 1000 ng/mL and ^{125}I-antiferritin Ab were added to 0.4 × 5-cm test tubes, followed by immediate immersion of an antiferritin-coated fiber into the test tubes. After a 10-min incubation, fibers were washed twice in PBS, placed in 12 × 75-mm test tubes, and counted in a gamma spectrometer to quantitate the amount of labeled Ab that was bound. Because the reaction must be terminated prior to signal measurement, only end-point analyses were compared to EWS.

Tests to quantitate the absolute binding capacity of the fiber were performed in the following manner: First, the fiber was incubated for several days with 1000 ng/mL ferritin. Following washing, fibers were reincubated with high concentrations of radioactive Ab for several days, washed, and counted. Control fibers, not exposed to Ag in the first-stage incubation, were processed in a like fashion to determine the extent of nonspecific binding. To assess the mole-binding ratio of Ag to labeled Ab, evaluation of loading Ag directly onto the fiber was made as follows: First, the fiber was allowed to react with high concentrations of ferritin. The ferritin content of the solution was measured both before and after inubation with the fiber in a commercial IRMA. Comparison of the absolute number of moles of Ag or labeled Ab bound to the fiber under these conditions were identical, indicating a 1:1 stoichiometric relationship between labeled Ab and Ag.

2.5.2. Evanescent-Wave Immunosensor (EWS) Format

When fibers were tested in an optical format, the fiber was inserted in a special flow-through chamber (5 cm length × 0.4 cm id) with two ports positioned on opposite ends perpendicular to the chamber. Tubing from the chamber was connected to a peristaltic pump to allow the sample/FITC–Ab mixture to flow into the chamber and around the fiber to initiate the reaction. The fiber was held in place by two specially fabricated silicone caps that prevented leakage of fluid from the chamber. For each data point, a fresh fiber was used.

This chamber assembly was secured on a carrier and the proximal end of the fiber aligned to an optical beam in a modified front-face epifluorometer developed by ORD, Inc. (Cambridge, MA). The characteristics of this instrument are described in the chapter by Lackie et al. in this volume. Excitation-light input and fluorescent-light output were conducted at the same proximal end of the fiber.

Analog signals from the solid-state detector were converted to digital output via a DASH-8 analog-to-digital converter (Metrabyte, Taunton, MA) coupled to an IBM-AT personal computer using appropriate software for data collection and reduction (Lab Notebook, Metrabyte). Assays were initiated by adding a sample to an equal volume of FITC–Ab in a test tube (0.7 mL total volume) and pumping it into the fiber chamber. Precise timing control was maintained by delaying intitiation of data collection until the sample chamber was completely filled with fluid. Integration time between data collection points was 10 s.

2.6. Data Processing

Since the primary immunoreaction between Ab coupled to fiber and Ag (ferritin)–FITC–Ab complex could be monitored directly, assays could be conducted in either a kinetic (rate) mode or to a timed end-point. In general, all signals were processed after subtracting the response of the first 10 s of reaction. Removal of this portion of the response eliminated nonspecific signals from the following sources:

1. Back-reflection fiber-end surface,
2. Background fluorescence from the fluorescein content of the Ab-conjugate solution prior to reaction and immunoconcentration,
3. Endogeneous background fluorescence of the serum sample, and
4. Fluorescence from nonspecific adsorption of proteins in the buffer and sample onto the fiber surface.

This last factor was more than 90% complete after the first 10 s of reaction. Removal of this early portion of the signal from rate determinations also eliminated falsely elevated initial rates, which resulted from residual convective or stirred-layer effects caused by sample injection. After the first 10 s of measurement, signal increases caused by concentration of fluorescent Ab in the evanescent zone were more uniform, approximating a strictly diffusion-limited process.

Data were processed in both rate and end-point modes. For rate analysis, slopes were derived for each standard and unknown analyte concentration by comparison of signal changes (Δ mV/ min) assessed at 10-s intervals from 20 s to 10 min, after subtraction of the initial 10-s measurement. End-points were computed by subtraction of 10-s signal from the total signal accumulated at a given time interval.

3. Characterization of the Fiber as a Solid-Phase Surface-Isotopic Model

3.1. Precision and Stability

Figure 2 demonstrates that, when utilizing 500-μm fibers containing chemically attached Ab in a single-step assay (simultaneous ferritin and labeled Ab addition) with serum-based samples, a precise standard curve could be generated throughout the clinically significant range. Precision of replicate samples was consistently within a coefficient of variation of <7%. High precision is essential for acceptance in quantitative clinical diagnostic systems for quantitative analysis. These results demonstrate that uniform protein coatings of fibers with retention of immunologic activity is achievable.

Further, as demonstrated in Fig. 3, the Ab-coated fibers maintained binding activity over long periods of storage. Concurrently prepared fibers, when tested periodically up to 5 mo after preparation, maintained up to 80% of their initial immunologic activity when stored in buffer at 4°C. Testing of fibers for this demonstration was done only at 200 ng/mL ferritin in 0.1% BSA–PBS buffer. This concentration was the midpoint of a typical ferritin standard curve.

3.2. Parameters Affecting Accumulation of Labeled Ab–Ag Complex on Fibers

3.2.1. Labeled-Antibody Concentration

Since one of the prime objectives in immunosensor development is to configure the system to accumulate as much specific signal (labeled tag) in as short a time as possible, modeling experiments in the isotopic mode were used to evaluate the parameters that affected the magnitude of labeled Ab–ferritin (Ag) complex accumulation on the solid phase. Since immunometric assay systems use excess reagents, in terms of both the solid-phase primary Ab and the labeled Ab species, the influence of increasing the tag concentration was investigated.

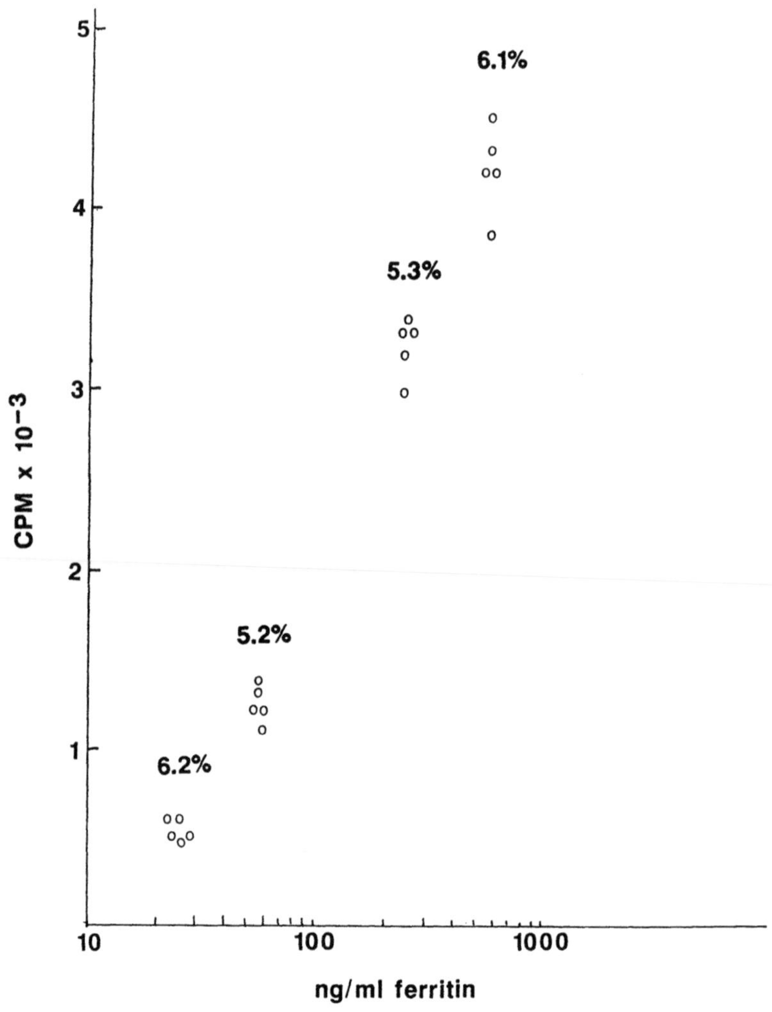

Fig. 2. Radioactive isotope modeling of ferritin standard curve on 500 μm diameter optical fibers containing Ab chemically attached to ferritin. Serum-based ferritin standards from 20–1000 ng/mL were diluted with an equal volume of ^{125}I-antiferritin Ab, mixed and transferred to 0.4 × 5-cm micro test tubes. Fibers were immediately immersed in the tubes. Reactions were at ambient temperature. Following incubation for 10 min, fibers were rinsed by dipping in a beaker of PBS, transferred to 12 × 75-mm plastic test tubes, and counted for bound radioactive Ab in a gamma spectrometer. Five replicates were determined for each data point. Numbers above each set of data are CVs.

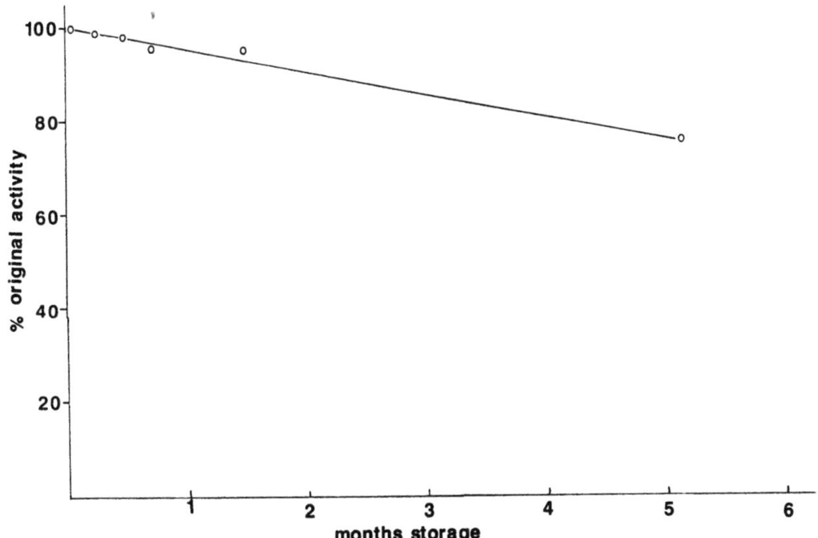

Fig. 3. Long-term storage stability of Ab-coated fibers. Antiferritin-Ab-coated fibers were stored in a batch in a beaker of 0.1% BSA–PBS at 4°C. At periodic intervals, fibers were tested at a ferritin concentration of 200 ng/mL. Each data point was measured in triplicate. From the specific activity of the labeled Ab, the cpm bound was converted to moles of labeled Ab bound and compared to values for freshly prepared fibers.

As seen in Fig. 4, increasing the tag concentration served to increase the signal at the high concentrations of the standard curve. This resulted in eliminating a problem that is inherent in immunometric assay systems, namely a progressive reduction in binding to formed labeled Ab–Ag complex at higher concentrations of Ag to give signal intensities less than those for very low concentrations of Ag. This is termed the "high-dose hook effect." Without proper modulation of reactants, erroneously low concentrations will be measured for high-concentration samples. Theoretical aspects of labeled Ab immunoassay systems have been addressed by Kemp et al. *(29)*.

A partial explanation of this phenomenon lies in the fact that there is a competition for the solid-phase Ab-binding sites between free Ag and the Ag in the labeled Ab–Ag complexes in solution. If the

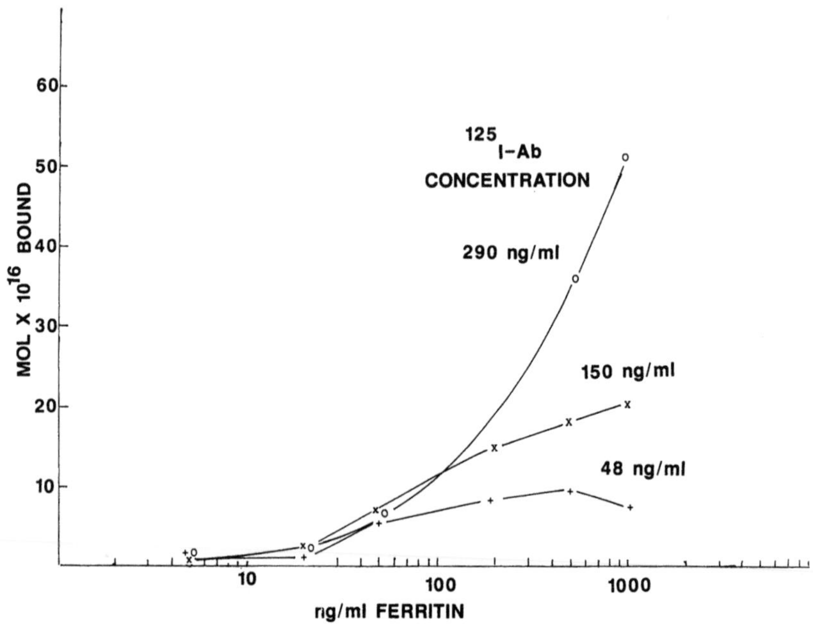

Fig. 4. Effect of increased labeled Ab concentration on the ferritin standard curve. Results are standard curves for 15 various final concentrations (48–290 ng/mL) of ^{125}I-labeled Ab.

amount of labeled Ab is less than that of the highest concentration of Ag, there will be a progressive increase in the proportion of free Ag in solution vs Ag-labeled Ab complexes and thus an increase in the proportion of free Ag binding to the solid phase.

One method of overcoming this effect is to increase the concentration of solid phase Ab to a level in excess of the total Ag concentration. In systems that use multiparticulate solid phases (beads and particles), this is easily accomplished by using more beads. However, the fiberoptic solid phase is constrained by the surface area of a single piece of fiber. Use of more highly purified Abs on the solid phase may be one way of generating an excess reagent system; however, the approach selected here is to increase the final concentration of labeled Ab to a number of molecules equal to or in excess of the total number of molecules of the highest concentration of Ag. For the fiberoptic system, this final concentration was 500 ng/mL ($3.1 \times 10^{-9}M$) FITC-labeled monoclonal antiferritin Ab. This corresponds to the 1000 ng/mL ferritin standard (500 ng/mL in the final assay mixture, or $1.1 \times 1^{-9}M$) and represented an excess ratio of 2.8:1(moles Ab:moles Ag), ensuring that all Ag in the reaction mixture is bound to labeled Ab. When the upper range of Ag to be measured is at a lower concentration, proportionately less labeled Ab is required in the test system.

3.2.2. Diffusional Constraints

Since the fiber has a fixed surface area, other than increasing the number of molecules of active Ab per unit of area, any additional parameters that could enhance the reaction rate are controlled by modulation of solution properties. Because of the particular geometry of the fiber and the fact that one of the reactants is immobilied, the concentration of labeled Ab can be maximized to make the assay pseudo-first-order. Once the system becomes first-order and strictly dependent on analyte concentration, the rate becomes diffusion-limited in nature. Signal (labeled-Ab) accumulation becomes dependent on diffusion of Ag-labeled Ab complexes through the solution and onto the solid phase. Unlike microbead or particle solid-phase immunoassay systems, in which the microdispersion of particles approximates solution kinetics and minimizes diffusion-path length, diffusion distances in the fiber system are large.

The influence of solution viscosity on accumulation of radioactive signal is shown in Fig. 5. Here, serum at 50% of the total reaction volume reduced tagged-complex accumulation at the fiber after incubation at room temperature for 1 min with the antiferritin fiber. This reduction of binding, as compared to that with the less viscous 0.1% BSA–PBS buffer-based ferritin standards, is most marked at higher analyte concentrations. When the tests were conducted at 37°C, this effect could be somewhat overcome, since there was a doubling of signal accumulation over the same time period for the assessed concentration range (data not shown). Use of 8% (w/v) BSA, which approximates the protein content and viscosity of serum, demonstrated a similar reduction in tag-complex accumulation, suggesting that suppression of binding was not attributable to factors in the serum.

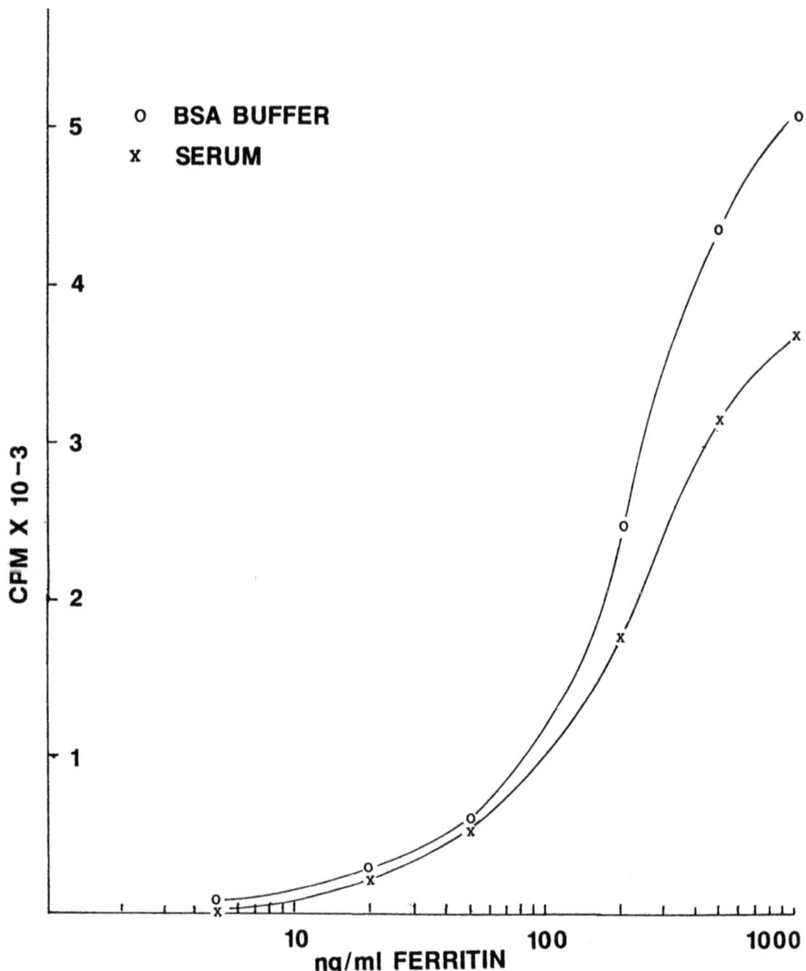

Fig. 5. Influence of high serum-protein concentration (solution viscosity) on accumulation of ^{125}I–Ab–ferritin complexes on fiber surface at room temperature. Ferritin standards for buffer (0.1% BSA in PBS) were replaced by serum-based standards of higher viscosity. Tagged Ab concentration was 250 ng/mL monoclonal Ab equivalent final concentration (50:50 dilution of sample and tag Ab).

Although dilution of the total serum protein content would be desirable to reduce the plasma protein diffusional constraint, this must be balanced by providing sufficient analyte concentration at the low end to allow significant discrimination from zero. Since the prototype optical instrumentation did not have temperature-control capability, the serum content was set at 50% of the total reaction volume. When coupled with an optimal concentration of tagged Ab species, the 50% serum value still provided signal discrimination from zero at the low analyte concentration of $1.2 \times 10^{-11}M$.

Final assay optimization prior to work in the optical EWS format included adjustment of labeled Ab to 500 ng/mL monoclonal Ab, adjustment of serum-sample volume to 50% of the total reaction mixture, and a standardized reaction time of 10 min with simultaneous addition of sample and labeled Ab. Results of emulation experiments suggested the use of 37°C as the better reaction temperature; although this was not included in the EWS prototype system because of hardware limitations.

3.3. Fiber Binding Capacity

The amount of labeled antiferritin Ab accumulated under actual assay conditions was estimated from conversion of cpm bound to moles of labeled Ab based on the specific activity of the radioiodinated Ab. When results were compared to equilibrium binding of labeled Ab and Ag (as described in the Materials and Methods section), they suggested that, under assay conditions, less than 10% of the active immobilized Ab was used to bind Ag, even at the highest Ag concentration. This ensured that, relative to the system developed, the solid-phase binding capacity was not being exceeded (Table 3).

Studies of ferritin-active Ab sites on the fiber at equilibrium, relative to total immobilized ^{125}I-IgG suggested that only 0.6% of the total coupled Ab protein was Ag-active. Antigen-specific titration of these Abs in solution prior to immobilization on the solid phase demonstrated that 0.5% of the total IgG population of the starting fraction was antiferritin Ab. Thus there appeared to be no loss of the Ab ability to bind ferritin after immobilization onto the fiber. These results suggest that use of more highly specific monoclonal Ab for immobilization should increase the total labeled complex bound to the fiber as a result of the increased density of active Ab immobilized per unit area and could correspondingly reduce the requirements for labeled Ab in the system.

Table 3

^{125}I-Antiferritin–Ferritin Complex Binding to Optical Fibers

Ferritin, ng/mL	Reaction concentration, ng/mL	Initial ferritin concentration, $\times 10^{-11}M$	^{125}I bound to Ab,[a] $\times 10^{-16}M$	Effective concentration,[b] $\times 10^{-8}M$	Apparent fold concentration	Total fiber capacity utilized, %[c]	Total serum volume completely extracted of ferritin, μL
6	3	0.67	0.56	2.37	3537	0.09	4.2
30	15	3.30	3.30	13.94	4224	0.54	5.0
64	32	7.10	8.50	36.52	5143	1.40	6.0
250	125	28.0	28.10	118.70	4239	4.70	5.0
640	320	71.0	49.30	208.30	2933	8.10	3.5
1100	550	120.0	58.60	248.0	2067	9.70	2.4

[a]Assay is for 10 min of total reaction time. Independent studies (as described in the Materials and Methods section) demonstrated equimolar (1:1) interaction of Ab and Ag.

[b]The length of a "sandwich" of Ab-ferritin-tagged Ab is assumed to be 30 nm. The differential volume of the fiber + "sandwich" cylinder and the fiber alone would be 2.36×10^{-9} L. This would be the volume into which all bound Ag would be concentrated during the reaction with the fiber.

[c]Based on an equilibrium binding capacity of 6.02×10^{-14} mol.

Even with the use of more highly purified Ab, there are limits to the magnitude of binding of labeled complex that could be expected, since the total surface area would still be unchanged (78 mm^2). Studies with fibers of progressively larger diameter, which thereby progressively increased the total surface area exposed to solution, resulted, at equilibrium, in proportional increases in total signal accumulated when all other factors remained constant. However, under the reduced reaction time, subequilibrium conditions were required as a criterion for rapid analysis. Accumulation of labeled complexes was less than half of the value expected based on the proportional increase in surface area. This again pointed to diffusional limitations in consequence of the cylindrical geometry of the optic fiber.

A solid phase with larger surface area would provide higher probability for molecular interaction and, thus, increased mass accumulation, but this has to be balanced by the optical constraints imposed when using fibers of larger diameter. These factors include the need for larger optical components, such as lenses and detectors, in order to fill the fiber for excitation completely and to capture fully the emitted fluorescent signal from the fiber surface. Although the ideal situation would be to maximize both surface density of active Ab and total surface area, limitations in the availability of large quantities of monoclonal Ab required for fiber coating, and restrictions in optical instrumentation limited the present studies to fibers 500 μm in diameter.

Finally, mass-accumulation studies were used to estimate, under actual assay conditions, the minimum sample volume required (prior to dilution) to conduct an assay. This was based on analysis of the volume of serum that had to be totally depleted of Ag in order to be concentrated on the fiber. These results are shown in Table 3 and indicate that the fiber has a mass-concentrating ability ranging from 2000- to 5000-fold, depending on the analyte concentration. Serum volumes of <6 μL were required. Calculations relative to the extent of Ag concentration assumed that the labeled complex was concentrated into the volume of a cylinder surrounding the fiber with a differential radius of 3 nm. This radius was estimated based on the longitudinal length of an IgG molecule being 10 nm and the diameter of a spherical ferritin molecule being 10 nm *(30)*. If loaded onto the fiber in an "end-on" conformation, the length of an Ab–ferritin-labeled Ab complex would be 3 nm. The differential volume of cylinders with radii r_1 (equal to the radius of the fiber [0.25 mm]) and r_2 (equal to the radius of the fiber plus stacked immune complex [0.250030]), would yield a volume of 2.36×10^{-9} L in which all the labeled complex would be contained. This volume is based on a fiber length of 50 mm.

Allowing for a 10-fold excess of Ag, to maintain a reasonable concentration gradient allowing for high reaction rates of complex accumulation according to the law of mass action, less than 60 µL of serum sample would be required in a 10-min assay system. Correspondingly less Ag would be required if the reactions were terminated at shorter time intervals, although less total signal would be accumulated. Small volume requirements are essential for use in immunosensor test systems in which capillary pipet-drawn blood samples may be employed. Because of fabrication limitations, the prototype instrumentation and assemblies used in this work required 700 µL of fluid volume (350 µL serum), far in excess of the predicted requirements.

4. Fiberoptic Evanescent-Wave Immunoassay (EWS)

4.1. End-Point Assay

Results of a typical end-point assay under conditions described earlier and in the legend are shown in Fig. 6. Each bar represents the total signal accumulated after 10 min of reaction time after subtraction of the signal at 10 s. A different fiber was used for each data point, emphasizing the single-use nature of these sensors. The hatched portion of each bar represents the residual nonspecific signal accumulated after 10 min of reaction minus that from the first 10 s. Comparing the zero Ag concentration with all components added (serum and FITC-Ab) and the zero concentration without added FITC-Ab, this nonspecific signal can be seen to be caused by residual adsorption of plasma and buffer proteins onto the fiber surface, rather than by nonspecific adsorption of tag Ab.

Unlike the radioimmunoassay system, reactions were monitored instantly and continuously after injection of the sample and labeled Ab mixture into the sample chamber surrounding the fiber. There were no additional washing and processing steps required.

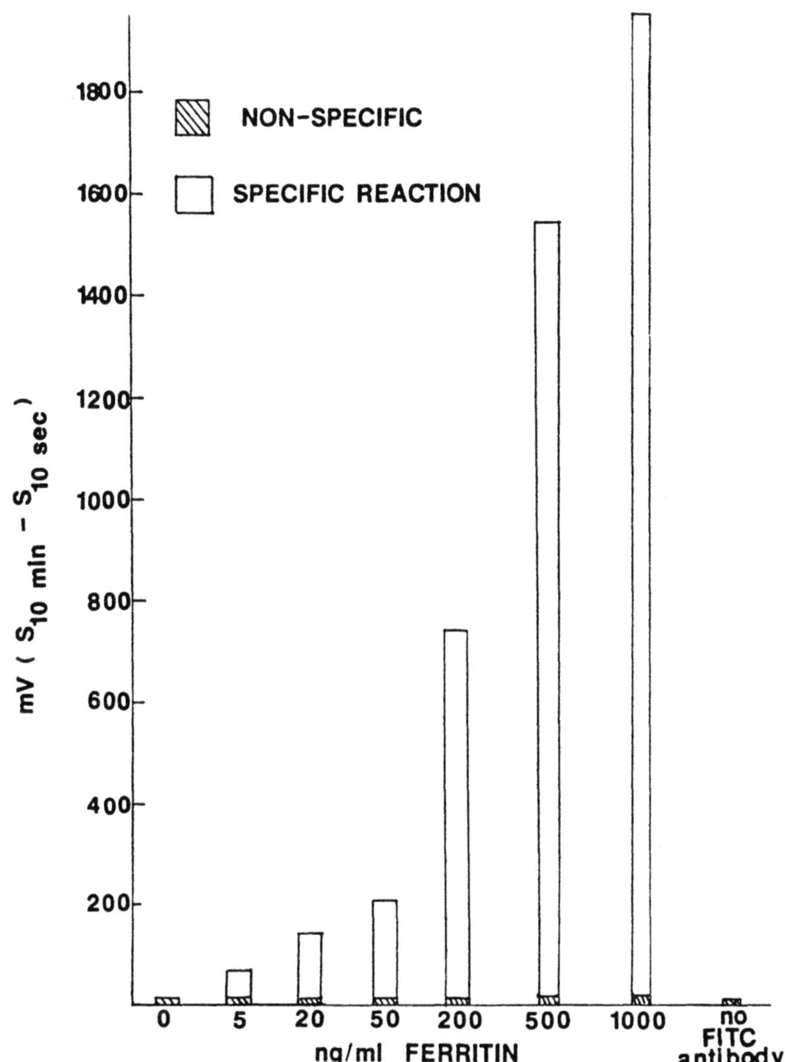

Fig. 6. Ten-minute end-point optical immunoassay for ferritin with serum-based standards. Utilizing FITC-labeled antiferritin Ab adjusted to a final concentration of 500 ng/mL monoclonal equivalent, fibers were monitored with the evanescent wave for signal accumulation at various concentrations of ferritin. The accumulated millivolt (mV) signal after the first 10 s of reaction was subtracted to remove nonspecific response. The residual nonspecific signal (hatched bar) was caused primarily by further adsorption of nonspecific proteins (*see* text).

4.2. Rate Analysis

Because the system functions in an essentially homogeneous fashion (nonseparation), direct analyses of the rate of signal accumulation (fluorescent Ab concentration) vs analyte concentration could be evaluated. This is shown in Fig. 7 for a total reaction time of 5 min. By keeping the tag concentration high, pseudo-first-order rate kinetics were maintained, even at high analyte concentration, for at least 2 min. Deviation from linearity is highlighted by the dashed lines at high concentrations of ferritin. Beyond 2 min, the extraction of Ag-labeled Ab complexes by the fiber results in removal of tagged Ab from solution. Once this concentration is no longer in excess, loss of first-order kinetics ensues. As expected, this is first evidenced at the higher Ag concentrations.

Examination of the system over the first 60 s of reaction is shown in Fig. 8. Rates in this time frame were found to be proportional to Ag concentration in as little as 40 s after reaction initiation. Determination of the slope for each ferritin concentration (Δ mV/min) resulted in the standard curve shown in Fig. 9. The time interval for accession of data is currently limited to the noise of the system (± 2.5 mV). This noise is imposed by the resolution of the A/D convertor and general system characteristics of the optics, detector, and lamp. Use of A/D conversion of higher resolution and more efficient coupling and detection of high-angle light entering into and emerging from the fiber should result in signal enhancement, thereby providing measurement of higher precision in shorter time intervals. Likewise, increasing the temperature at which reactions are run would enhance the magnitude of signal development in shorter times. Given these present limitations, the system was still repeatedly able to discriminate $6.0 \times 10^{-12}M$ ferritin from a zero dose in as little as 40 s at a signal:background ratio of 2.5:1. This concentration of ferritin is the actual amount in the reaction mixture after the 1:2 dilution of sample.

The signal:background ratios in both the rate and end-point modes, when compared to a commercial radioimmunoassay kit for ferritin, far exceeded the results of the commercial immunometric procedure. This is shown in Fig. 10. Values used for background determination were based on signal accumulation at 0 ng/mL ferritin concentration

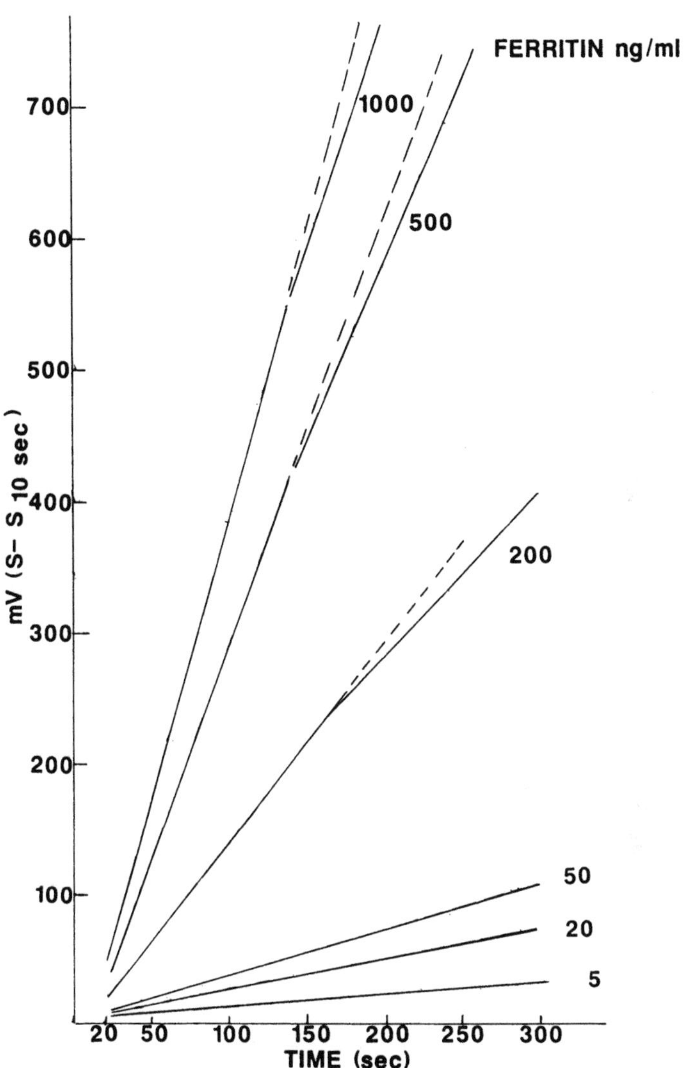

Fig. 7. Real-time reaction-rate monitoring of ferritin fiberoptic immunoassay. Various concentrations of ferritin in serum were mixed with an equal volume of FITC-labeled antiferritin Ab and injected into the sensor chamber. Rates of fluorescent signal accumulation resulting from binding of labeled complex to antiferritin Ab immobilized on the fiber surface were monitored for a period of 5 min. Each line represents a separate sensor monitored at a different ferritin concentration. Dashed lines emphasize deviation from linearity after the first 2 min of reaction.

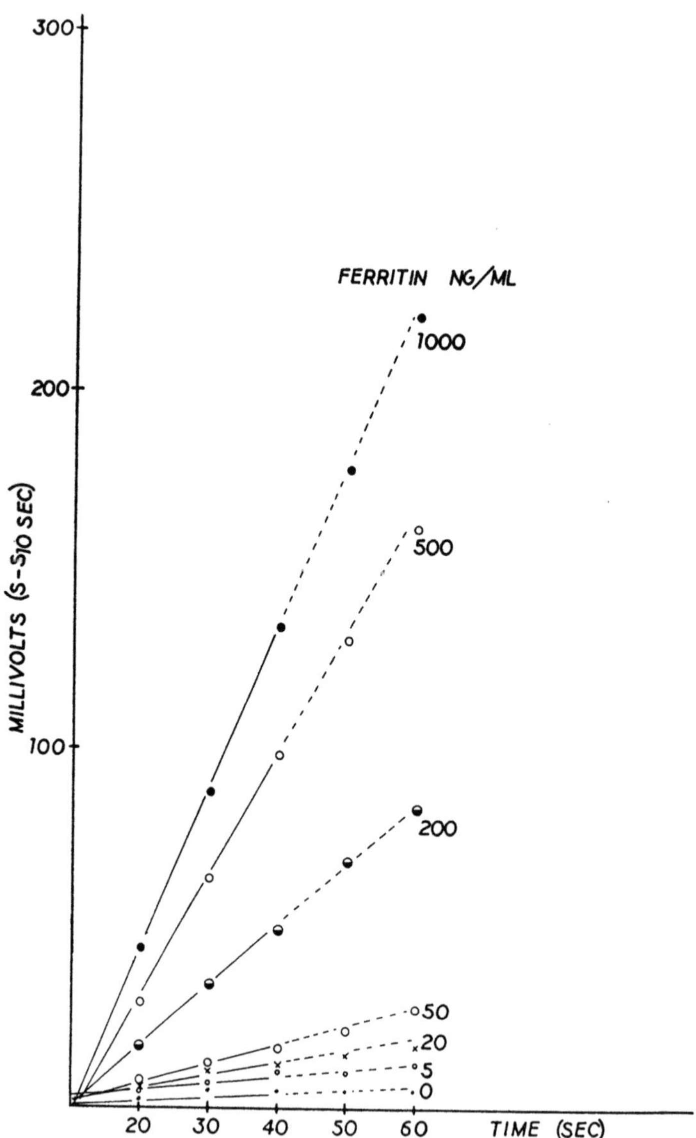

Fig. 8. Reaction rates over the first minute. Data are from Fig. 7 with expansion of the y axis to demonstrate the linear relationship. Reactions are monitored directly after sample addition. From ref. *24*, p. 462.

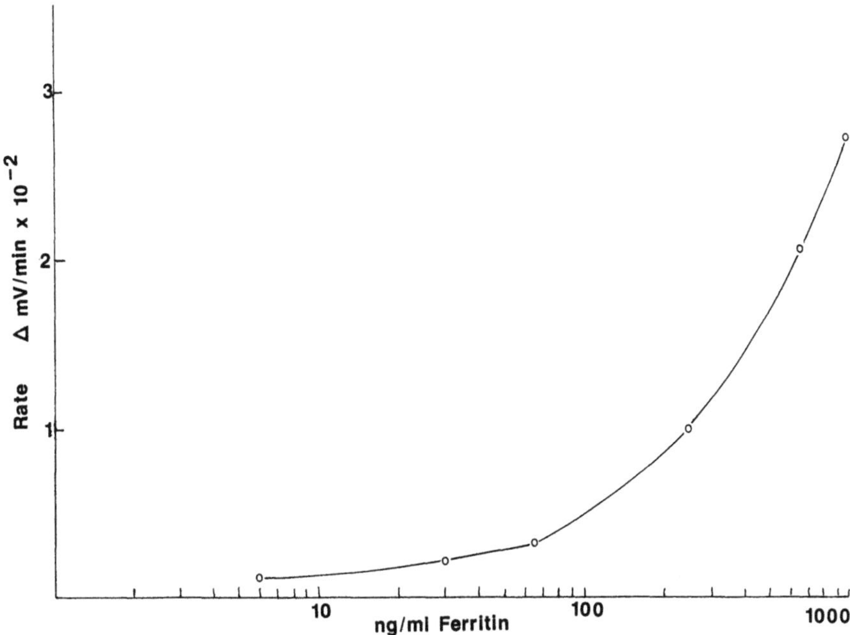

Fig. 9. Standard calibration curve for ferritin based on the reaction rate for the first 40 s (plot of slopes of Fig. 8 data for the first 40 s).

(mV for EWS or radioactive cpm for the IRMA). The low signal:background ratios in IRMA are caused by high background cpm from trapping of radioactive Ab by the solid-phase glass particles. Total assay time to run the IRMA assay included a $2\frac{1}{2}$ hr incubation plus processing time for two separate reagent additions as well as washing, centrifuging, decanting, and counting the sedimented pellet. Total EWS assay times are as indicated following addition of sample and fluorescent tag to the fiber. These results demonstrate the extreme speed of the EWS system as compared to current technology. As can be seen from Fig. 10, the use of the end-point mode in EWS can be advantageous in terms of increased sensitivity, since the proportion of specific accumulation of signal relative to nonspecific adsorption (background) is magnified over longer time intervals.

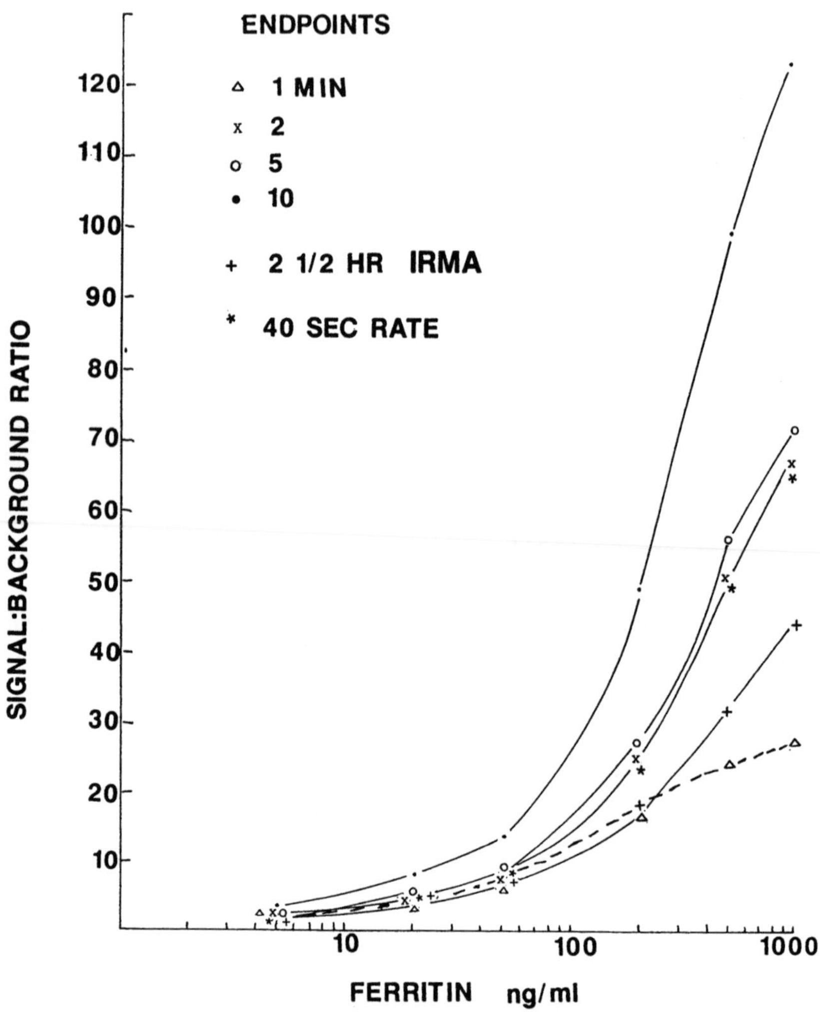

Fig. 10. Signal:background ratios for an EWS, 40 s rate; EWS end-points at 1, 2, 5, and 10 min, and a 2 1/2 h commercial solid-phase immunometric assay for ferritin. Backgrounds were determined on data from the 0 ng/mL (ferritin-stripped human serum) standard at identical time intervals as the various concentrations of ferritin. Background for the commercial radioassay (IRMA) was based on cpm bound at 0 ng/mL; in EWS, background was based on the millivolt signal accumulated with this standard. Evaluations were made with serum-based standards.

Table 4
Fiber-to-Fiber Precision
of Ferritin Optical Immunosensor, 5-Min End-Point

Serum concentration,	Total signal, mV			
ng/mL ferritin	*N*	$\overline{X}$	SD	% CV
50	8	576	32	5.6
1000	12	1025	52	5.0

4.3. Precision

Signal:background ratios seen at 40 s in the rate mode were comparable to those found with 2- or 5-min end-point EWS analyses. However, because of detection constraints of the prototype system, as described above, high precision cannot yet be achieved at the low concentrations in the rate mode. This results from the small signal increments over short time intervals at the low analyte concentrations. Potential precision of the EWS system therefore has been assessed using the end-point signals at 5 min. As shown in Table 4, assessment of total signal coefficient of variation (CV), including all nonspecific signal from fiber background and serum components, resulted in CVs at high and low serum-ferritin concentrations of <6%. Comparable results were achieved when 2-min end-points were utilized.

Further evidence of the reproducibility of the entire system (fiber and optical detection system) was confirmed by the ability of the system to reproduce standard curves for ferritin over long periods of time. As shown in Fig. 11, replicate standard curves for ferritin from 5 ng/mL to 1000 ng/mL run 2 wk apart gave virtually superimposable results, strongly supporting the reproducibility of the system. Since immunoassay is predicated on referencing unknown samples to a stored standard curve or to one that is run simultaneously with the unknown samples, the overall implication of these findings is that development of a commercial system requiring infrequent calibration and possessing long-term stability is feasible with this technology.

In order to develop an overall system with high precision, it is essential that the error contributed by each individual component be

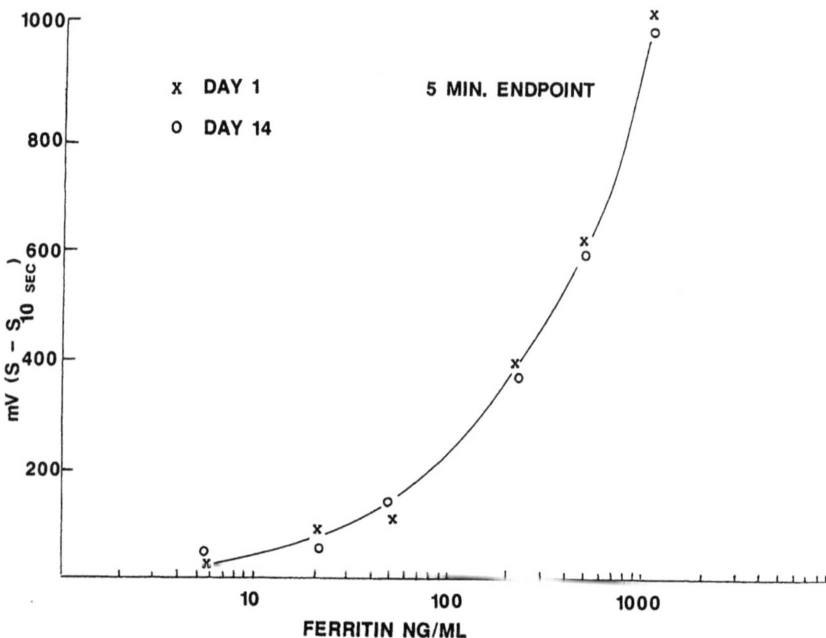

Fig. 11. Standard curve stability. The potential for long-term stability of both reagents and instrumentation is demonstrated by the reproducibility of standard curves prepared 2 wk apart with identical lots of fibers, serum calibrators, and labeled antibody. Millivolt data were obtained at a 5-min end-point after subtraction of signal at the first 10 s.

thoroughly understood. Variables, and thus potential sources of error, are listed in Table 5.

4.4. Comparison of Optical and Isotopic Systems

Since the EWS system is a solid-phase immunoassay and signals developed are dependent on accumulation of labeled Ab–Ag complexes on the solid-phase, experiments were undertaken to relate fluorescent-signal accumulation with radioactive Ab bound to the fiber under identical assay conditions. Since signal is collected in entirely different forms by these two approaches, comparison of the numerical relationship of accumulated molecules of tagged Ab, either isotopically or fluorescently labeled, should help provide an understanding of the optical characteristics of evanescent wave sensors. In order to deter-

Table 5
System Variables in Fiberoptic Immunoassay

Sensor Requirements	
Materials	**Chemistry**
Diameter variations	Silane coating
End-surface polishing and defects	Ab coating
Fiber-to-fiber variation	Methods of application
in optical-transmission properties	Stability of coatings
Optimum lengths and diameters	Optical influences of coatings

Instrument and Sensor Assembly Requirements
Methods of holding sensors
Alignment of sensor to optical beam
Sample introduction to sensor chamber
Maximizing input of high-angle light
Precise timing of assay reactions
Lamp and detector noise
Optimization of interference filter
Sources of sensor noise
Sources and control of nonspecific fluorescence

mine the moles of fluorescent Ab bound to the fiber as determined under evanescent wave sensing conditions, fluorescent signals from bound Ag-labeled Ab complex were related to fluorescent signals of solutions of fluorescein dye measured on fibers to which no Ab had been chemically coupled and to which, presumably, no dye had adsorbed. These sensors can detect solution fluorescence concentration, as any fluorophore brought into the evanescent zone can be excited and will emit light regardless of whether it is physically bound to the fiber surface.

Signals found in the ferritin EWS at various analyte concentrations were related to a standard concentration curve of FITC dye as shown in Fig. 12. As seen from the curve, the lowest detectable concentration of FITC dye was $5 \times 10^{-9}M$. This concentration was comparable to the specific signal accumulated by $6.0 \times 10^{-12}M$ ferritin at 2 min. The fact that the equivalent fluorescein dye concentrations at

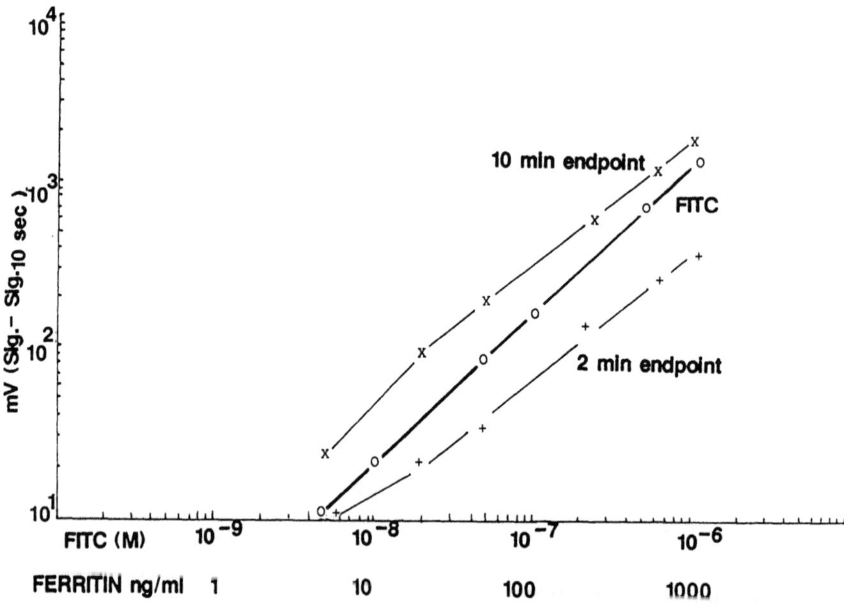

Fig. 12. Comparison of ferritin optical-immunoassay fluorescent signals to those for free fluorescein dye in solution. The fluorescent signal at a given analyte concentration was first related to the equivalent signal generated by a concentration of free FITC dye in solution. Dye concentrations were determined with fibers to which no Ab had been immobilized or adsorbed. Ferritin values shown are the levels of the initial serum sample; actual concentrations in the assay system are one-half of those shown. As described in the text, with assumptions regarding the thickness of the evanescent zone and the total number of dye molecules contained within it, the number of fluorescent Ab–ferritin complexes present on the surface of the fiber after reaction for 2 and 10 min can be estimated.

each ferritin level measured exceed the initial analyte concentration by three orders of magnitude again demonstrates the tremendous mass-concentrating capability of the fiber.

To relate the molar concentration of equivalent FITC dye at each ferritin standard to the number of moles of bound FITC-labeled Ab, a simple set of first approximation assumptions were chosen. First the number of moles of FITC detected by the sensor was assumed to be fully contained in the volume defined by the depth of penetration (d_p)

of the evanescent wave. This distance is defined as the length from the fiber surface in which the electromagnetic field amplitude decays to $1/e$ of the surface intensity. Its derivation and analysis are discussed in the chapters by Love et al. and Lackie et al. in this volume. At an excitation wavelength of 500 nm, d_p has been determined to be approx 100 nm. The differential volume of this cylinder of molecules sheathing the fiber would thus be 7.85×10^{-9} L. Calculations are similar to those described in Section 3.3.

The next assumptions for this analysis were that all molecules were excited by equal field strength, regardless of their distance from the fiber surface, and that their subsequent fluorescent emissions were tunneled with equal efficiency back into the fiber. A further assumption in development of the relationship of free-dye concentration to flourophore conjugated to Ab was that the relative quantum yield was the same for both free FITC dye and FITC covalently conjugated to antiferritin Ab. In fact, there is a 50% reduction in observed fluorescence between free dye and fluorescein conjugated to protein via thiocarbamyl linkages *(31,32)*. Given these assumptions, the number of molecules of FITC dye in solution that were excited could be related directly to the bound FITC–Ab tag and compared to the amount of radioactive Ab found bound to the fiber under identical assay conditions. These results are summarized in Table 6.

Comparison of the number of moles of radioactive Ab bound vs the apparent number of moles of FITC coupled to Ab are in remarkably good agreement, being on the order of 10^{-16} mol bound per fiber by both radioactive and fluorescent determinations. The moles of labeled Ab in both systems increase proportionately based on increasing analyte concentration. These findings confirm the validity of EWS measurements in quantitatively evaluating the amount of labeled-tag accumulation on the fiber in Ag–fluorescent Ab complexes and conclusively demonstrate the ability of the technology to measure mass accumulation on a very specialized solid-phase immunoadsorbent utilizing the evanescent wave as a source of excitation and detection.

Based on this comparison, the apparent F:P ratio for labeled Ab is on the order of 1–2 fluoresceins/molecule Ab. Given the 50% reduction in fluorescence intensity of bound protein vs free dye, this is equivalent

Table 6
Comparison of Fiber Binding
of Radioactive and Fluoresceinated Antiferritin–Ferritin Complexes

Serum ferritin, ng/mL[a]	FITC equivalent, $\times 10^{-8}M$[b]	FITC in evanescent zone (F), $\times 10^{-16}M$[c]	^{125}I bound to Ab, (P), $\times 10^{-16}M$	Apparent F:P ratio
6	1.3	1.02	0.56	1.80
30	7	5.49	3.30	1.67
64	12	9.42	8.50	1.10
250	40	31.40	28.10	1.20
640	80	62.80	49.30	1.27
1100	140	109.90	58.60	1.87

[a]Values have been computed in a commercial IRMA for ferritin. The actual concentration in EWS was 1:2 as a result of 50:50 dilution of serum with fluorescent Ab mixture.

[b]Values have been determined from extrapolation of corrected end-point millivolt signals from a 10-min assay at each concentration to a calibration curve prepared with FITC dye.

[c]Volume of the evanescent zone based on an evanescent-wave d_p of 100 nm. Differential calculation of the fiber plus evanescent-wave vol minus the fiber-cylinder vol is 7.85×10^{-9} L.

to 2–4 fluoresceins/molecule Ab. This is somewhat lower than the calculated value, 8–10, derived from spectrophotometric measurement of absorption by the conjugate at both 280 and 495 nm, and is most likely attributable to the intramolecular quenching demonstrated when multiple fluorophores form a complex with a single protein. Studies on multifluoresceinated IgE molecules have demonstrated that addition of fluorescein molecules beyond 3.5 mol/mol protein will result in a reduction of relative fluorescence intensity of each bound fluorescein molecule *(32)*. For example, if the ratio of fluorescein to protein is 10:1, the relative intensity per mole of fluorescein is only 56% of that seen at a ratio of 3.5:1.

Another possible reason for this discrepancy is that the assumptions chosen for comparison include the assumption that all the FITC molecules measured in solutions of the dye exposed to the fiber are contained within the distance of a single d_p. In fact, the energy excitation level at this distance is still approx 36% of that found at the fiber surface and is thus capable of exciting flourophores in solution at a greater distance than the 100 nm chosen as the radius of the sheath surrounding the fiber. Obviously molecules in this outer zone would exhibit some fractional level of excitation relative to those found at the fiber surface, and thus the total number of FITC molecules used in the calculations would be erroneously low.

Using an expanded evanescent zone would increase the total evanescent volume and likewise the total number of moles of FITC utilized in calculations. This in turn would increase the apparent F:P ratio of labeled Ab to be more in line with the F:P ratio calculated by conventional means (optical absorbance at 280 and 495 nm). Complete assessment of all the fluorescent dye molecules excited with equal power, and thus the true distance of the evanescent zone, requires a weighted integral evaluation, fully accounting for the exponential reduction in field intensity with distance from the fiber surface. Assuming that all fluorophores in immunoassay are held at a constant distance from the fiber (30 nm, based on the intermolecular diameter of a "sandwich" complex), the field intensity of that distance should be 80–90% of that found at the fiber surface.

4.5. Accuracy

As a final test of the system's feasibility as a quantitative immunoassay system, various commercial serum-based controls and patient-serum samples were tested in a commercial IRMA assay for ferritin and by fiberoptic methodology. These studies were conducted to assess the accuracy of ferritin concentration values generated in the EWS technology with a commercially accepted and validated procedure. For the EWS system, ferritin standards calibrated against reference preparations, were used to construct a calibration curve. Patient and control sera were then tested and their signal values (mV) converted to serum ferritin-concentration values via extrapolation from the calibration curve.

As shown in Table 7, correlations of both control and patient sera were excellent over a concentration range of 15–450 ng/mL. The slopes of the regression line were close to one for the control sera, indicating

 Bluestein et al.

Table 7
Correlation of FOIA with Radioimmunoassay

Sample	N	Concentration range, ferritin ng/mL	r	Slope
Control sera	31	50–400	0.97	1.08
Patient sera	23	15–450	0.89	0.80

no system bias in the measurements relative to the commercial IRMA technique. The slight negative bias of the patient EWS values relative to IRMA was found on closer evaluation to be caused by extensive hemolysis found in a number of patient samples.

Excessive chromogens, if present in patient samples, can exert a quenching effect on observed fluorescence if there is spectral overlap of the chromogen with either the excitation or emission wavelength of the fiber-bound fluorophore. This inner filter effect *(14)* observed with hemoglobin is a result of the high molar extinction coefficient of heme at about 525 nm, where the fluorescent emission of fluorescein is maximal. Other chromogens, such as bilirubin (whose absorption maximum is at 440 nm, well below the excitation and emission wavelengths of fluorescein), did not exhibit any quenching effects. Potential problems with serum-based optical interferents can be virtually eliminated if, in optimization of assay systems, care is taken to minimize the volume of sample aliquot utilized. In the described system, serum hemoglobin demonstrated no negative effects at concentrations up to 50 mg/dL, at which concentration gross hemolysis is clearly visible to the unaided eye.

Because of the short path length determined by the evanescent-wave zone, the fiberoptic system is relatively insensitive to inner filter effects, wherein a fluorescent emission is directly attenuated by a closely adjacent fluorophore (nonradiative energy attenuation). In standard solution fluorimetry, in which path lengths are large, the cumulative effects of energy absorption are demonstrated by a marked reduction in observed signal, especially at high chromophore concentration. This is aptly demonstrated in Fig. 13: Reduction in observed fluorescent

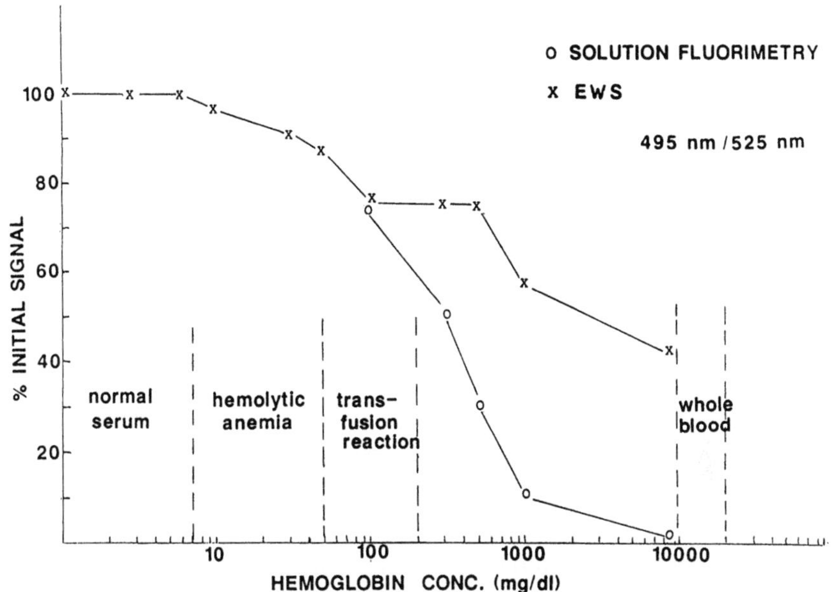

Fig. 13. Hemoglobin interference quenching effects. Various concentrations of human hemoglobin were prepared from crystallized hemoglobin powder (Sigma) in 0.1% BSA and mixed 1:2 with $10^{-6}M$ sodium fluorescein. Values shown for hemoglobin are concentrations of the original stocks prior to mixing with fluorescent dye (1:2 final dilution). The extent of fluorescence quenching at each concentration level was evaluated by comparison to the fluorescent signal of fluorescent dye in the absence of any added hemoglobin. Differences in the extent of fluorescence quenching in standard solution fluorimetry vs evanescent-wave sensing were made by reading the same solutions on a standard solution fluorimeter and the fiberoptic (EWS) instrumentation. A clean fiber without Ab coating was used in these evaluations.

signal from $10^{-6}M$ fluorescein dye in the presence of increasing concentrations of soluble hemoglobin was measured in both a standard solution fluorometer and via the EWS. At concentrations approximating whole-blood hemoglobin levels (10,000 mg/dL), conventional fluorimetry showed complete attenuation of fluorescence, whereas almost 50% of the original fluorescent signal was still detectable in the EWS test system when a non-Ab-coated fiber was used as the test sensor. These findings enhance the potential of using EWS technology to make direct readings of immunoanalytes in whole blood, although such experimentation was not undertaken in the present studies. Work thus far has been restricted to measurements in either serum or plasma samples.

5. Summary and Discussion

The recent emergence of hybridoma-based technology, with its ability to tailor-make antibodies of high specificity directed against analytes present in low concentration in the blood, has opened the domain of immunoassay into fields of medicine in which it was previously not utilized. Immunoanalyte analysis has proved beneficial not only for primary disease diagnosis, but also for screening, monitoring, and risk assessment. Extended use of the technology is currently limited by the lack of a rapid, easy-to-use delivery system.

The results presented here demonstrate the feasibility of utilizing fiberoptic EWS for quantitative measurement of medically important analytes. Unlike current immunodiagnostic procedures, EWS technology can be considered characteristic of a biosensor format. These characteristics include simplicity of both process and instrumentation, rapid turn-around time from sample input to answer output, and reagents containing only one component.

No user- or instrument-mediated separation of bound and free labeled tag is required prior to final signal readout. In a conventional nonisotopic heterogeneous immunoassay, separation requirements dictate that assays be conducted in four separate stages:

1. Primary incubation of antigen, solid-phase Ab, and tagged Ab;
2. Separation of bound from free solid-phase tag;
3. Initiation of a secondary reaction by addition of color or fluorescent signal-developing substrates; and
4. Signal readout in appropriate instrumentation.

In EWS, because of the unique phenomenon of a restricted light zone in which fluorophores can be excited and detected, primary Ag–Ab reactions can be monitored directly as they form and concentrate on the fiber. Thus, the multistep protocols required in other technologies occur simultaneously in EWS.

The ability for real-time measurement reduces to a time scale of seconds the determination of ultralow concentrations of high-mol-wt analytes, which require hours to determine with today's technology. With turn-around time minimized and a simple operational format requiring less user skill and attention, immunoassay-based technology is now free to enter alternate sites and provide rapid information to clinicians and medical personnel directly involved with patient care. These areas include patient bedside, emergency rooms, ambulances, operating rooms, doctor's offices, and community health centers.

The results of previous studies and ongoing work in our laboratories suggest that EWS will be able to meet the stringent requirements of medical diagnostic tests, including (1) sensitivity equivalent to radioimmunoassay procedures, (2) accuracy, (3) precision, and (4) stability. Because the system is a one-step protocol, EWS can be formatted as a very simple delivery system. This trend in turn will result in enhanced patient management and more cost-effective delivery of medical services while reducing the trauma of uncertainty experienced by the patient waiting for test results.

References

1. Hunter, W. M. (1978) Radioimmunoassay, in *Handbook of Experimental Immunology*, (Weir, D. M., ed.), Blackwell Scientific, London, chap. 14.
2. Avrameas, S. (1981) Heterogeneous enzyme immunoassays, in *Immunoassays for the Eighties* (Voller, A., Bartlett, A., and Bidwell, D., eds.), MTP Press, Lancaster, PA, p. 85..
3. Dandliker, W. B., Kelly, R. J., Dandliker, J., Farquhar, J., and Levin, J. (1973) Fluorescence polarization immunoassay, theory and experimental method. *Immunochemistry* **7**, 219.
4. Rupchock, P., Sommer, R., Greenquist, A, Tybach, R., Walter, B., and Zipp, A (1985) Dry reagent strips used for determination of theophylline in serum. *Clin. Chem.* **31**, 737.
5. Rubinstein, K. E., Schneider, R. S., and Ullman, E. F. (1972) Homogeneous enzyme immunoassays. *Biochem. Biophys. Res. Commun.* **47**, 846.
6. North, J. R. (1985) Immunosensors: Antibody based biosensors. *Trends in Biotechnol.* **3**, 180.

7. Sepaniak, M. J. (1986) Fiberoptic sensors for bioanalytical measurements. *Clin. Chem.* **32,** 1041.

8. Seitz, W. R. (1984) Chemical sensors based on fiber optics. *Anal. Chem.* **56,** 16A.

9. Schultz, J. S. (1982) Optical sensor of plasma constituents. US Patent 4,344,438.

10. Angel, S. M. (1987) Optrodes: Chemically selective fiberoptic sensors. *Spectroscopy* **2,** 38.

11. Harrick, N. H. (1967) *Internal Reflection Spectroscopy* (Interscience, New York).

12. Hirschfeld, T. B. (1965) Total reflection fluorescence. *Can. J. Spectroscopy* **10,** 128.

13. Glass, T. R., Lackie, S., and Hirschfeld, T. (1987) Effect of numerical aperture on signal level in cylindrical waveguide evanescent fluorosensors. *Appl. Opt.* **26,** 2181.

14. Wiechelman, K. J. (1986) Empirical correction equation for the fluorescence inner filter effect. *Am. Lab.* **18,** 49.

15. Kronick, M. N. and Little, W. A. (1975) A new immunoassay based on fluorescence excitation by internal reflection spectroscopy. *J. Immunol. Methods* **8,** 235.

16. Sutherland, R. M., Dahne, C., Place, J. F., and Ringrose, A. R. (1984) Immunoassays at a quartz liquid interface: Theory, instrumentation and preliminary application to the fluorescent immunoassay of human immunoglobulin G. *J. Immunol. Methods* **74,** 235.

17. Badley, R. A., Drake, R. A. L., Shanks, I. A., Smith, A. M., and Stephenson, P. R. (1987) Optical biosensors for immunoassays: The fluorescence capillary fill device. *Philos. Trans. R. Soc. Lond.* **316,** 143.

18. Hirschfeld, T. B. and Block, M. J. (1984) Fluorescent immunoassay employing optical fiber in capillary tube. US Patent 4,447,546.

19. Block, M. J. and Hirschfeld, T B. (1984) Assay apparatus and method, US Patent 4,558,014.

20. Block, M. J. and Hirschfeld, T. B. (1986) Apparatus including optical fiber for fluorescence immunoassay. US Patent 4,582,809.

21. Andrade, J. D., and Vanwagenen, R. (1983) Process for conducting fluorescence immunoassays without added labels and employing attenuated internal reflection. US Patent 4,368,047.

22. Sutherland, R., Dahne, C., and Place, J. (1984) Preliminary results obtained with a no label homogeneous optical immunoassay for human immunoglobulin G. *Anal. Lett.* **17,** 43.

23. Bluestein, B. I. (1987) Optical response immunosensors—instant quantitative immunoassay technology for clinical diagnosis. *Clin. Chem.* **33,** 1061.

24. Slovacek, R., Bluestein, B., Craig, M., Urciuoli, C., Stundtner, L., Lee, M., Walczak, I., Love, W., and Cook, T. (1987) Optical immunosensors, in

Proceedings of the Symposium on Chemical Sensors (Turner, D., ed.), The Electrochemical Society, **PV-87-9**, 456. Paper originally presented at the 1987 Fall Meeting of The Electrochemical Society Inc., held in Honolulu, HI.

25. Weetall, H. H. (1976) Covalent methods for inorganic support material. *Methods Enzymol.* **44**, 134.
26. Lowry, O. H., Roseborough, N. J., Farr, A. L., and Randall, R. J. (1951) Protein measurement with the folin–phenol reagent, *J. Biol. Chem.* **193**, 265.
27. Greenwood, F. C., Hunter, W. M., and Glover, J. S. (1963) The preparation of ^{131}I-labeled human growth hormone of high specific radioactivity. *Biochem. J.* **89**, 114.
28. Kawamura, A. (1969) Preparation of materials, in *Fluorescent Antibody Techniques and Their Applications* (Kawamura, A., ed.), University Park, Baltimore, MD, p. 33.
29. Kemp, H. A., Woodhead, J. S., and Rhys, J. (1984) Labeled antibody immunoassays, in *Practical Immunoassay* (Butt, W. R., ed.), Marcel Dekker, NY, p. 179.
30. Rodbard, D. and Chrambach, A. (1971) Estimation of molecular radius free mobility and valence using polyacrylamide gel electrophoresis. *Anal. Biochem.* **40**, 95.
31. Watt, R. M. and Voss, E. W. (1984) Affinity labeling of antifluorescyl antibodies, in *Fluorescein Hapten: An Immunological Probe* (Voss, E. W., ed.), CRC, Boca Raton, p. 177.
32. Rowley, G. L., Henriksson, T., Louie, A., Nguyen, P. H, Kramer, M., Derbailan, G., and Kameda, N. (1987) Sensitive fluoroimmunoassays for ferritin and IgE. *Clin. Chem.* **33**, 1563.

Instrumentation for Cylindrical Waveguide Evanescent Fluorosensors

Steve J. Lackie, Thomas R. Glass, and Myron J. Block

1. Introduction

This chapter contains a description of the physical principles of a cylindrical evanescent-wave fluorosensor and a discussion of some practical considerations for use of the sensor. Evanescent fluorosensors are based on the principle of total reflection fluorescence, first described by Hirschfeld *(1)*, wherein fluorescence is excited by the evanescent wave on or near the boundary of a totally reflecting element, such as a prism or waveguide. Early investigators *(2–4)* used evanescent excitation, but collected fluorescent emission in free propagation. Block and Hirschfeld *(5–7)* first proposed evanescent excitation and evanescent collection (i.e., collection of the fluorescence that is tunneled back into trapped modes of the waveguide) using an optical fiber or rod as the sensing element. Subsequently, Sutherland et al. in 1984 *(4,8)* at Battelle and Andrade et al. in 1985 *(9)* at the University of Utah described similar research approaches. The topic of evanescent-wave theory is reviewed by Harrick *(10)* (good introduction) and supplemented by Carniglia, Mandel, and Drexhage's treatment *(11)*.

Using an optical fiber or rod as the total reflection fluorescence element and collecting the fluorescence tunneling back into the fiber,

Biosensors with Fiberoptics Eds.: Wise and Wingard ©1991 The Humana Press Inc.

as diagrammed in Fig. 1, has a number of advantages over other approaches:

1. The fluorescence intensity for radiation tunneled into the waveguide is enhanced by as much as two orders of magnitude over normal propagation, such as in a standard 90° fluorimeter *(12–14)*. This advantage applies to planar and cylindrical waveguides and to prisms as long as tunneled, or evanescent, fluorescence is collected.

2. The fiber or rod, as shown by Block and Hirschfeld *(5–7)*, has an additional advantage over planar waveguides or prisms. If the total optical absorption is much less than the incident power, as is usually the case, all the fluorescent molecules along the waveguide receive essentially the same excitation power. This uniform power presentation, in conjunction with fluorescence tunneling into bound modes of the fiber, brings about efficient transmission of the signal to the fiber face for detection. The result is higher brightness of the observed fluorescence, with the observed signal proportional to the length of fiber used.

3. As discussed by Hirschfeld and Block *(5–7)* collection of the fluorescence at the proximal end of the fiber, into which the excitation is coupled, reduces the background compared to detection of fluorescence at the distal end of the fiber *(4)*.

4. The fiber geometry, in conjunction with a capillary, lends itself to an assay format with no requirement for volume measurement *(5,7)*.

This last advantage may prove particularly important, since it removes operator skill-intensive steps from the fluorescent immunoassay procedure.

2. Evanescent Waves

The theory is described elsewhere in this book; only key points are highlighted here. An optical waveguide typically consists of a transparent core surrounded by cladding of a lower refractive index. The light is retained inside the core by total internal reflection, but the internally reflected light actually penetrates a small distance into the material of lower refracive index (i.e., the cladding). The penetrating light, called the evanescent wave, has an electric field amplitude that decreases exponentially (for a particular mode or angle of incidence)

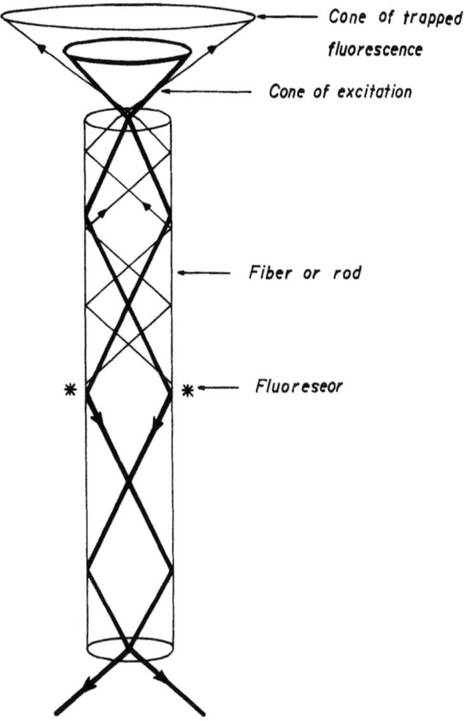

Fig. 1. Fluorophor compounds ("fluorescers") outside the waveguide are excited by the evanescent wave and emit light that becomes trapped within the waveguide. Here the waveguide can be a quartz rod.

with increasing distance into the cladding medium, which is optically less dense.

In an evanescent-wave sensor, the sample solution is the "cladding" and the exponential decrease in amplitude (or intensity) results in preferential excitation of those fluorophores that are very close to the waveguide (core) boundary, thus optically separating the fluorophores bound to the waveguide from fluorophores in the bulk of the sample solution. This separation of fluorophores is further enhanced by the efficiency of coupling of fluorescent light back into the waveguide, which also decreases with distance. The primary advantage of an evanescent-fluorescence approach in a biosensor is that physical separation of bound from unbound labeled analyte is not needed, since optical separation is accomplished automatically through the use of the evanescent wave. This holds true even for small sample volumes because of the very small effective-penetration depth of the evanescent wave.

In order to quantify separation of fluorophores bound to the waveguide from those not bound, it is necessary to calculate the relative signal as a function of distance from the fiber. The separation can be conveniently (and approximately) referred to in terms of an effective-penetration depth; however, this should not be confused with the commonly used evanescent-penetration distance, d_p, which does not include a factor for the efficiency of fluorescence tunneling into trapped modes in the waveguide. A complete expression for the effective penetration depth of an evanescent fluorosensor can be derived in the following manner:

Given a waveguide (fiber core) with refractive index n_1 and a sample of refractive index n_2, the electric field amplitude in the sample decays exponentially with distance (*10*), as follows:

$$E = E_0 \exp\{-[2\pi\,(\cos^2\alpha - n^2_{21})^{1/2}Z/\lambda]\} \tag{1}$$

where E = electric field amplitude a distance Z from the fiber–sample interface, E_0 = electric field amplitude at $Z = 0$, $n_{21} = n_2/n_1$, α = angle of incidence measured relative to a tangent plane to the interface, λ = wavelength of excitation light, Z = distance from interface, and the evanescent penetration depth d_p is defined by Eq. (2):

$$d_p = \lambda/2\pi\,(\cos^2\alpha - n^2_{21})^{1/2} \tag{2}$$

The rate of this decay ($1/d_p$) is a strong function of the incident angle, or mode, of the internally reflected ray. The rate becomes smaller, and hence the evanescent-penetration depth becomes larger, as the incident angle approaches the critical angle. In addition to d_p, the strength of the interaction of the ray with the sample is also influenced by E_0 (which is a function of incident angle) and relative indices, and incident polarization. For polarizations parallel and perpendicular to the plane of incidence, the electric field amplitude in the rarer media, assuming unit-amplitude incident waves, is given by Harrick as

$$E_\perp = [2 \sin\,\alpha/(1 - n^2_{21})^{1/2}] \exp\{-[2\pi\,(\cos^2\alpha - n^2_{21})^{1/2}Z/\lambda]\} \tag{3}$$

$$E_\parallel = \{4 \sin^2 \alpha \, (2 \cos^2\alpha - n^2_{21})/(1 - n^2_{21})$$
$$[(1 + n^2_{21}) \cos^2\alpha - n^2_{21}]\}^{1/2} \exp \{-[2\pi \, (\cos^2\alpha - n^2_{21})^{1/2} Z/\lambda]\} \qquad (4)$$

where $E_\perp$ = electric field amplitude perpendicular to plane of incidence, and $E_\parallel$ = electric field amplitude parallel to plane of incidence.

The strength of the excitation is governed by the intensity, E^2. By reciprocity *(10,15)*, the trapped fluorescence has the same angular distribution and dependence on index, wavelength, and distance.

In real systems, it is unlikely that all modes will be equally excited. If the excitation source is incandescent (or of any type that can be treated as lambertian) and is imaged onto the fiber face, then from basic radiometry *(15)*, E_0 has an additional angle-dependent factor, given by

$$E_0 \propto \cos \alpha \sin \alpha \qquad (5)$$

Note that, for light sources other than lambertian (e.g., a laser), this factor, and hence E_0, changes.

For a lambertian source, the total electric-field intensity in the sample can be calculated using Eqs. (1) and (3–5) *(10)*. In all real waveguide devices, there is a range of angles described by Eq. 6:

$$\alpha_{max} = \cos^{-1} (n_2/n_1) \qquad (6)$$

Integration of the electric-field intensity $E_\perp^2$ and $E_\parallel^2$ over α gives the excitation as a function of distance for the two extremes of polarization. In a similar fashion, the relative fluorescent return can be obtained by integrating Eqs. (3) and (4). If the light is assumed to be randomly polarized, it can be averaged for the two polarizations to get the total fluorescent return. Then multiplying the excitation by the collection gives the average signal as a function of distance from the fiber. The predicted relative-signal strength is shown in Fig. 2 as a function of distance into the cladding.

It is important to remember that the separation of bound from unbound fluorophores is the real issue and that there is no true cutoff distance beyond which a fluorophore cannot be seen. Given a sample size and the distance from the fiber to the bound fluorophores, Fig. 2 can be used to determine the effective separation of bound from unbound material by calculating the relative signals expected from each.

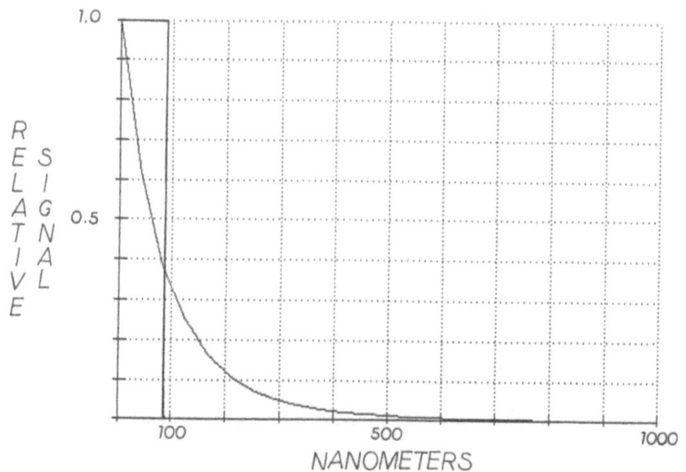

Fig. 2. Relative signal strength vs distance from fiber into cladding (average of the two polarizations).

The same evanescent phenomena responsible for separating bound from unbound analyte also greatly reduces interference from endogenous sample fluorescence. Because of this, even whole blood can be used as a sample without significant interference problems.

3. Importance of Numerical Aperture

The numerical aperture (NA) of the sensor is determined by the refractive indices of the fiber and sample.

$$NA = (n^2_1 - n^2_2)^{1/2} \tag{7}$$

If the NA is limited to less than this, either by the optical system or by a fiber coating, a dramatic decrease in signal strength results. A semiquantitative description of how fluorescent signal relates to NA is presented elsewhere *(16)*. We found that if the NA is less than that calculated using Eq. (8), the fluorescent signal will vary approximately as NA^9.

In the chapter by Love et al. in this book, the theory is developed in a more comprehensive and quantitative manner, but reaches a similar conclusion regarding the importance of NA. Love et al. conclude that the fluorescent signal varies as NA^8 with the exponent increasing somewhat beyond 8 as $\sin \alpha_{max}$ approaches the NA of the fiber.

In any case, the sensor NA, as set by the optical indices of both the sample and the fiber, must be maintained in the length of fiber between the sample and the proximal face of the fiber to prevent loss of signal. To accomplish this, the medium surrounding the fiber must have an optical index less than or equal to the optical index of the sample (commonly 1.33–1.35 for aqueous samples). Air, with an optical index of approx 1.0, does not limit the NA; however, there is a need to support the fiber, to hold the proximal face of the fiber in proper alignment with the optical system, and to seal the ends of the sample chamber. Since there are no available materials with the needed properties (transparent solids with an optical index ≤ 1.33), other methods have been devised.

A method of supporting and aligning the proximal end of the fiber while preserving the NA is to use an alignment ring to hold the proximal face of the fiber. The ring is fixed at the correct focal position relative to the objective lens. The clear aperture of the ring, which is the same size as the optical excitation spot, is slightly smaller than the fiber diameter. As can be seen in Fig. 3, the ring can contact the fiber at a distance L from the proximal end of the fiber without interfering with any light ray entering at less than the maximum acceptance angle according to Eq. (8):

$$L = [(D - R)\, n_2/2\, NA] \tag{8}$$

where L = distance from proximal face at which the ring can contact the fiber, D = fiber diameter, and R = diameter of the clear aperture of the ring. The alignment ring makes it easy to change fibers because the fiber is guided into alignment as it is inserted.

A drawback to the use of an alignment ring is that it does not allow a liquid seal to the fiber, thus complicating the sample handling. This problem can be overcome by integrating a transparent seal with the proximal end of the fiber. The excitation energy can then be coupled through the seal with no internal reflections and into the fiber at the maximum NA (Fig. 4). In a modification of the system diagrammed in Fig. 4, the seal can also serve as a lens (Fig. 5). Using the seal as a lens

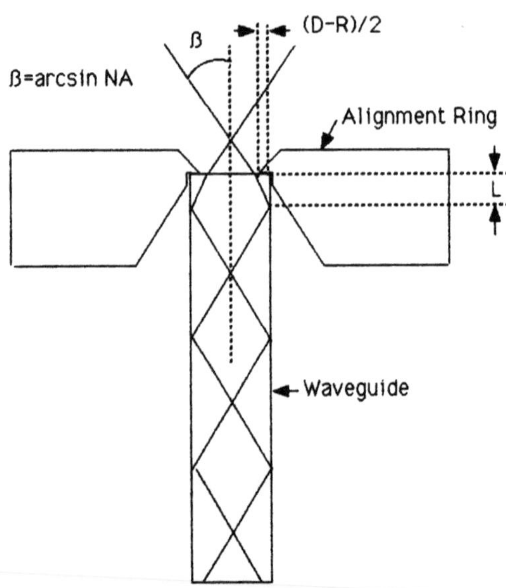

Fig. 3. The alignment ring can contact the fiber without frustrating the total reflection of the excitation.

provides the added advantage of increasing the allowable alignment tolerance for the fiber–seal assembly.

We have constructed injection-molded polystyrene fiber–seal assemblies. In the case of a material with a relatively high refractive index, such as polystyrene, the lens also eases implementation of the high-NA optical system that the fiber requires (the NA of a polystyrene fiber in water is 0.88).

4. Why Cylindrical Waveguides?

Evanescent fluorescent assays have been performed with both planar *(2,3)* and cylindrical *(9)* waveguides. The cylindrical-waveguide geometry has several advantages over that of the planar form, the greatest being better sensitivity. This stems from the fact that, in the cylindrical geometry, the light is contained in two dimensions rather than one, as in the planar case. To illustrate the difference that the geometry makes, consider two sensors with the same active area: one cylindrical and the

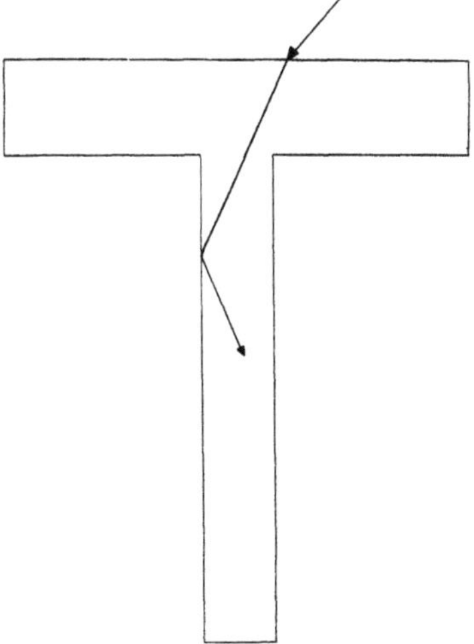

Fig. 4. Fiber with proximal seal.

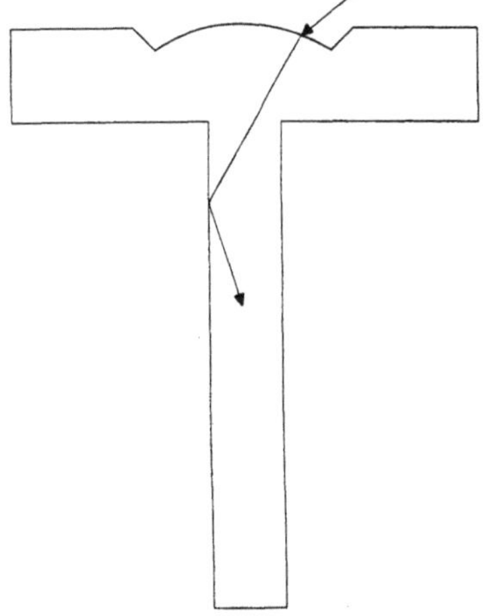

Fig. 5. Fiber with proximal seal and lens.

other planar. Further, assume that the active-area length is the same for both sensors, so the circumference of the cylindrical sensor equals the width of the active area on the planar sensor.

In the case of the planar waveguide, excitation light on the proximal face is accepted for all angles up to the maximum acceptance angle in the y dimension (determined by the refractive indices of the waveguide and sample). In the x dimension there is no internal reflection, and the acceptance angle can be as large as 180°. The lack of guidance in the x dimension may result in a reduction in excitation power with distance from the proximal face. Likewise, the returning fluorescence is collected over a range of angles that is position-dependent and is smaller for fluorophores further from the proximal face. This causes a reduction in the signal level from fluorophores further away from the proximal face, making the signal level a function of fluorophore position.

In the cylindrical geometry, the proximal face is a circle that accepts excitation light from all angles up to the maximum acceptance angle in both the x and the y dimensions. The fluorescence emission is confined in both the x and y dimensions; therefore, the signal is not dependent on fluorophore position.

In any practical planar geometry, the source and detector size (because of the proximal face size) need to be larger than in an equivalent cylindrical geometry, requiring increased power (for the larger source) and generally resulting in higher noise (from the larger detector area).

5. Optical System

The optical system guides the excitation light from the source, through the excitation filter, and into the fiber. It also must guide the fluorescence from the fiber, through the fluorescence filter, and onto the detector.

Different layouts of fiberoptic-sensor readout instruments have been proposed *(6,17)*. The most important difference is whether the fluorescence is collected at the proximal or the distal end (relative to the excitation) of the fiber. Since the fluorescence photons in the waveguide have no preferential direction, the signals that exit opposite ends of the fiber are equal in power. On the other hand, approx 90% of the excitation light launched into the waveguide exits the waveguide at the distal end. Therefore the signal-to-background ratio is lower for proximal- than for distal-end collection.

In addition to a performance advantage, there are practical advantages to proximal-end collection. Having the distal end free makes the installation and alignment of the sensor easier, since the distal end is accessible for removal and replacement. If both the input and output of the waveguide are through the proximal face, only that face needs to be focused to the instrument. Another advantage is that the distal face is not used for optical transmission; therefore, it does not need to have an optical finish, which reduces the cost.

An example of an optical system that accomplishes the above is the collimated-beam system shown schematically in Fig. 6. Collimation allows easy and efficient use of interference filters for excitation and fluorescence wavelength selection, and makes it simple to use a dichroic beam splitter for separating the excitation light from the fluorescent light. The collimated beam is also insensitive to length (neglecting vigneting) and is therefore easier to align. The objective lens (the lens closest to the proximal end of the fiber) is spaced a distance equal to its focal length from the proximal face of the fiber. To accommodate the maximum NA of the fiber sensor it must have a clear optical aperture (C) of:

$$C = 2\{r + [f_0\,NA/(1 - NA^2)^{1/2}]\} \tag{9}$$

where r = radius of fiber, f_0 = focal length of objective lens, and NA = maximum numerical aperture as determined by fiber and sample refractive indices. The collimated beam has a divergence half-angle (θ) given by Eq. (10):

$$\theta = \tan^{-1}(r/f_0) \tag{10}$$

and, at a distance f_0 from the lens, has a beam diameter (B) (Eq. [11]) of

$$B = 2f_0\,NA/(1 - NA^2)^{1/2} \tag{11}$$

so the clear aperture (A) of the lens at the other end of the collimated beam must be at least that given by Eq. (12):

$$A = 2\{[f_0\,NA/(1 - NA^2)^{1/2}] + (r/f_0)\,(1 - f_0)\} \tag{12}$$

where l = optical path length of collimated beam, as measured from the objective lens.

Any elements placed in the collimated beam, such as filters or dichroic mirrors, must also have clear apertures of at least the size given by Eq. (12).

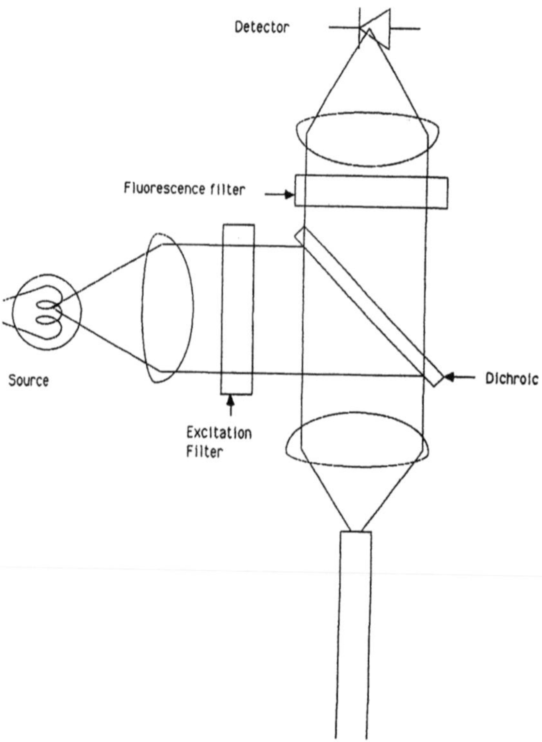

Fig. 6. Light from the source is collimated, spectrally filtered, and focused onto the fiber (waveguide) face. Fluorescence returning from the fiber is collimated, filtered, and focused onto the detector.

The image formed by the second lens has, from Eq. (13), a diameter of I:

$$I = (2r)\,(F/f_0) \tag{13}$$

where I = image diameter and F = the focal length of the second lens. The second-lens image has a numerical aperture (NA_I) given by Eq. (14):

$$NA_I = \tan^{-1}\left\{[f_0\,NA/F\,(1-NA^2)^{1/2}] + (r/f_0\,F)\,(1-f_0)\right\} \tag{14}$$

It should be noted that the other lenses can be "slower" than the objective lens if they have a longer focal length than the objective lens. However, this results in magnification of the fiber-face image, which requires a larger detector or source.

In some cases (for example, a fiber with a very high refractive index) a lens that is fast enough to accommodate the NA at the fiber face may not be available. In such cases, integrating a lens with the fiber as described earlier can reduce the speed requirements of the other lenses. The "fiber" lens and the "objective" lens can then be treated as a multielement objective lens in the previous analysis. Another approach to high-NA requirements is to use a tapered element *(18)* as an "NA transformer" (Fig. 7) to reduce the NA to a value that can be accommodated by a conventional optical system. The tapered element is a transparent solid of conical section, normally with a taper angle <5°. The change in NA achieved is equal to the ratio of the areas of the two ends of the tapered element. In use, the objective lens is focused on the larger end of the tapered element and the small end of the tapered element is positioned next to the proximal end of the fiber.

6. Excitation System

The excitation system is designed to provide excitation light to the proximal face of the fiber, with the cone of excitation light having an angle as close as possible to the maximum acceptance angle determined by the optical refractive indices of the fiber and sample. Therefore both the size of the excitation spot and the cone angle are determined by the fiber, and the rest of the optical components are chosen to accommodate the fiber parameters. A throughput calculation is a very useful tool to aid in component selection.

Throughput is defined here as the sine of the cone half–angle times the diameter of the spot. This is useful because the angle of the cone of excitation light and the diameter of the excitation spot are inversely related to each other by the optical magnification in such a way that the throughput is a constant. In practice, the throughput needed at the fiber is determined by multiplying the fiber diameter by its NA. This gives the throughput needed at all points in the optical system to accommodate the fiber.

Calculating the throughput of a source (its emitting diameter times the sine of the half–angle of the cone of light that can be collected from it) and comparing it to the throughput needed at the fiber allows quick and easy determination of how efficiently the source can be used. A

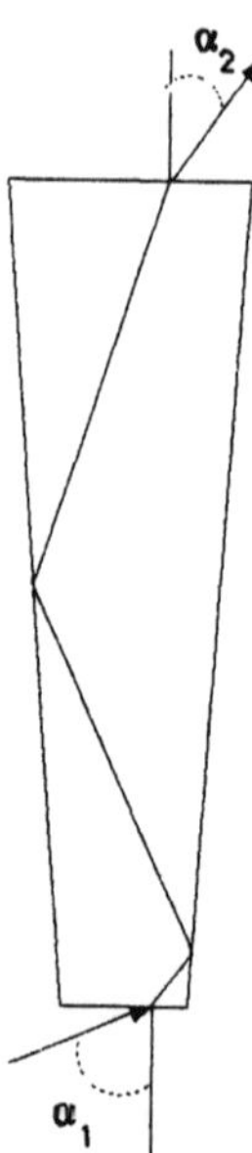

Fig. 7. A tapered waveguide can act as an NA transformer.

throughput lower than that of the fiber will not be able to fill the fiber completely, whereas a throughput higher than that of the fiber cannot be completely utilized and would be an inefficient match. For example, consider the selection of a tungsten halogen incandescent lamp. Since all such lamps operate at approximately the same color temperature (3200K) and emit over the same range of angles, lamp throughput determines the size of filament (and hence the wattage) needed to match the fiber throughput. In the ORD Inc. instrument with a quartz fiber, we use a small 4-W tungsten halogen lamp to match fiber throughput. A tungsten halogen lamp with a higher wattage (larger area) would increase only the power-supply requirements and heat load, without increasing the excitation power into the fiber.

There are many other considerations in the source selection, since a single source is not best for all applications. Trade-offs must be made among excitation power, physical size of the source, cost, power-supply restrictions, and amount of heat generated by the source.

Lasers generally give very high powers of excitation and, because of their narrow output spectrum, can be used with a wider bandpass and a more transmissive fluorescence filter. However, laser sources are also expensive, large, have high input-power requirements, and cannot be conveniently adapted to different excitation wavelengths.

Arc sources also give very high powers of excitation and can be adapted to different excitation wavelengths by using filters. However, because arc sources are broad-band sources, they require much better (and less transmissive) filters than narrow-band sources to prevent reflected and scattered excitation light from reaching the detector. Arc sources also are expensive, generate a lot of heat, and have high input-power requirements.

Incandescent sources are low in cost, small in size, have moderate input-power requirements, exhibit moderate heat generation, and are broad-band sources allowing adaptation to different excitation wavelengths by a filter change. They do, however, generally give lower excitation powers than a laser or an arc source and, because they are broad-band sources, require better filtering than do narrow-band sources.

LEDs are the lowest in cost, produce the least amount of heat, require the lowest amount of input power, and are physically the smallest sources available. The main disadvantage of LED sources is that they are available only at discrete wavelengths, which do not always match well with dye-excitation wavelengths. Also, the LEDs at the shorter wavelengths needed for most dyes have very low output power. One potentially interesting combination being explored in our lab is the use of a phycobiliprotein dye, such as allophycocyanine, in conjunction with a powerful GaAlAs LED source emitting light at 660 nm.

7. Detection System

The output from an evanescent fluorosensor consists of the fluorescent signal plus background wavelengths. This sum plays a critical role in determining the signal-to-noise performance of an optimized detector system. In particular, if a shot-noise limited-silicon photodiode detector system can be used, it will give better signal-to-noise performance than will a photomultiplier tube (PMT). To explain this, consider the following description of the noise sources for both PMTs and solid-state photodiodes:

In the photodiodes, the shot noise is proportional to the square root of the current. So, if the current increases, the shot noise also increases. Above some limiting value of current, the shot noise becomes larger than the other, constant, noise terms. At this value of current, the measurement of current is said to be "shot-noise-limited." Although shot noise increases with increasing current, it does not increase as fast as the current (because shot noise is proportional to the square root of the current). Therefore, in a shot-noise-limited measurement, an increase in current improves the signal-to-noise ratio.

For a PMT device the total noise is the sum of two terms, the shot noise of the signal and the noise of the dark current. These two noise sources are added as uncorrelated sources (e.g., the square root of the sum of the squares). Optimum measurement is achieved when the shot noise of the signal current is the dominant noise source. For a typical PMT at room temperature, this is the case for optical input to a lower limit of about 10^{-14} W. If needed, this can be improved to still lower inputs by cooling the PMT to reduce the dark-current noise.

For a silicon photodiode also, the total noise is the sum of two noise terms, the shot noise of the signal current and the Johnson noise of its shunt resistance, with the Johnson current noise $I_{(RMS\ noise)}$ of a resistance given by Eq. (15):

$$I_{(RMS\ noise)} = (4\ k\ TB/R)^{1/2} \tag{15}$$

where k = Boltzmann constant, T = temperature in degrees Kelvin, B = bandwidth in hertz, and R = resistance in ohms.

As with the PMT, optimum performance occurs when the shot noise of the signal dominates. The shunt resistance of the photodiode determines the optical level above which shot noise dominates. The shunt resistance is proportional to the size of the active area of the photodiode and is also a function of temperature.

Because silicon photodiodes have a higher quantum yield (typically 75%, compared to 10–15% for a PMT), shot noise is a smaller percentage of the signal for a photodiode than for a PMT when both are exposed to the same light intensity. In other words, the signal-to-noise ratio is higher for a silicon photodiode than for a PMT whenever shot noise is the dominant noise term.

Figure 8 shows the signal-to-noise ratio as a function of incident power for a typical PMT and for a photodiode with a 1.6 mm active area, both at room temperature. There is an incident power at which the signal-to-noise ratios of the photodiode and the PMT are equal. At higher incident power levels, the signal-to-noise ratio of the photodiode is better than that of the PMT. The equality point occurs approximately at the input level, where the photodiode signal shot noise is equal to the Johnson current noise of its shunt resistance. This equality point can be lowered by reducing the active area of or by cooling the photodiode. Cooling the photodiode reduces the Johnson noise (a) because the Johnson current noise is proportional to the square root of temperature and (b) because the shunt resistance increases as the photodiode is cooled, thus reducing the Johnson current noise.

In order to maintain the signal-to-noise advantage of the silicon photodiode, the preamplifier used with it must have an input current noise less than the photodiode Johnson current noise. This will prevent the preamplifier from significantly degrading the performance of the photodiode. Conventional photodiode preamplifier circuitry uses an op-amp transresistance circuit (shown in Fig. 9), which is a current-to-voltage conversion amplifier, the conversion gain being set by the value of the feedback resistor. The input current noise of this circuit is the sum of two uncorrelated terms, the Johnson current noise of the feedback resistor, and the operational amplifier input current noise. Recent advances in op-amp design have lowered the input current noise of better field effect transistor (FET) input op-amps to below that for most photodiode Johnson noise.

In order for the Johnson current noise of the feedback resistor to be lower than the Johnson current noise of the photodiode, the feedback resistance must be higher than the photodiode shunt resistance. For a photodiode with a relatively small active area, such as the Hamamatsu (Hamamatsu City, Japan) s-1087-01 (1.6 mm^2 area), the shunt resistance is quite high ($3 \times 10^{10}\ \Omega$). Using this high a feedback resistor limits the response speed of the circuit as a result of the RC network formed by the feedback resistor and the stray capacitance. For

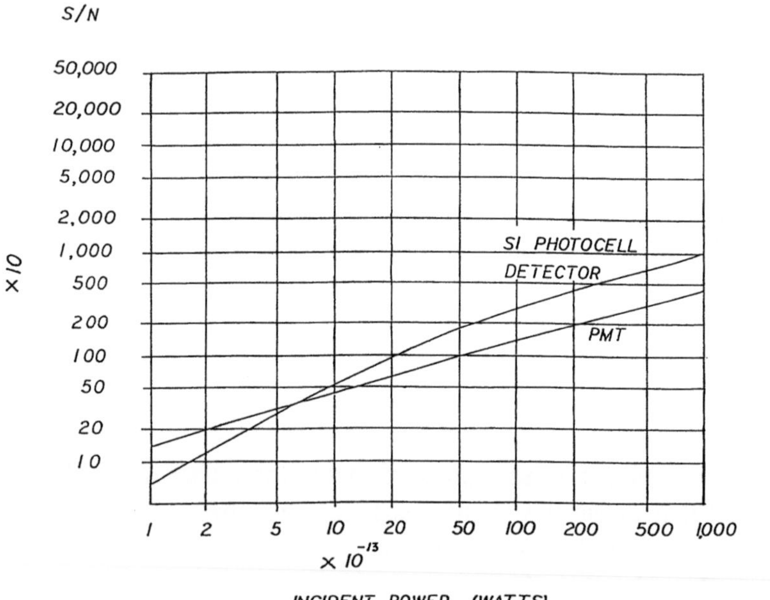

Fig. 8. Signal-to-noise ratio for a PMT and a photodiode detector system at room temperature.

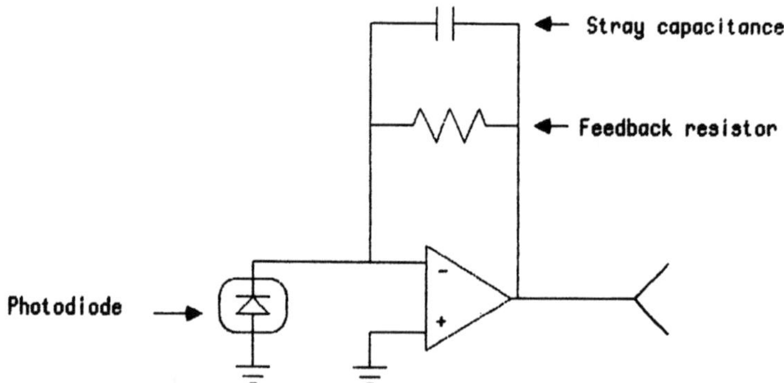

Fig. 9. A typical transimpedance amplifier for use with a silicon photodiode.

example, a 10^{11}-Ω feedback resistor with a stray capacitance of 1 pF limits the response speed to about 1.6 Hz. The response speed can be increased by reducing the value of the feedback resistor; however, this would increase the Johnson current noise of the resistor. A 1.6 Hz bandwidth is fast enough to follow the binding curves for all assays done at present. However, the response is not fast enough to allow the excitation light to be chopped, or modulated, in order to improve DC stability. We have found the DC stability of the feedback resistor to limit the accuracy of this detector system to about 1%.

8. Filters

The optical filters define the excitation spectra seen by the fluorophores and limit the spectra seen by the detector. Ideally the filters should have 100% transmittance in the passband and 0% transmittance at all other wavelengths. Real filters always have some attenuation in the passband and operates some transmittance outside the passband. How close to ideally a filter determines the performance of the instrument and is important in filter selection. Other considerations in filter selection are cost, size, and throughput.

Both monochromators and dielectric filters can be obtained with adequate throughput for use with present sensors. Monochromators have the advantage of being able to change the passband wavelength. This is useful in a laboratory instrument that is to be used with different dyes. The monochromators can be reset to accommodate a new dye, rather than replacing the filters as would be necessary with an interference-filter setup. The monochromators also allow wavelength sweeping for determination of the fluorescence spectra of a compound, which is not possible with an interference-filter-based instrument. However, monochromators with adequate throughput are large, heavy, and expensive, whereas interference filters are small, lightweight, and inexpensive.

In selecting interference filters, the goal is to find the filter and dichroic set that give the highest signal-to-noise ratio. The general approach is to calculate the signal and background levels for a particular filter set and then to calculate the signal-to-noise ratio for each set.

A computer model, developed in our laboratory for performing these calculations, uses the ASYST™ language (Macmillan and Co.), a FORTH offshoot. The light source, dye emission, dye absorption, and filter transmission, represented as spectral curves, are digitized and placed into arrays. Other variables that are not wavelength dependent are represented as scalars. In developing this program, transmission curves for the excitation and emission filters, as well as the dichroic beam splitter, were taken as representative of typical interference filters and were obtained by comparison of actual filter-transmission curves. When LED sources were investigated, the emission curves were individually constructed from data provided by the manufacturer. Incandescent light sources were modeled with a black-body curve at the appropriate color temperature for the lamp being modeled. The spectral curve for photodiode response was constructed from the manufacturer's data. Dye absorbance and emission curves were obtained from standard reference works, and several scalars were developed to represent factors that are not frequency-dependent. These include percentage of fiber face covered by the filament image (for incandescent sources), geometric coupling efficiency (for LEDs), reflection from fiber face, background fluorescence, absorption, quantum efficiency, concentration of the dye, and detector current noise.

The calculations that the model performs are done in the following manner: The fluorescent signal is computed by multiplying the source power spectrum by the source geometric coupling efficiency (in the LED case), the transmission curve of the excitation filter, the spectral reflection of the dichroic beam splitter, and the spectral absorbance characteristic of the dye. The resulting product is integrated over wavelength to determine an "effective" excitation power. The total power of the fluorescent response is proportional to the integrated effective-excitation power. The constant of proportionality takes into account such factors as the concentration of dye, the active area of the fiber, the efficiency of coupling excitation light to the dye, and the efficiency of the emitted fluorescence tunneling back into the fiber. The weighted fluorescent response, i.e., the signal spectrum, is then multiplied by the spectral transmission curve of the dichroic beam splitter, the fluorescence filter, and finally the spectral sensitivity of the detector. The integral over wavelength of the resulting distribution is the predicted signal output of the system.

Before the model can predict sensitivity, it must calculate system noise. Noise in the system consists of detector noise, which is fixed for a given detector, and shot noise of background, which must be computed from the background level for each filter set.

For modeling purposes, we assumed reflection of 4% of the excitation power at the fiber face and applied the same computations, i.e., filtering through the dichroic and fluorescence filter, as we applied to the fluorescence exiting the fiber, in order to get the background for excitation wavelength light. For background from fluorescence, we first measure the excitation and emission spectra of the background fluorescence spectra, then calculate a level by procedures similar to those used to obtain signal predictions. From the combined background, a noise value is calculated and a signal-to-noise ratio determined.

The model is operated as an iterative calculator. Working from a predetermined range of filter values, it steps through the values one at a time, performing the calculations and then saving the signal-to-noise ratio along with its corresponding filter values in an array. This array is then used to determine the optimum filter specifications and can also be used to see how a change in filter specifications affects the signal-to-noise ratio.

9. Optical Background

"Background" is the output reading caused by light reaching the detector that does not come from the fluorescence (signal) of bound fluorophores. If the background is constant over the period of time of an assay, it can be measured prior to the assay and subtracted from the final reading. However, any noise or variation of the background directly affects the accuracy of the assay. For example, if the background changes for any reason during the assay, an error is introduced by subtracting the background level measured before the assay.

The noise on the background is at least the fundamental, or shot-type noise. Shot noise is proportional to the square root of the background level and, at high background levels, may be the dominant noise term. Reducing the background level can reduce the system noise whenever a background-level-dependent noise source, such as shot noise, dominates.

The sources of background can be divided into two categories: background at the excitation wavelength and background at the fluorescence (signal) wavelength. Background at the excitation wavelength can be light that has been reflected or scattered back towards the detector and "leaked" past the filters. This part of the background is strongly dependent on the filter parameters and on the level of reflected and scattered light. The majority of the reflected light comes from the proximal and distal faces of the fiber, and scattering can be from imperfections in the waveguide, imperfections at the waveguide surface, or elements in the optical system of the instrument. The scattering is collected much more efficiently in and on the waveguide than elsewhere in the system (such as at the lenses) and is therefore the source of the majority of that part of the background caused by scattering.

The portion of the background at the fluorescence wavelength is attributable to a shift in the wavelength of the excitation light, either by fluorescence or by Raman scatter. Fluorescence may be caused by impurities in and on the optical elements in the instrument, impurities in and on the waveguide, or optical coupling to the fluorophores in the bulk of the solution. Fluorescence in and on the optical elements is dominated by the fluorescence of a layer in the dielectric filters; therefore, the direction in which the filters are placed in the instrument affects the background level. Fluorescence from impurities in and on the waveguide is very efficiently collected, and even small amounts of contamination can result in large background levels. Scattering both in and on the waveguide redirects a fraction of the excitation light into the bulk of the sample. Bulk fluorescence thus excited can then be coupled back into the fiber via these same scatterers (Fig. 10). We have found that much of this scattering takes place in the section of the waveguide between the proximal face and the sample. Therefore, coupling to the bulk can be reduced by placing a cladding on the waveguide to strip the very-high-angle rays that would otherwise enter the bulk. The coupling efficiency to and from the bulk is very low, but the effect can cause a significant background level as a result of the large bulk volume in comparison to the evanescent volume.

We have previously derived a relationship between the signal level from an evanescent fluorosensor and the NA of the optical system *(16)*, approximated by $S \approx NA^9$. A more significant relationship is how the signal-to-noise ratio depends on NA. For this, variation of noise with NA must be known for a particular instrument.

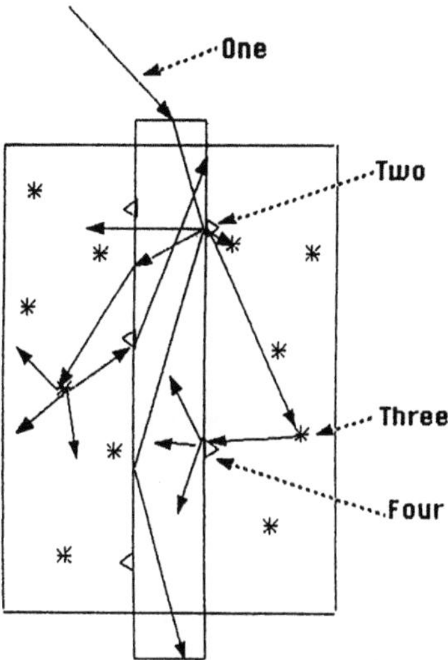

Fig. 10. Background. Excitation light (One) can be coupled out of the fiber through scattering centers (Two) and excite bulk fluorescence (Three). Some of the fluorescence is then coupled back into the fiber by scattering (Four).

10. Sensitivity

Our biosensor is designed for chemical-recognition assays, such as immunoassay. As early as 1985, the ORD Inc. instrument achieved picomolar detection of bound antigen (oxytocin) in a very small observed volume (results presented at Pittsburgh Conference in 1985 and 1986). This is very different from detecting low concentrations in large sample volumes, in which the number of molecules detected can be large. In 1985, detection of oxytocin was reported at 10^{-12} mol/L as a trace in a background solution of 10^{-8} mol/L fluorescein. Independent work with our instrument at Allied Corporation (Morristown, NJ) demonstrated detection of <1000 Y-pestis bacteria. Performing fluoroimmunoassay without a separation step and rejection of the fluorescent background of the sample was the goal in developing our method, and this goal has been achieved.

11. Multiplex Operation

Evanescent fluorosensors are compatible with the use of multiple tags having different fluorescent wavelengths, enabling a "multiplex" mode of operation. By multiplexing we mean different, correlatable measurements on the same sample, *in the same microenvironment,* and at the same time. It is not merely adding additional or parallel sensors in the same sample.

The most obvious application of multiplex detection is the simultaneous measurement of multiple analytes. This also could be done with parallel sensors and thus does not reveal the true power of multiplexing. Making the measurements in the same microenvironment enables additional correlations, making multiplexing a better way to provide internal references and standards than taking parallel measurements.

Of most interest are those applications of multiplexing that provide capabilities beyond those achievable in a single-channel method. For example, by labeling the immobilized phase with a different tag than that used for the assay, an internal quantification of the immobilized phase is obtained. Uncertainty in the amount of immobilized antibody is a noise source in assays *(19)*. Multiplexing can provide a means of reducing this noise and thus improving sensitivity in some systems.

In sandwich assays, nonspecific binding limits ultimate sensitivity. Multiplexing provides a means of correcting for nonspecific binding and, therefore, increases the sensitivity limit beyond that achievable in any single-channel system. We do this by using a physically similar but immunologically irrelevant antibody that is tagged differently than the analyte antibody and measuring its nonspecific binding. This is related to the nonspecific binding of the analyte antibody.

Multiplexing also can be used for ultraspecificity or to reject crossreactivity by the simultaneous use of different monoclonal antibodies against the same antigen *(20)*.

12. Applications

We have developed prototype fiberoptic evanescent-sensor readout instruments for NASA, the US Army, the US Air Force, Allied Signal, Ciba Corning Diagnostics Corporation, Lawrence Livermore National Laboratories, Fujirebio Corporation (Tokyo, Japan), the UK Ministry of Defence, the Defence Research Establishment (Ralston Alberta, Canada), the University of Wisconsin, the University of Maryland, and the University of Texas.

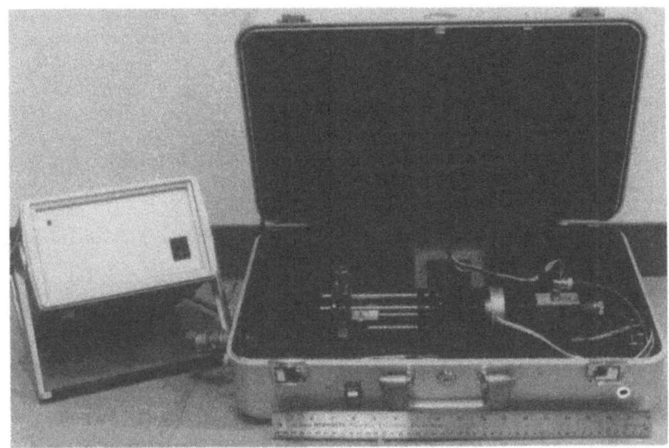

Fig. 11. ORD's briefcase-sized fiberoptic fluorimeter with power supply.

The optical system of these prototype units is of briefcase size, with a collimated beam that uses a dichroic beam splitter to separate the excitation from the detection wavelengths. A shutter in the detection system prevents saturation of the detector while the case is open. Figure 11 shows the briefcase-size instrument and its power supply. The power supply includes the electronics to control the shutter as well as regulate the power for the source and detector. A miniature, hand-held-size instrument with a similar optical layout also has been tested and includes the optics as well as the source, optical filters, detector, and detector electronics. It does not include the power supply or a readout.

The instrument should be applicable for specific binding systems, including antibodies, DNA probes, and receptors. Because the method is rapid, eliminates the need for sample preparation, and eliminates skill-intensive steps, it is expected to find quantitative-assay applications in clinical offices and military field operations. A preliminary report on work with our instrument has appeared *(21)*. Work is being pursued on a receptor-based assay, immunoassay of toxins and pathogens, detection of effluent from an affinity column, measurement of enzyme inhibition, screening of bloodbank blood for specific components, viral detection, detection of the AIDS antibody, and immunoassay for drugs of abuse.

Acknowledgments

We wish to acknowledge the tremendous technical contributions of Tomas Hirschfeld.

We also wish to thank Mark K. Mondol for technical support and stimulating discussion and to acknowledge the support of the US Army CRDEC, NASA, US Army USAMRIID, and NIH.

References

1. Hirschfeld, T. B. (1965) Total reflection fluorescence. *J. Can. Spectrosc.* 126.
2. Kronick, M. N. and Little, W. A. (1973) A new fluorescent immunoassay. *Bull. Am. Phys. Soc.* **18,** 782.
3. Kronick, M. N. and Little, W. A. (1975) A new immunoassay based on fluorescence excitation by internal reflection spectroscopy. *J. Immunol. Methods* **8,** 235.
4. Sutherland, R. M., Dahne, C., Slovacek, R., and Bluestein, B. (1987) Interface immunoassays using the evanescent wave, in *Nonisotopic Immunoassays* (Ngo, T. T., ed.) Plenum, New York.
5. Hirschfeld, T. B. and Block, M. J. (1986) Apparatus including optical fiber for fluorescence immunoassay. US Patent 4,582,809.
6. Hirschfeld, T. B. and Block, M. J. (1984) Assay apparatus and method. US Patent 4,558,014.
7. Hirschfeld, T. B. and Block, M. J. (1984) Fluorescent immunoassay employing optical fiber in capillary tube. US Patent 4,447,546.
8. Sutherland, R. M. and Dahne, C. (1984) Optical detection of antibody-antigen reactions at a glass liquid interface. *Clin. Chem.* **30,** 1533.
9. Andrade, J. D., Van Wagenan, R. A., Gregonis, D. E., Newby, K., and Lin, J. N. (1985) Remote fiber-optic biosensors based on evanescent-excited fluoroimmunoassay: Concept and progress. *IEEE Trans. Electron. Devices* **32,** 1175.
10. Harrick, N. J. (1967) *Internal Reflection Spectroscopy* (Interscience, New York) Chap. 2.
11. Carniglia, C. K., Mandel, L., and Drexhage, H. (1972) Absorption and emission of evanescent photons. *J. Opt. Soc. Am.* **62,** 479.
12. Lee, E. H., Benner, R. E., Fenn, J. B., and Chang, R. K. (1976) Angular distribution of fluorescence from liquids and monodispersed spheres by evanescent wave excitation. *Appl. Opt.* **18,** 862.
13. Holland, W. R. and Hall, D. G. (1985) Waveguide mode enhancement of molecular fluorescence *Opt. Lett.* **10,** 414.

14. Holland, W. R. and Hall, D. G. (1987) Method and system for the enhancement of fluorescence. US Patent 4,649,280
15. Boyd, R. W. (1983) *Radiometry and the Detection of Optical Radiation* (Wiley, New York) Chap. 5.
16. Glass, T., Lackie, S., and Hirschfeld, T. (1987) Effect of numerical aperture on signal level in cylindrical waveguide evanescent fluorosensors. *Appl. Opt.* **26,** 2181.
17. International Patent WO 83/01112 (1983).
18. Hirschfeld, T. B. (1987) Apparatus for improving the numerical aperture at the input of a fiber optic device. US Patent 4,654,532.
19. Jackson, T. M., et al. Immunoassays for clinical chemistry. (Hunter and Carrie, eds.), Churchill Livingstone, Edinburg, p. 557.
20. Khosravi, M. J. and Sudbury, R. (1989) Ultra-specific time-resolved immunofluorometric assay of latrotin in serum. *Clin. Chem.* **35,** 2251.
21. Bluestein, B. (1987) Optical response immunosensors—instant quantitative immunoassay technology for clinical diagnosis. *Clin. Chem.* **33,** 1061.

Immunoassay Kinetics at Continuous Surfaces

*John F. Place, Ranald M. Sutherland,
Andrew Riley, and Ciaran Mangan*

1. Introduction

The key to immunoassay success is analytical sensitivity and specificity, which in turn lead to clinical utility. The immunoassay reagent, the antibody, has the characteristic of being able to bind tightly and relatively specifically to the invading agent (the antigen) and thereafter initiate a series of events terminating in the biological neutralization of the invading agent. It is this binding event that gives immunoassays their importance in clinical medicine. The reaction between the selected antibody and its target antigen is generally highly specific, rapid, and effectively irreversible. Thus, using antibodies as reagents in a testing system should allow the specific detection of relatively small amounts of antigen from mixtures of generically similar materials. For example, thyroid stimulating hormone can be determined at 10^{-12} mol/L in the presence of greater than approx 10^8-fold other proteins that are essentially physically identical *(1)*.

Two factors distinguish immunoassays from color reactions with reagents used to determine higher concentrations of analytes in biological samples: *First,* antibodies do not have intrinsically specific physical characteristics that allow rapid detection in the presence of factorially greater concentrations of physically similar proteins. Therefore, antibodies usually must be labeled—for example, with an isotope, enzyme, or fluorescent tag—to allow specific signal detection. *Second,* immunoassays rely on a separation system to distinguish labeled antibody (or antigen) bound to the antigen (or antibody) from labeled antibody (or antigen) in the free form. When the signal from a bound label can be directly distinguished from the signal from a free label, no physical separation step is necessary, as in nonseparation or homogeneous assays. However, often the bound label must be physically separated from the free label prior to signal measurement, as in separation or heterogeneous assays.

A heterogeneous immunoassay thus relies on at least two steps, the first being the reaction between analyte and specific reagent (antibody) and the second being the physical separation of bound from free label. These steps have fundamentally different requirements, which affect assay design. For example, the primary reaction requires intimate contact of analyte and antibody and is best carried out in solution, whereas the separation step requires the opposite, i.e., the bound fraction must be completely removed from contact with the free fraction. In heterogeneous assays, specific reagents are often immobilized on a carrier matrix. This is done for two reasons: *First,* to allow primary separation of analyte from potentially interfering complexes in the sample and, *second,* to provide a practical method for separation of the bound and free labels for measurement.

As soon as a reagent is immobilized on a surface, the kinetics of the reaction are changed. In addition to possible changes in the reagent itself (reaction rate constants) as a result of attachment, the surface properties of the support matrix (surface charge, hydrophobicity) also affect the primary reaction kinetics. The most obvious change is the limitation of movement of reagent molecules as a result of the spatial geometry of the reaction, e.g., the spatial angle of approach of analyte to the surface. Thus, although immobilization may be convenient for separation, it does not provide ideal conditions for the primary reaction between analyte and specific reagent.

This limitation of immobilization was recognized early in immunoassay development, and techniques for overcoming the disadvantage were considered. Many immunoassays rely on the use of very small particles as the immobilization matrix, with these particles providing the opportunity for separation (centrifugation, magnetic attraction) without seriously affecting the primary reaction rate, since they provide a large and well-dispersed surface for reaction with the analyte. The use of such small particles requires technical skill, in that it is difficult to separate the particles from the liquid reaction medium, and complete separation is a prerequisite of good assay precision. If the size of the individual particles is increased to improve the convenience of separation, it becomes necessary to agitate the mixture for reaction with the smaller surface area.

Because they require physical separation of signal from background interferences stemming from the sample, heterogeneous assays are usually more sensitive. However, this physical separation step is technically demanding, time-consuming, and probably the greatest source of imprecision in most immunoassays. For this reason, a number of approaches have been adopted in attempting to carry out immunoassays without physical separation of antibody-bound from nonbound materials. The general aim of these homogeneous systems is to retain all the advantages of the antibody as a reagent (specificity, sensitivity) without the cumbersome separation step present in heterogeneous assays (Table 1).

The current trend toward decentralization of testing has given rise to the development of immunoassays for use not only in central laboratories, but also in nonlaboratory environments, such as the specialized outpatient clinic, the doctor's office, or even the home. Decentralization of testing emphasizes the need for ease of manipulation in testing protocols and requires the development of immunoassay systems with minimal instrumentation. In microanalytical systems, including immunoassays, such convenient systems are often based on the use of continuous surfaces for immobilization.

With the growing interest in biosensors, it is unlikely that conventional spectroscopy techniques will provide the necessary sensitivity or specificity for microanalytical immunoassays. Sensitive biosensors with immobilization of specific reagents at the transducing interface are envisioned, with the transducer providing an electrical (or optical) signal for further processing, and the interface necessarily being a continuous surface. Immunoassays at continuous surfaces present a

Table 1
Optical Homogeneous Immunoassays: A Classification

Approach	Principle	Examples
1	Change in label activity	EMIT™ TDX™ *(2)* CEDIA™ *(3)*
2	Immune complex formation, direct	Nephelometry Radial immunodiffusion
3	Immune complex formation, indirect	IMPACT™ Latex agglutination
4	Immunoassays on a continuous surface	*See* Table 2

novel assay format, with the antibody fixed to the surface of the sensing device and the antibody–antigen reaction monitored as binding occurs. This approach obviates the necessity of a distinct separation step, and can therefore be regarded as "homogeneous." This chapter is concerned mainly with the kinetics of such an immunoassay.

2. Abbreviations

Ag	antigen
Ab	antibody
Ag–Ab	antigen–antibody complex
k_1	binding rate constant, $cm^3 \times mol^{-1} \times s^{-1}$
k_{-1}	dissociation rate constant, s^{-1}
β	diffusion boundary layer thickness, cm
D_j	diffusion coefficient for species j, cm^2/s
t_d	approximate time-scale of diffusion-controlled assay, s
$[Ag–Ab]_s$	surface concentration of complex, mol/cm^2
$t_{1/2}$	equilibrium half-life of binding reaction controlled assay, s
$J_j(x,t)$	flux of species j at distance $= x$, time $= t$; $mol \times s^{-1} \times cm^{-2}$
x	distance, cm
t	time, s
Ω	an operator whose exact form depends on the geometry of the system. Here it is always a distance operator. cm^{-1}

V	solution velocity, cm/s
$C_j(x,t)$	concentration of species j, mol/cm^3
ZF/RT	a constant, V
Z	electronic charge on diffusing species
F	Faraday
R	gas constant
T	absolute temperature
ϕ	electrical potential
k_f	effective forward reaction rate constant, cm$^3 \times$ mol$^{-1} \times$ s^{-1}
$[Ag]^t_s$	surface concentration of antigen at time t, mol/cm^3
$[Ab]^0_s$	initial surface concentration of antibody, mol/cm^2
$[Ag]^0_b$	initial bulk concentration of antigen, mol/cm^3
D_{Ag}	diffusion coefficient of antigen, cm^2/s
Γ	time constant for assay, according to Stenburg *(11)*, s
ζ	dimensionless time
erfc	error function
K	equilibrium constant for assay, cm^3/mol
b	channel width, cm
α	shear rate, s^{-1}
v	fluid velocity, cm/s^1
k_{sp}	wave vector for dispersion of surface plasmons, Hz $\times$ m$^{-1} \times$ s
W	frequency of the incident light, Hz
c	speed of light, m/s
ε	dielectric constant of the dielectric outside of the metal, C^2 dyne^{-1} cm^{-2}
ε_m	dielectric constant of the metal, C^2 dyne^{-1} cm^{-2}
k_x	wave vector of light in air/glass, Hz $\times$ m$^{-1} \times$ s
θ	angle of incident light $^\circ$

3. The Continuous Surface

The amount of reagent that can be immobilized on a surface is a physical limitation of this system. For example, IgG antibodies a monolayer thick are limited to about 1 µg/cm^2, i.e., the amount of IgG on 1 cm^2 of surface area is equivalent to the immunoglobulin G (IgG) in 1 mL of solution at 10^{-8} mol IgG/L. This appears to be more than adequate for an assay, especially since the concentration of analyte in a 1-mL sample is often many orders of magnitude smaller. However, the major problem in such a system is ensuring adequate contact between the analyte and the immobilized reagent. Without this, the time required for completion of the reaction will be very long.

Methods of avoiding or circumventing these problems include:

1. Increased surface area relative to the volume of the analyte solution, for example, using convoluted surfaces, sponges, or microporous matrices. However, this often requires a trade-off against separation convenience.
2. Use of labels that provide amplification of signal to increase the potential sensitivity of the assay; or
3. Kinetic measurements that increase the overall speed of the assay.

A major factor in the use of continuous surfaces is the static layer of liquid at the surface, through which the analyte must diffuse to reach the surface. It has been estimated that, even with vigorous agitation, this surface layer is about 100 μm thick *(5)*. Although the diffusion coefficient does not decrease proportionally with mol wt (but rather in proportion to the square or cube root), diffusion through this layer still represents a formidable barrier to rapid reaction.

4. Effects of Diffusion at Continuous Surfaces
4.1. Introduction

All the systems discussed rely on components in solution contacting a continuous surface, the sensing element. Of the forces acting on the movement of molecules toward and away from surfaces, including diffusion, convection, and electrostatic migration, diffusion can play a dominant role.

In immunoassays, when an antigen (Ag) reacts with antibody (Ab) in a highly specific manner to form a relatively stable complex (Ag–Ab), diffusion can play an important role in determining overall reaction kinetics.

$$Ag + Ab \underset{k_{-1}}{\overset{k_1}{\longleftrightarrow}} Ag\text{–}Ab \qquad (1)$$

In solid-phase immunoassays, either the antigen or the antibody is initially immobilized, and the corresponding binding partner then must diffuse to the surface, where it reacts to form a complex. This discussion concerns systems in which one of the reactants is immobilized at a continuous surface that is planar or cylindrical.

In the vast majority of immunoassays used in routine diagnostics, it is assumed that equilibrium has been reached before the measurement is made. Although diffusion may have insignificant effects on the equilibrium position, it can significantly affect the initial kinetics and the overall timing of the assay.

4.2. Estimation of the Equilibrium Constant of Immunoassays

The problem in designing equilibrium-based immunoassays is how to tell when the binding reaction has reached equilibrium. A practical approach is to derive a way of estimating the equilibration time for the immunoassay and optimize the design accordingly. An estimate (within an order of magnitude) of the rates of diffusion and reaction can be obtained from a knowledge of the diffusion coefficient and the forward and reverse reaction rate constants. From the theory of diffusion, the diffusional effects are concentrated in a boundary layer of thickness β, which is a simple function of the diffusion coefficient (D) of species j (D_j) and the duration (t_d) of the diffusion process *(6)* as in Eq. (2):

$$\beta = (D_j t_d)^{1/2} \qquad (2)$$

In the case of immunoassays in which one of the reactants is immobilized at a continuous surface, β is equivalent to the dimensions of the reaction cell. Consequently, the time-scale of the diffusion process (t_d) is approximately equal to the time for diffusion to reach the physical limits of the reaction volume. By rearranging Eq. (2), t_d can be estimated:

$$t_d = \beta^2/D_j \qquad (3)$$

The binding kinetics of Ag–Ab reactions become very complex if more details than those given in Eq. (1) are included. These more com-

Table 2
Equilibration Times Estimated
from the Dissociation Constant of the Binding Reaction Using Eq. (5) *(4)*

	Dissociation constant, s^{-1}	Equilibrium half-life, s
Anti-ADHB	6.0×10^3 (low affinity)	11.0×10^{-4}
Antidigoxin	2.4×10^{-4} (high affinity)	2.9×10^3

plex situations have been extensively reviewed by Pecht and Lancet *(4)* and Mason and Williams *(7)*.

An approximation of the time for completion of a reaction under binding reaction control can be obtained from the dissociation constant of the binding reaction. The forward-reaction rate in most Ag–Ab reactions is usually many orders of magnitude faster than the reverse rate, and hence dissociation of the complex is generally the rate-determining step. Expressed mathematically, the rate of change in concentration of the complex is proportional to the dissociation rate constant in a first-order process:

$$d[\text{Ag–Ab}]/dt = -k_{-1}[\text{Ag–Ab}] \tag{4}$$

Integrating and assuming the Ag–Ab complex concentration ([Ag–Ab]) to have reached half its equilibrium value gives the half-life of the binding reaction:

$$t_{1/2} = 0.693/k_{-1} \tag{5}$$

This simple analysis shows for any first-order process that, to a reasonable approximation, the equilibration time is a function of the diffusion coefficient (D_j) of the diffusing species, the reaction cell dimensions (β), and the dissociation rate constant (k_{-1}). In this approximation, the equilibrium time is thus independent of the Ab or Ag concentrations. Using Eqs. (3) and (5), we can make some simple conclusions concerning assays controlled by diffusion or binding reaction rate.

For reactions in which the binding reaction is the rate-determining step, the time to equilibrium can be estimated (Table 2) using val-

Table 3
Effect of Mol Wt of Diffusing Species
on Equilibration Times for Diffusion
with a Cell Dimension of 1 mm Using Eq. (3)

Mol wt, daltons	Diffusion coefficient, m^2/s	Estimated equilibrium time, s
10^5	10^{-11}	10^5
10^2	10^{-10}	10^4

Table 4
Effect of Reaction Cell Size on Equilibration Time
for Diffusion with a Diffusion Coefficient
of 10^{-10} m^2/s Using Eq. (3)

Maximum reaction cell dimension, mm	Estimated equilibration time, s
0.1	10^2
1.0	10^4
10.0	10^6

ues for the dissociation rate constant of typical low- and high-affinity antibodies.

For reactions in which diffusion is the rate-determining step, the mol wt of the diffusing species may vary from 10^2 dalton for small drugs to 10^5 dalton for antibodies. In such cases, the diffusion coefficient varies by only approx one order of magnitude. The diffusion coefficient itself, therefore, does not much affect the equilibration time for most applications (Table 3). On the other hand (Table 4), it can be seen that the reaction cell dimension can have an important influence on the equilibration time.

A comparison of Tables 2–4 indicates that equilibration times for immunoassays at continuous surfaces usually are determined by diffusion rates. Only if the dissociation constant is very small (for high-affinity antibodies) and if the reaction cell dimension is very small (<1 mm) is the equilibration time of the immunoassay determined by the binding reaction kinetics.

4.3. Diffusion Rate or Binding Reaction Rate Determination of Overall Reaction Rate

A more thorough understanding of the kinetics may enable immunoassays to be designed based on rate measurements. Clearly, if rate measurements can be made during the initial phase of the reaction, there is no need to wait for equilibrium to occur. It might be possible to obtain the same sensitivity in a shorter time and to minimize the quantities of expensive reagents used in the immunoassay. In addition, analysis of the factors determining reaction rate may provide insight into how continuous surface immunoassays behave in a medium of unlimited reaction cell size, as used in continuous biosensing in vivo or in some bioreactors.

A first step is to ascertain whether it is the Ag–Ab reaction or diffusion of the analyte that is the rate-determining step in an immunoreaction at a surface. It is therefore important to analyze the effects of diffusion in greater depth, and for a kinetic assay, to determine at which stage diffusion becomes the rate-determining factor.

Two approaches for modeling diffusion problems (analytical and numerical) appear appropriate.

4.3.1. An Analytical Solution of the Model Diffusion Equations

Rigorous analytical treatment of the overall binding reaction incorporating diffusion is extremely difficult because of the large number of reaction steps, competing reactions, and complex boundary conditions. A simplified discussion outlining the physical nature of the diffusion problem and the equations describing the diffusion process are presented. No attempt is made to solve them for any specific situation, since this has been done elsewhere *(8–10)*. However, the equations are outlined here.

In a continuous surface immunoassay, one or more molecular species diffuse independently toward the surface, and all the binding reactions are assumed to take place at this surface. The system can thus be treated as a diffusion problem with special boundary conditions at the surface.

For any arbitrary geometry, the flux of species j to a surface is described by the Nernst-Planck equation as follows *(11)*:

$$J_j(x,t) = -D_j C_j \Omega - (ZF/RT) D_j C_j \Omega \phi + C_j V \qquad (6)$$

$$\text{Diffusion} \quad \text{Electrical} \quad \text{Convection}$$
$$\text{term} \quad \text{migration term} \quad \text{term}$$

The flux (J_j) has three components, one from diffusion caused by a concentration gradient, a component from migration in the presence of an electric field, and a component from either applied or natural convection in the solution. In immunoassays, the component from electrical migration can be neglected, since normally there is no applied electric field, and the solutions are buffered to reduce electrostatic effects. The convective component is usually minimal, since the solution volume is very small and analyte is usually not circulated through the cell chamber. For many purposes, the analyte can be considered to be a stagnant solution in which mass transport occurs solely by diffusional processes.

The system can be described in terms of a simplified version of the Nernst-Planck equation, also known as Fick's First Law of diffusion, which, when further simplified for linear diffusion in one dimension, becomes Eq. (7):

$$J_j(x,t) = -D_j \, dC_j(x,t)/dx \qquad (7)$$

It should be noted that the flux and concentration of species j are functions of time and location. Fick's First Law expresses the flux of a species as a function of its concentration gradient. Fick's Second Law describes the concentration change of a species as a function of time, and it is presented here in both its general form (Eq. [8]) and its simplified, linear diffusion version (Eq. [9]):

General form
$$dC_j/dt = D_j \Omega^2 C_j \qquad (8)$$

Linear form
$$dC_j(x,t)/dt = D_j \, d^2 C_j(x,t)/dx^2 \qquad (9)$$

The operator Ω^2 takes different forms according to the geometry of the diffusion process.

To determine the equilibrium conditions requires the solution of Fick's First and Second Laws according to a set of special boundary conditions (initial conditions) in the cell chamber at time $t = 0$:

1. Antigen, antibody concentration (assumed initially to be uniform in the reaction cell);
2. Cell geometrical considerations, i.e., maximum cell dimension in relation to the "diffusion length" β;
3. Boundary conditions at the surface: the rates of removal or production of antibody and antigen at the surface; the binding reaction kinetics;
4. Any other interfering reactions occurring in the solution or at the surface, i.e. nonspecific binding altering the concentration of available antibody, competition/removal of antigen in solution by unattached antibody, or multiple binding steps.

The simultaneous analytical solution of Eqs. (7) and (9) in accordance with these initial boundary conditions is very complex. Numerical methods often offer the only practical way of obtaining a solution and have the advantage that realistically complex systems can be considered. Various workers have attempted to find analytical solutions for the binding process incorporating diffusion *(12,13)*, but these have usually been limited to steady-state conditions. However, Stenberg et al. *(8)* recently described the initial binding kinetics of a generalized form of continuous surface immunoassay. An analytical solution was found assuming the following: Dissociation of the complex was negligible during the period of interest; or the assay was not saturated, and only a small fraction of the total number of binding sites was occupied.

The immunoassay analyzed by Stenberg et al. *(8)* is illustrated in Fig. 1. The antibody is the analyte to be determined, but for the sake of consistency, we will assume that the antigen is being determined. In the initial stage of the assay, very little of the antigen is bound. In addition, with a high concentration of binding sites the reverse reaction rate can be shown to be very small. The effective forward rate constant (k_f) can be simplified to

$$k_f = 2\,k_1 \tag{10}$$

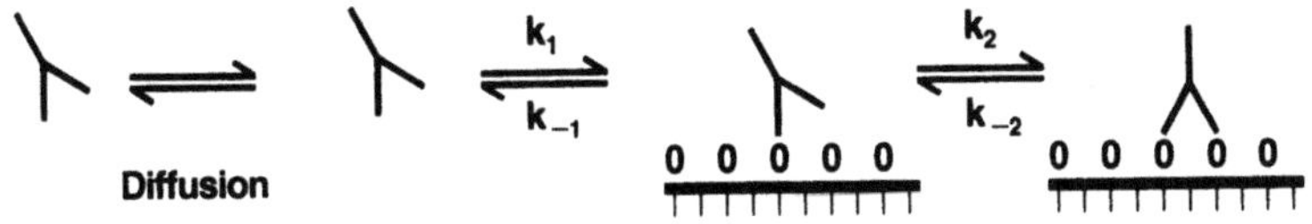

Fig. 1. Immunoassay systems analyzed by Stenberg et al. *(8)*. Diffusion is followed by two binding steps. Y represents the antibody antigen 0, which binds at two points to immobilized.

This means that the rate of formation of bound complex is equal to

$$d[\text{Ag--Ab}]_s/dt = k_f\,[\text{Ag}]^t_s\,[\text{Ab}]^0_s \tag{11}$$

where $[\text{Ag}]^t_s$ is the concentration of Ag at the surface as a function of time, and $[\text{Ab}]^0_s$ is the initial surface concentration of bound Ab (assumed to stay roughly constant with time, i.e., the Ab is in excess).

The rate of formation of the complex must be equal to the rate at which antigen diffuses to the surface.

$$d[\text{Ag--Ab}]_s/dt = D_{\text{Ag}}\,d[\text{Ag}]/dx$$
$$= k_f\,[\text{Ag}]^t_s\,[\text{Ab}]^0_s,\ \text{for } x = 0 \tag{12}$$

We now have the boundary condition that the initial concentration of antigen at the surface is equal to the bulk concentration, i.e.,

$$[\text{Ag}(x,t)] = [\text{Ag}]^0_s\ \text{for all } x,\ \text{at } t = 0 \tag{13}$$

Stenberg et al. solved these equations for a radial geometry by noting that the problem is analogous to that of heat flow from a sphere, solved by Carlslaw and Jaeger *(14)*. The solution is also equivalent to a plane surface when the radial dimension is infinitely large, in which case the concentration of complex on the surface as a function of time is

$$[\text{Ag--Ab}]_s = [\text{Ag}]^0_s\,(D_{\text{Ag}}\Gamma)^{1/2}\,\{2(\zeta/\pi)^{1/2} + \exp(\zeta)\,\text{erfc}\,(\zeta)^{1/2} - 1\} \tag{14}$$

where

$$\Gamma = D_{\text{Ag}}/(k_f\,[\text{Ab}]^0_s)^2 \tag{15}$$

is a time constant, and

$$\zeta = t/\Gamma \tag{16}$$

is dimensionless time.

To recapitulate, Eq. (14) describes the kinetics of the assay when both the binding reaction and diffusion are taken into account, assuming that dissociation of the complex is negligible and that binding saturation does not occur.

After an initial period, the kinetics are governed by diffusion. This rapidly becomes the case for planar surfaces. However, for spherical geometries, diffusion control is less important, since the spherical geometry allows greater access of the analyte to the surface. Under diffusion control, assuming very little complex dissociation and no saturation, the [Ag] at the surface is effectively zero at all times, as the forward binding reaction consumes all the antigen faster than it can diffuse to the surface. The kinetics of complex formation are then, by Eq. (17) for diffusion-limited kinetics,

$$[Ag\text{–}Ab]_s = [Ag]^0_s \, (D_j\Gamma)^{1/2} \, 2(\zeta/\pi)^{1/2} \tag{17}$$

which is the first term of Eq. (14).

The time constant Γ is an estimate of when the rate of diffusion dominates over the rate of binding. From Eq. (15) for the time constant Γ, it can be demonstrated that the onset of diffusional limitation occurs more rapidly if

1. k_f (forward reaction rate) is increased;
2. $[Ab]^0_s$ (initial surface concentration of antibody) is increased;
3. D_{Ag} (the diffusion coefficient of the antigen) is decreased.

There are two important conclusions to be drawn from Eqs. (14–16): *First,* regardless of whether the assay is under reaction or diffusional control, the rate is proportional to the initial bulk concentration of antigen. This means that kinetic measurements are a feasible method of measuring an unknown concentration of antigen. *Second,* there are no terms corresponding to cell dimension or volume in the equations governing the kinetics. Hence, the kinetics are not affected by an infinite vs a micro cell volume. Conversely, the kinetic equations as derived above do not accurately describe the assay rate when the effects of diffusional limitations are felt at the observed reaction cell boundaries. Diffusion effects are observed within a distance β from the interface after time t, according to Eq. (2).

The distance β can be one of several physical limitations. It can be the radius of the reaction pipe or vessel when there is no flow or mixing. If the analyte is flowing past the surface, then the critical distance is equivalent to the thickness of the stagnant diffusion layer that always exists on a reaction surface. The thickness of this diffusion layer is governed solely by the flow velocity and the viscosity of the analyte.

4.3.2. Numerical Solutions
of the Model Diffusion Equations

An accurate simulation of reaction and diffusion kinetics requires the simultaneous solution of a large number of equations and boundary conditions. It is a relatively simple procedure to set up a discrete model of the immunoassay, which can then be made as complex as required. Although boundary conditions still have to be present in a numerical model, they can be made more realistic.

For example, the effects of competition between labeled and nonlabeled diffusion species and nonspecific binding can be included. The advantages of making an accurate simulation of the evolution of the assay with time offsets the extra work required to build and test such a model, which can be adapted to many different situations and a solution found regardless of the degree of complexity.

Electrochemistry at solid–liquid interfaces is a very close analogy to the problem of immunoassay kinetics. Many numerical models already exist for solving complex electrochemical reactions incorporating diffusion *(11,15)*; these possibly can be modified to model immunoassays.

To date, there are few published numerical solutions of the diffusion problem of continuous surface immunoassays; however, there is an extensive literature pertaining to enzyme kinetics, in which numerical models have been used to describe complex reaction schemes *(16–18)*.

Caras et al. *(19)* recently described a time-dependent numerical solution of the kinetics of enzyme reactions at a field effect transistor incorporating the effects of diffusion. They describe the conditions in which an enzyme system obeying Michaelis-Menten kinetics is either reaction or diffusion rate limited.

4.4. A General Description
of a Surface Immunoassay Under Reaction
or Diffusion Rate Control

The exact analytical solution of the overall kinetics of an immunoassay at a surface incorporating both diffusion and binding reaction kinetics is very complex, but schematic diagrams are presented showing how the concentrations of free Ag, free Ab, and AgAb change with time or position.

The two situations considered are immunoassay under binding reaction control and that under diffusion (of Ag) control. In each situation, several possible boundary conditions can affect the concentration profiles. These include:

1. Whether Ag or Ab is in excess (usually Ag will be in excess);
2. The relative rates of k_{-1} and k_1;
3. The absolute values of k_{-1} and k_1.

When the immunoassay is under binding reaction-rate control and the conditions are as described in Eqs. (3) and (5), the diffusion of Ag to the surface occurs faster than Ag is consumed by the binding reaction. Hence, there will be no diffusion gradient in the cell vol during the assay (Figs. 2 and 3). In most assays the cell vol is finite, i.e., there is no flow of analyte through the cell. If the antigen is greatly in excess, then there will be no effective change in bulk Ag concentration (Fig. 3). If this is not the case, the bulk Ag concentration will decrease with time at a rate determined by the binding reaction kinetics. If $k_1 \gg k_{-1}$ and the antibody is in excess (the assay is far from saturation), then the kinetics of the process may be as represented in Figs. 4 and 5 and as described by Eq. (14).

When diffusion becomes rate controlling, the forward binding reaction rate is very much greater than the rate of diffusion of Ag to the biosensor surface. In this case, the surface concentration of Ag is effectively zero, since Ag is consumed at a greater rate than mass transport can supply it. The reverse reaction is very slow; hence, dissociation of the complex does not create significant amounts of free Ag. Complex dissociation becomes important as the binding reaction approaches

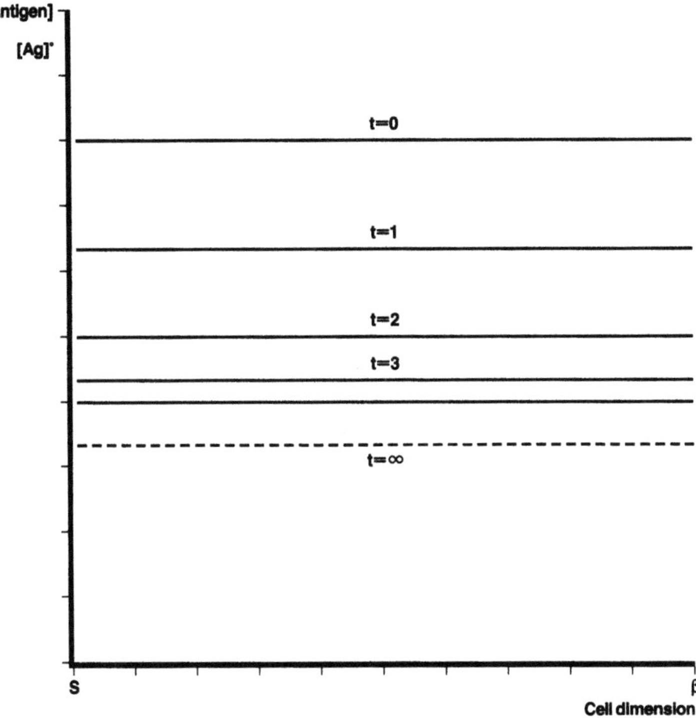

Fig. 2. Under binding as the rate-controlling step, the free antigen concentration $[Ag]^0$ is shown as a function of distance from surface at different times (t) during the reaction. The boundary of the cell dimension (β) is to the far right. At the surface, antigen concentration at $t = 0$, $[Ag]^0$, is smaller than the antibody concentration, $[Ab]^0_s$. *See also* Eq. (14), which defines the surface-bound concentration of complex $[Ag–Ab]_s$.

completion. The kinetics may be as represented in Fig. 6 and as described by Eq. (17).

The initial concentration profiles for Ag in solution are governed by diffusion when the surface concentration of Ag is zero. In this time period, the overall reaction rate is under diffusional control, and the concentration of Ab decreases and that of complex increases as $(t_d)^{1/2}$. Eventually, the kinetics diverge from this behavior as the reverse reaction begins to influence the kinetics.

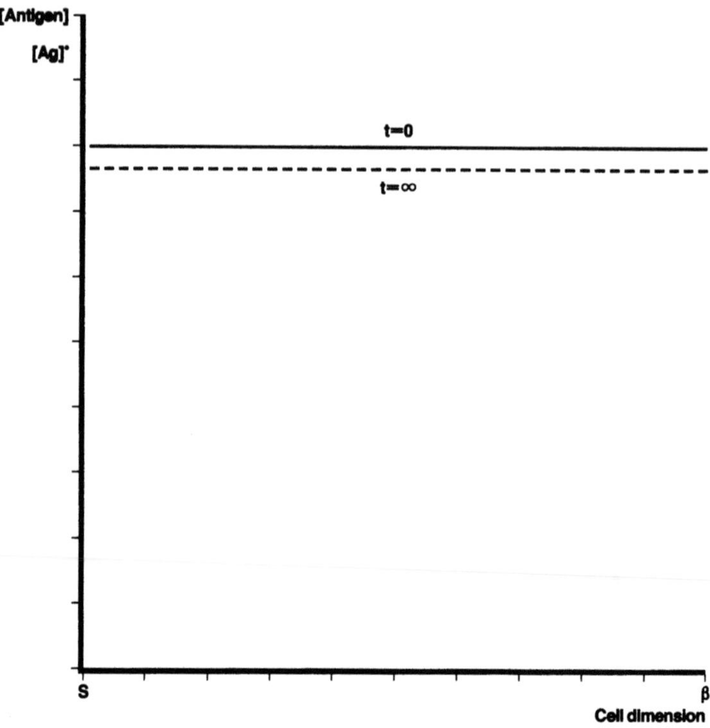

Fig. 3. An illustration of free antigen concentration $[Ag]^0$ as a function of distance from the surface at different times (t) during the reaction. The conditions are the same as those for Fig. 2, except that $[Ag]^0 > [Ab]^0_s$.

The situation has been analyzed in detail by Thompson et al. *(20)* using an earlier result by Reinmuth *(21)*. The rate constants k_1 and k_{-1} are assumed to be sufficiently large that equilibrium is readily established at the surface. The surface coverage of complex is assumed to approximate a Langmuir adsorption isotherm,

$$[Ag\text{–}Ab]/[Ab]^0_s = [Ag]_s K/(1 + [Ag]_s K) \qquad (18)$$

where K is the equilibrium constant. The rate of complex formation now depends solely on the diffusion of antigen to the surface. Reinmuth *(21)* proceeded to use a numerical method to derive an approximate solution for linear diffusion to a plane surface.

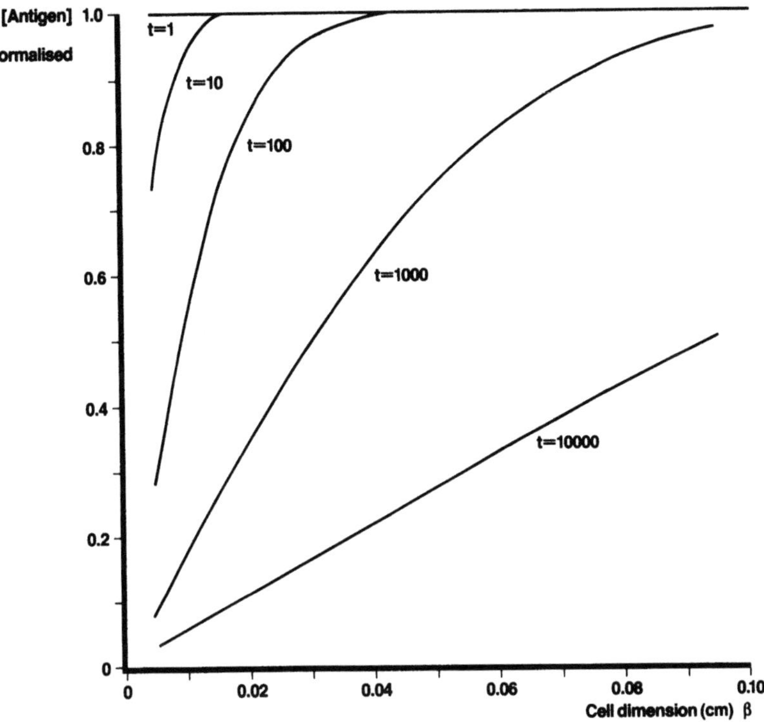

Fig. 4. For diffusional rate-limiting conditions, variation of free antigen concentration (normalized to 1.00 at $t = 0$) is shown as a function of distance from the surface at different times (t) during the reaction. $\beta = 0.1$ cm; $D = 10^{-6}$ cm^2/s; t given in seconds. Lines calculated according to Bard and Faulkner *(11)*:

$$[Ag]^t_x = [Ag]^0_x \, \text{erfc} \, \{x/2 \, (D_j t)^{1/2}\}$$

Antigen profiles show that effects of cell boundary (β) start to be seen at $t = {\sim}1000$ s.

4.5. Real Time Immunoassay Biosensors: The Effect of Diffusion

Present immunoassays are generally based on equilibrium measurements, and many involve washing steps. These features require that immunoassays be used as single-use devices, incapable of consecutive measurements at a rate that might be described as operating in real time. This "one-use" nature, coupled with slow response times,

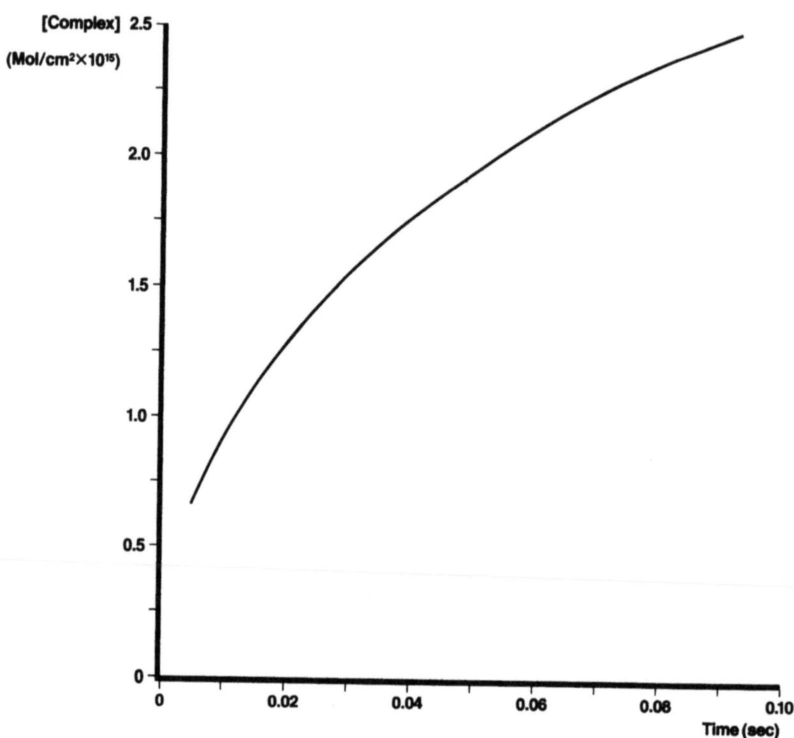

Fig. 5. For binding rate-limiting conditions, per *(8)*, variation of complex concentration at the surface, $[Ag-Ab]_s$, is shown as a function of time, according to Eq. (14). $k_1 = 10^{10}$ cm^3 mol^{-1}s^{-1}; $D = 10^{-6}$ cm^2/s; $[Ab]^0_s = 10^{-13}$ mol/cm^2; $[Ag]^0 = 10^{-11}$ mol/cm^3. The surface antibody is far from saturation, and the small effect of dissociation of the complex is ignored. The time constant according to Eq. (15) is 0.25 s. After 0.1 s, the reaction is controlled by diffusion.

tends to limit the application of immunoassay biosensors in biotechnology and clinical medicine for which rapid continuous measurements are required. Ideally, one would like to be able to plug or implant a biosensor directly into a bioreactor or a patient and obtain a continuous, accurate, and quantitative measurement of the analyte. However, the sampling vol in each case is effectively infinite, and, in addition, the flow of analyte past the biosensor may not be well defined. Both of these factors have an influence on how much of the reacting species reaches the active surface of

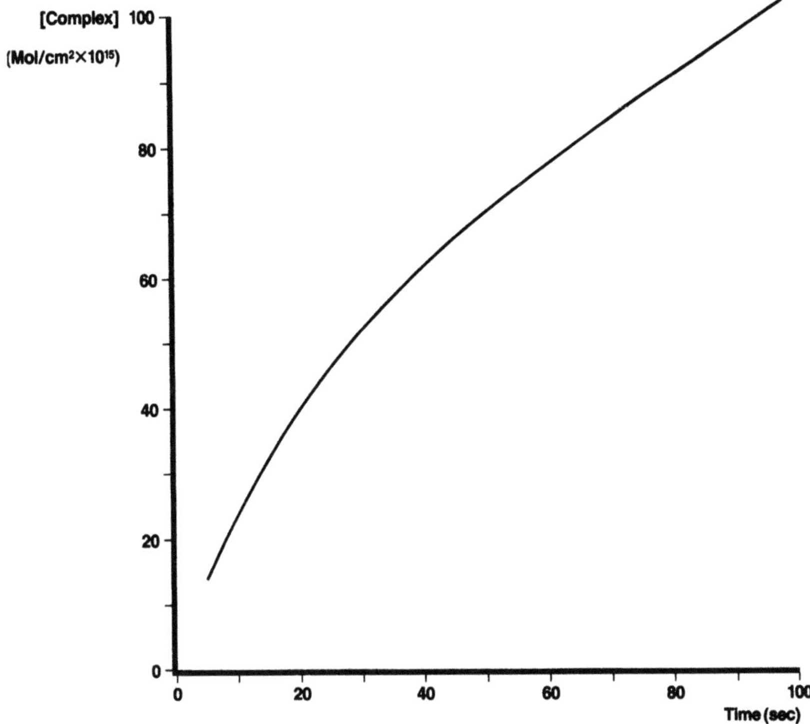

Fig. 6. For diffusion as the rate-limiting step, variation of complex concentration at the surface $[AgAb]_s$ is shown as a function of time, according to Eq. (17). Conditions are the same as in Fig. 5.

the biosensor, and how fast it does so. The major concerns are how this affects both the equilibrium position and the overall binding reaction kinetics.

To have a practical response time, a biosensor needs to utilize a binding reaction with a fairly large dissociation constant (low-affinity antibody). This is because the sensor may be required to measure a decrease, as well as an increase, in analyte concentration. The response of the biosensor, therefore, depends on diffusion of the analyte away from the surface and on eventual convection or mixing to expose the sensor to the new analyte concentration.

It has already been shown that kinetic methods are feasible for measuring unknown concentrations of analyte, provided that the effects of reaction cell boundaries do not play a role, but it is not known under what situations small cells might be used.

4.6. The Limit of Reaction Cell Boundaries

At the surface there is always a static liquid layer through which transport is effected only by diffusion. The thickness of this diffusion layer may be reduced by agitation to perhaps 10 μm, and the overall dependence on diffusion can be reduced further by increasing the surface area. However, this is usually not practical for continuous-surface immunoassays.

In a stirred bioreactor, the stagnant layer may be on the order of 1 mm; hence, the time-window is 10^5 s. In the case of a planar surface placed in a large reaction cell with no circulation of the analyte, natural convection establishes an average diffusion-boundary-layer thickness of 0.5–1 mm. However, since the convection process may not be well controlled, the thickness of the diffusion layer may randomly fluctuate both spatially and temporally.

In reaction cells with dimensions <0.5 mm, there is little convective mass transport. This is the case for continuous surface immunoassays in which there is no forced flow of analyte over the substrate surface and the cell vol is very small.

The forced flow of analyte greatly affects the diffusion process at a surface. To date, this has not been utilized in continuous surface immunoassays, primarily because it requires more complex equipment and larger amounts of expensive reagents. However, hydrodynamic control may offer some advantages, in particular, when on-line continuous action biosensors are being designed. There is much experience in the use of hydrodynamics in the construction of enzyme-based, on-line sensors that is directly applicable to continuous surface immunoassays *(22)*.

The effects of fluid flow on protein adsorption have been extensively studied *(23,24)*. Using total internal reflection fluorescence, protein solutions were made to flow through a rectangular flow channel situated in the base of a silica prism. Fully developed laminar flow was assumed to exist in the channel, creating a diffusion boundary layer of thickness β, given by

$$\beta = b \, (D_f/\alpha \, b^2)^{1/3} \tag{19}$$

where b is the channel width, v is fluid velocity, and $\alpha = 2v/b$ is the shear rate.

4.7. Conclusions

The use of equilibrium measurements thus seems to have little hope of forming the basis of a real time biosensor for measurements in infinite reaction vol. By contrast, kinetic measurements have the advantage of speed and are less affected by reaction cell vol. To assess the feasibility of using kinetic measurements requires a time-dependent solution of the overall assay kinetics, incorporating both diffusion and the binding kinetics.

Although diffusion plays a very important role in surface reactions, other factors also play a part at the cellular or molecular level *(25,26)*. Prior calculations of the effects of diffusion tended to ignore these factors and to use spherical models as a basis. This may be quite far from reality for macromolecules such as nucleic acids, enzymes, and membrane structures *(26)*. The physical nature, such as hydrophobicity or electrostatic charge, and dimensions of the reacting molecules and the surface seem to have an important influence on effective reaction rates at surfaces *(27)*. The ability of some molecules to attach nonspecifically to a part of the target molecule and to then migrate within the potential well of electrostatic interactions may be a natural method for overcoming the barrier of diffusional resistance, making multimolecular reaction processes at low concentrations more efficient *(26)*.

5. Optical Techniques Applied to Immunoassays at a Continuous Surface

Many optical techniques are available to utilize thin films on continuous surfaces *(28,29)*. To carry out immunological binding reactions at such surfaces imposes a number of requirements.

1. The technique must retain the sensitivity and specificity of immunoassays and should be chosen to maximize the potential advantages of this binding reaction.

2. Ideal optical techniques should allow monitoring of the binding reactions between the immobilized component and the solution component as the reaction occurs, with minimal interference from other solution components.
3. The optical technique should rely on components and hardware that are chemically amenable to immobilization of proteins to their sensing surface, that are chemically stable in physiological buffer, and that have the potential for manufacture in large quantities to allow general use.
4. The technique should give advantages over current methods, for example, allowing use of immunoassays in sites different from presently used sites, or allowing complete automation of immunoassays in clinical chemistry laboratories in which high-throughput multichannel analyzers deal with the majority of the workload.

To date, most published work has been directed at using optical techniques for simple immunoassay systems that may be used in decentralized testing. However, future lines of research may extend the potential of these technologies into the central laboratory.

5.1. Optical Techniques
for Continuous Surface Reactions

Several optical techniques can be categorized according to how the light energy impinges on the surface (e.g., external vs internal reflection, as discussed in earlier chapters) or by the method of detecting the surface reaction with labeled or unlabeled reagent systems. A partial list of the different techniques is given in Table 5.

Total internal reflection fluorescence (TIRF) can be used to monitor surface reactions between two transparent media of different refractive indices *(43, 44)*. The electromagnetic waveform called the evanescent wave is generated essentially at the reflecting surface in the medium of lower refraction index, as discussed in several other chapters and elsewhere *(29,41)*.

The fluorescence emitted at the interface is detected with the detector either at a right angle to the interface or in line with the primary light beam. There is an enhancement that theoretically predicts in line fluorescence intensity to be 50 times higher than that emitted at right angles to the waveguide. This has been verified experimentally *(45,46)*.

Table 5
Interface Immunoassays

Detection principle	Device design	Reference
TIRF[a]	Planar, quartz slide, single reflection	(30)
TIRF	Planar, quartz slide, single reflection	(20)
TIRF	Planar, quartz slide, single reflection	(31)
TIRF	Planar, quartz slide, multiple reflection	(32)
TIRF	Quartz optical fiber, multiple reflection	(33)
TIRF	Quartz optical fiber, multiple reflection	(34)
TIRF	Quartz optical fiber, multiple reflection	(35)
TIRF	Planar glass-slide waveguide	(36)
SPR[b]	Silver on quartz prism, single reflection	(37)
SPR	Optical grating	(38)
Scattering	Indium on quartz slide	(39)
Ellipsometry	Quartz slide	(40)

[a]TIRF, total internal reflection fluorescence (41).
[b]SPR, surface plasmon resonance (42).

Using our own research as an example, the critical factors in designing an assay system based on TIRF are waveguide choice, materials choice, incorporation into a disposable device, optical system design layout, and instrument concept.

TIRF can be applied to hapten measurement as well as to large antigens. Kronick and Little (30) first used a labeled antihapten Ab and indirectly immobilized the hapten conjugate. On addition of a sample containing free hapten, the labeled Ab was positioned between the solid-phase hapten and the free hapten, thus producing the type of dose–response curve seen with a radio immunoassay (RIA) or limited reagent assay *(47)*. Our group first applied TIRF to the "two-site" or "sandwich" fluorescent immunoassay, whereby a partially purified sheep antiserum to human IgG was immobilized to the solid phase (a quartz slide) and reacted with a human IgG standard in physiological buffer. Subsequently, an alternative highly purified fluorescein (FITC)-labeled rabbit antihuman IgG was reacted with the immunologically immobilized Ag, and this reaction monitored kinetically. Dose–response curves could be constructed either as a function of the absolute change in fluorescence signal in an endpoint fashion or by estimating the rate of change of signal during the first few minutes of the reaction *(32)*.

A further example illustrates the characteristics of TIRF immunoassays (Fig. 7). Here, a limited reagent assay is illustrated for the measurement of human IgG. The Ag is precoated onto the waveguide surface, and the sample or standard is premixed with the FITC-labeled Ab and then introduced into the system. As with the first example (*see above*), the labeled Ab partitions between the solution phase Ag and Ag (pre-)immobilized onto the optically sensitive surface. Binding curves show smaller signals with increasing concentration of solution phase Ag. When plotted as the absolute change vs Ag concentration (Fig. 8), the dose–response curve is similar to that found in an RIA.

The choice of waveguide format and its incorporation into a disposable device is an area that has resulted, and will result, in many developments. Waveguides are currently being used in two major formats, both based on multiple internal reflection: flat plates (e.g., microscope slide) and solid cylinders (fiberoptics). Both designs have positive and negative aspects and implications in instrumentation design and assay performance. The planar waveguide can be readily handled and manufactured as part of a disposable device (e.g., as one or more walls of a plastic cuvet). Coupling light into and recovering fluorescent light out of such a device can be readily achieved by designing the appropriate prism as part of the disposable unit, and the complete device may be made by injection molding, thus enabling relatively cheap mass manufacture. Such a design is illustrated in Fig. 9.

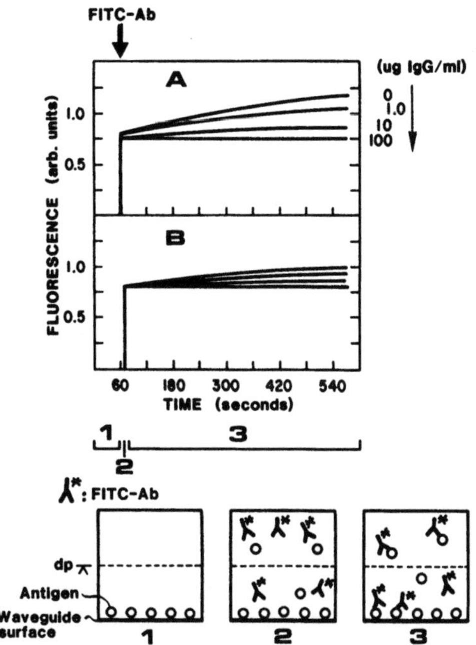

Fig. 7. TIRF immunoassay for human IgG, binding curves of FITC antihuman IgG to immobilized Ag in the presence of standard Ag solution. (1) Ag-coated waveguide; (2) addition of specific FITC-Ab and sample Ag; (3) competitive binding reaction of FITC-Ab for free and bound Ag. (A) Measuring tunneled fluorescence; (B) Measuring right-angle fluorescence (*see text* for explanation). dp = depth of penetration of evanescent detection; FITC-Ab = fluorescein isothiocyanate conjugated Ab.

The optical fiber design of fused quartz fibers to which Ab or Ag can be attached via conventional coupling chemistries is discussed extensively in other chapters. For plastic fibers, new chemistries have to be developed to allow efficient attachment of Ab. Also, there is a physical limit to how thin planar waveguides made of injection-molded plastic can be molded (~0.5–1.0 mm), and thus a smaller number of internal reflections is possible as compared to optical fibers. The technology associated with fiberoptic design for the telecommunication industry allows the manufacture of far thinner waveguides using quartz (~10–500 µm) with a potential for a factorial increase in assay sensitivity.

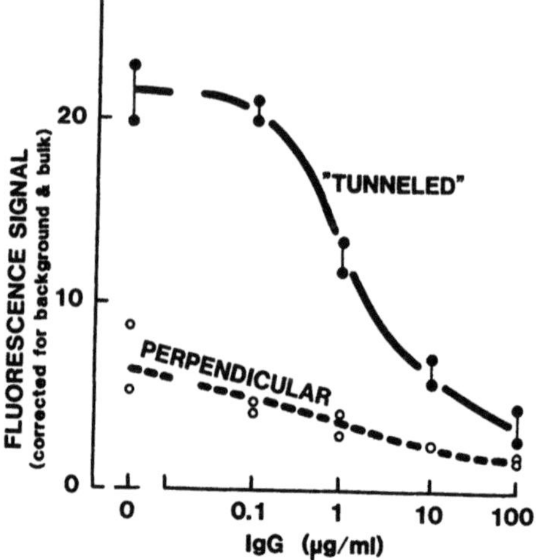

Fig. 8. Experimental results from TIRF immunoassay for human IgG (*see also* Fig. 7) plotted as change in fluorescence vs concentration of antigen. The signal has been corrected for background fluorescence and washed to remove the effect of fluorescence in the bulk solution.

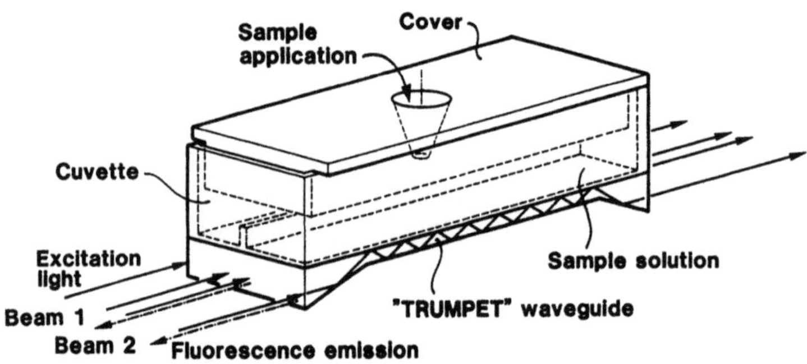

Fig. 9. Injection-molded planar surface device for immunoassays at surfaces by TIRF (*see text* for details).

Furthermore, almost all of the optically sensitive surface of the fiber would be in contact with the sample solution, whereas with the planar waveguide, probably only one surface would be in contact. Most of the potential problems with optical fibers are related to practical aspects of manufacturing and handling such a device.

5.2. Surface Plasmon Resonance (SPR)

Investigation of the optical properties of metals requires the preparation of clean surfaces and leads to the conclusion that the interaction of light and metals can be used as a tool for investigation of the electronic structure of metals and alloys, and for nondestructive studies of surfaces, interfaces, and very thin superficial layers *(48)*.

Such optical techniques include ellipsometry, multiple internal reflection spectroscopy, differential reflectivity, and surface plasmon resonance (SPR) excitation. SPR excitation was found to be one of the techniques most sensitive to surface and interface effects, and also offered a sound theoretical basis for data analysis.

SPR is a well-established concept *(42)*, but it is new in the area of sensing. The promising possibilities of its use in solid-phase optical immunoassays have only recently been demonstrated. The SPR phenomenon can be described in terms of a collective oscillation in the free-electron plasma at a metal boundary. These oscillations can be produced by an electric field, e.g., from an electron.

The plasmons represent the quanta of the oscillations of surface charges that are produced by exterior electrical fields in the boundary. The dispersion relation for a nonradiative plasmon is given by Eq. (20) *(49)*:

$$k_{sp} = W/c \, (1/\varepsilon + 1/\varepsilon_m) \tag{20}$$

where ε_m is the real part of the dielectric constant of the metal at the given frequency (W), c is the velocity of light, and ε is the dielectric constant for the medium outside the metal. The frequencies of interest are those at which ε_m is negative. If light is incident to the surface with an angle θ, then the wave vector of the light parallel to the surface is given by Eq. (21) *(50)*, with θ being the angle of the incident light.

$$k_x = W/c \, (\varepsilon) \sin \theta \tag{21}$$

From Eqs. (20) and (21), for a negative value of ε_m, the parallel wave vectors for the incident light and the surface plasmon are not equal for any values of W and θ. However, if the metal is in the form of a thin film on, e.g., a glass substrate, then the dielectric constants on the two sides of the metal are different, i.e., ε is larger in Eq. (21) than in Eq.

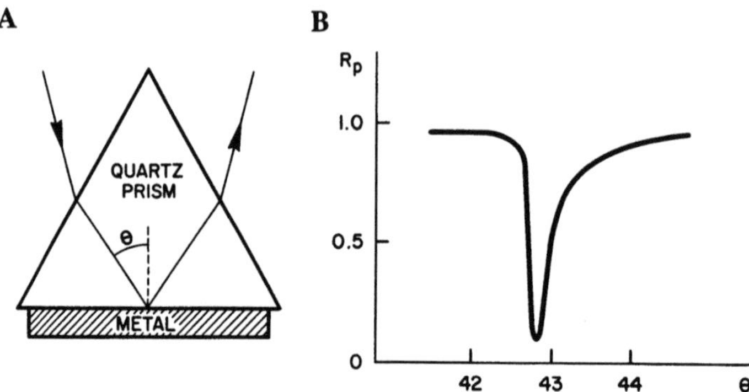

Fig. 10. (A) Arrangement for optical excitation of surface plasmons. (B) Reflectivity (R_p) vs angle of incidence (θ) for surface plasmon resonance.

(20). Surface plasmons can be excited by electron beams or by light. However, they cannot normally be excited optically in a metal dielectric interface, because energy and momentum conservation cannot be obtained simultaneously, i.e., dispersion relations for surface plasmons and incident light do not match at any frequency. This can be overcome with the Kretschmann arrangement shown in Fig. 10A. A beam of p-polarized light is passed through a prism onto the metal surface. At a certain angle of incidence, the parallel component of the light momentum matches the dispersion for surface plasmons at the opposite surface of the metal film, i.e., Eqs. (20) and (21) match each other. With a given thickness of the film, electromagnetic coupling between the interfaces gives rise to SPR and, thus, alteration of the light reflection, as shown in Fig. 10B.

The phenomenon can be explained mathematically if SPR is considered as a surface charge density wave at a metal surface. Two types of surface plasmons can be distinguished: radiative and nonradiative.

A radiative surface plasmon is one that could be excited directly by incident radiation. However, because the real part of the surface plasmon wave vector propagating along a metal dielectric interface is greater than the wave vector of a free photon in the dielectric, the SPR cannot be excited by direct coupling with incident radiation. To overcome this, the SPR is generated by coupling the evanescent tail exhibited in total internal reflection from a prism across the interface to excite the SPR on a silver film. This is called a nonradiative surface plasmon and is used for sensing applications.

In the case of glass/metal and metal/air or metal/liquid interfaces, an angle of incidence exists for which Eqs. (20) and (21) can be matched. At this angle, SPR occurs. This resonance is experimentally observed as a very sharp minimum of light reflectance when the angle of incidence is varied. The resonance angle is very sensitive to variations in the refractive index of the medium just outside the metal film. For example, if the medium outside the silver film is changed from air ($n = 1$) to water ($n = 1.33$), then the resonance angle for helium–neon light changes from 43 to 68° *(48)*. Since the electric field arising from surface resonance probes the medium within only a few thousand Angstroms of the metal surface, the resonance is very sensitive to variation in the thickness and refractive index of thin films on this surface. By choosing an angle halfway from the reflectance minimum and measuring the intensity of the reflected light at that constant angle, changes of about 10^{-5} in the refractive index are readily detected.

SPR has only recently been demonstrated as a new optical technique for use in immunoassays. Flanagan and Pantell *(51)* carried out experiments using human serum albumin as the antigen. The initial layer of Ag was 60 Å thick, but increased to 220 Å following reaction with the Ab. This corresponds to up to four layers of IgG, which could be a result of the nonspecificity of adsorbed Ab. The principle of detection has also been shown on the model system IgG/anti-IgG, in which a bulk concentration of <2 µg/mL of anti-IgG could be detected in a few seconds *(37)*. In these experiments, a glass slide with a 600-Å-thick silver film was used as one well of a 2-mL flow cell through which aqueous solutions were introduced. Figure 11 shows the reflectance vs the angle of incidence for several materials. For the metal/water interface alone, the angle of incidence at which resonance occurs was 68°. The illustrated curves show the resulting shifts in this resonance angle obtained when human IgG molecules bind to the silver surface and rabbit antihuman IgG molecules bind to the initial IgG layer.

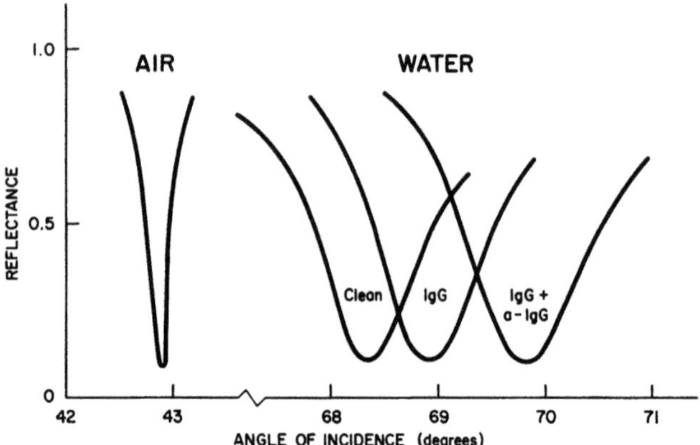

Fig. 11. Resonance curves for a 600-Å silver film in air and in water. Adsorption of an IgG layer causes the curve to shift to a larger angle of incidence. Additional adsorption of a-IgG gives rise to a further shift.

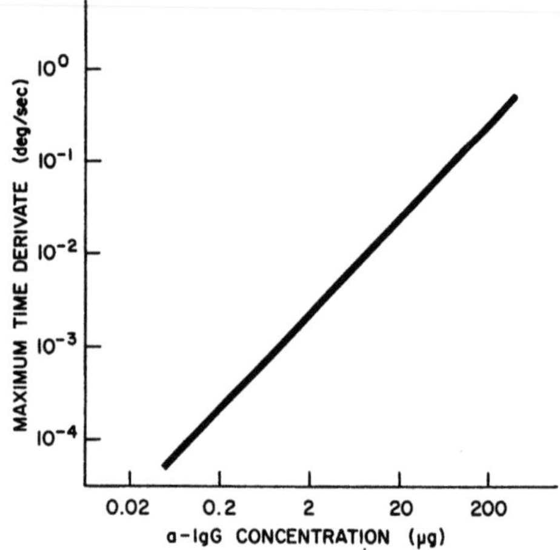

Fig. 12. Maximum derivative of change in angle with time vs a-IgG concentration (μg/mL) in the added solution.

Figure 12 shows a dose–response curve for antihuman IgG. The maximum time derivative needed for the resonance shift to occur for the addition of three different concentrations of anti-IgG gives a linear correlation between Ab concentration and resonance-angle shift. The data apparently show selective Ab reactions with a high sensitivity.

In its present format, this system is more of an experimental instrument than a sensor. However, the measurements could be automated using a differential amplifier and a peak holding display. Also, by use of a semiconductor laser or even a light-emitting diode (LED), miniaturization should be possible, allowing a small sensor to be formed. There are two major problems associated with this system of measurement. The first involves the effect of aqueous buffer salts on the metal film. Tarnishing of a silver film may occur in consequence of oxidants in the solution. This can drastically affect the optical quality of the surface and in some cases make it insensitive to a resonance shift. This problem might be alleviated by depositing a very thin layer of silica over the metal film, which could afford a certain degree of physical protection from the buffer but might also reduce the sensitivity of the system. An alternative might be the use of other reflective metals, particularly gold, which is more stable in oxidant-containing buffer solutions.

The second problem is that Ab binding reactions and enzyme activity reactions at surfaces are not readily reversible, and the reagents used may affect the reagent–surface binding, especially in the case of metals. Metal films must therefore be replaced between each measurement, making the system inherently expensive. One recently demonstrated way to overcome this problem is to use lamellar plastics onto which a single metallized diffraction grating has been deposited. Ab specific for a viral, bacterial, or fungal Ag can be coated on the grating. The sample is then applied, and Ag binds to the surface *(52)*. The slide is then rinsed and inserted into the readout instrument. The measurement technique uses essentially the same principle as described before. However, instead of taking the resonance angle halfway from the reflective minimum and measuring the intensity of the reflected light at this constant angle, the readout is done on the spectrum of light reflected from the metallized grating and shifts in the absorption band in the spectrum are calculated from changes in the dielectric constant above the metal surface. These changes are caused by the effect on the SPR shift of binding of Ag to the surface. The instrument then converts the reading of the k_x range in position of the dark band into a statement of concentration—in this case, that of the infectious organism *(52)*.

The potential for SPR and other surface sensing devices in immunoassays is based on their ability to monitor surface reactions without a major influence from the bulk of the solution. The examples above demonstrate the basic feasibility of the SPR concept, which shows promising simplicity and sensitivity. However, some obstacles still have to be overcome before the full potential of this system is realized in both research and commercial environments. A major problem is still the nonspecific binding that occurs in immunoassays. SPR is essentially a measure of effective layer thickness and refractive index; it does not have the potential to measure selectively (for example, at a characteristic wavelength). Nonspecific binding can therefore change the parameters detected in SPR, making it difficult to differentiate specific binding events. Conventional solid-phase immunoassays use procedures to minimize nonspecific binding—for example, surface treatment with blocking agents. These techniques need to be adapted to the surface treatment of SPR sensors.

The system described above, using lamellar plastics, lends itself to low-cost, high-volume manufacture and ease of operation; recently, improvements have been made in its sensitivity by binding a fluorescently labeled antibody to the surface of the grating. When an assay is carried out for the corresponding Ag, it is found that the SPR induced by incident light that is resonant with the grating surface has a profound effect on the behavior of the dye molecule. Light is emitted at a specific angle instead of reradiating uniformly in all directions, as expected from conventional considerations.

With fluorescence phenomena, there is also the benefit that measurements of the output radiation are taken at a wavelength different from that of the input radiation. The system also allows higher probability of a fluorochrome interaction, thus enhancing the excitation frequency.

The principle of signal generation in SPR appears to be different from that of conventional internal reflection methods. The theoretical basis and the practical results obtained with SPR suggest that the evanescent wave generated at the metal/liquid interface and used in measurement has an electromagnetic field strength about 80 times that of the internal reflection field.

When a fluorescent marker is used in SPR, it may be possible to generate an enhanced fluorescent signal relative to TIRF. As mentioned earlier, it is known that the fluorescent signal emitted in TIRF is tunneled back into the waveguide, but there is apparently no published evidence that, in addition to the conventional right-angled fluorescent detection, there is any tunneling of the fluorescent signal in SPR.

6. Discussion and Conclusions

Immunoassays using solid-phase immobilized reagents are well-established and sensitive methods for microanalysis of low-concentration analytes. Immunoassays at continuous surfaces seem to represent a compromise approach that has certain practical advantages. It is argued that continuous surface immobilization increases handling convenience, which is an essential feature of successful decentralized testing. At the same time, the continuous surface is a necessary part of the biosensor approach that measures reactions over a period of time.

The price paid for this convenience might be the restriction in surface area available for reaction and the reaction-rate limitation represented by the diffusional barrier at a surface. However, it appears that we are only beginning to understand the importance of surface phenomena in assay design, and that these limitations may not be as restrictive as was previously thought.

Among the important factors that may allow improvements in the efficiency of continuous surface immunoassays are

1. Genetic engineering of the Ab to suit the application and the surface;
2. Surface structures and physical characteristics that enhance the attachment of the analyte to the surface immobilized Ab, including the conditions of attachment of the Ab itself;
3. Use of signals that are particularly suited to the surface reaction;
4. Measurement of rates of reaction rather than concentrations at equilibrium.

The study of reactions at the cell surface shows that continuous surface reactions proceed by efficient methods, and suggests that modeling of such systems may allow the development of simpler and more sensitive immunoassay devices. The function of living organisms apparently does not suffer unduly from what would appear to be a severe limitation of the speed of reaction at a surface caused by diffusional barriers. Not only are successive reactions probably spatially organized at the surface, but also it appears that for larger molecules (e.g., nucleic acids), there are biophysical phenomena that allow these molecules to establish a foothold on the surface and subsequently react specifically in an efficient manner.

Optical techniques for measurement at continuous surfaces have proved to be quite sensitive, and there are still avenues to be explored to improve these techniques further. The techniques described have the ability to monitor surface reactions without major influence from the bulk of the solution (i.e., an *in situ* separation). In this way, since they do not require a separation step, they can be regarded as homogeneous assays.

SPR appears to have the potential for measurements more sensitive by one or two orders of magnitude than conventional internal reflection techniques. However, this method is less selective and cannot measure changes in specific wavelengths of interest for signal selectivity.

Hopefully, the results of research and development of continuous surface immunoassays will appear in the form of practical applications for decentralized testing in the near future. The possibilities for improving these systems do not appear to be exhausted, and more sensitive systems can be expected from more intensive research and development at the molecular level.

Acknowledgment

Our thanks to Jacqueline Boëque for completing the difficult task of preparing the original text of this chapter for publication.

References

1. Lawson, N., Mike, N., Wilson, R., and Pandov, H. (1986) Assessment of time-resolved fluoroimmunoassay for thyrotropin in routine clinical practice. *Clin. Chem.* **32,** 684–686.
2. Dandliker, W. B., Hsu, M.-L., and Vanderlaan, W. P. (1980) Fluorescence polarization immuno/receptor assays, in *Immunoassays, Clinical Laboratory*

Techniques for the 1980's (Nakamura, R. M., Dito, W. R., and Tucker, E. S., eds.), Alan R. Liss, New York, pp. 65–88.

3. Henderson, D. R., Friedman, S. B., Harris, J. O., Manning, W. B., and Zoccoli, A. (1986) CEDIA, a new homogeneous immunoassay system. *Clin. Chem.* **32,** 1637–1641.

4. Pecht, I. and Lancet, D. (1977) Kinetics of antibody–hapten interactions. *Mol. Biol. Biochem. Biophys.* **24,** 306–338.

5. Axén, R. E., Kaj, G. L., and Björkman, R. Method, porous matrix and device for analytical biospecific affinity reactions. International Patent Application PCT/SE84/00033.

6. Crank, J. (1975) *Mathematics of Diffusion*, 2d Ed. Clarendon, Oxford, UK.

7. Mason, D. W. and Williams, A. F. (1980) The kinetics of antibody binding to membrane antigens in solution and at the cell surface. *Biochem. J.* **187,** 1–20.

8. Stenberg, M., Stiblert, L., and Nygren, H. (1986) External diffusion in solid-phase immunoassays. *J. Theor. Biol.* **120,** 129–140.

9. Stenberg, M. and Nygren, H. (1985) A diffusion limited reaction theory for a solid-phase immunoassay. *J. Theor. Biol.* **113,** 589–597.

10. Schaaf, P. and Dejardin, P. (1987) Coupling between interfacial protein adsorption and bulk diffusion. A numerical study. *Coll. Surf.* **24,** 239.

11. Bard, A. J. and Faulkner, L. R. (1980) *Electrochemical Methods,* Wiley, New York.

12. Jerne, N. K., Henry, C., Nordin, A. A., Fuji, H., Koros, A. M. C., and Lefkovits, I. (1974) Plaque forming cells: Methodology and theory. *Transplant. Rev.* **18,** 130–191.

13. Wank, S. A., DeLisi, C., and Metzger, H. (1983) Analysis of the rate-limiting step in a ligand-cell receptor interaction: The immunoglobulin E system. *Biochemistry* **22,** 954–959.

14. Carlslaw, H. S. and Jaeger, J. C. (1959) *Conduction of Heat in Solids*, 2d Ed., Clarendon, Oxford, UK.

15. Feldberg, S. W. (1969) Digital simulation: A general method for solving electrochemical diffusion-kinetic problems. *Electroanal. Chem.* **3,** 199–296.

16. Tran-Minh, C. and Broun, G. (1975) Construction and study of electrodes using cross-linked enzymes. *Anal. Chem.* **47,** 1359–1364.

17. Mell, L. D. and Maloy, J. T. (1975) A model for the amperometric enzyme electrode obtained through digital simulation and applied to the immobilized glucose oxidase system. *Anal. Chem.* **47,** 229–307.

18. Brady, J. E. and Carr, P. W. (1980) Theoretical evaluation of the steady-state response of potentiometric electrodes. *Anal. Chem.* **52,** 977–980.

19. Caras, S. D., Janata, J., Saupe, D., and Scmitt, K. (1985) pH-based enzyme potentiometric sensors. *Anal. Chem.* **57,** 1917–1920.

20. Thompson, M., Dhaliwal, G. K., Anthur, C. L., and Calabrese, G. S. (1987) The potential of the bulk acoustic wave device as a liquid-phase immunosensor. *IEEE Trans. Ultrasonics, Ferroelectrics and Frequency Control* **34,** 127–135.

21. Reinmuth, W. H. (1961) Diffusion to a plane with Langmuirian adsorption. *J. Phys. Chem.* **65,** 473–476.

22. Carr, P. W. and Bowers, L. D. (1980) *Immobilized Enzymes in Analytical and Clinical Chemistry*, Wiley, New York.

23. Watkins, R. W. and Robertson, C. R. (1917) A total internal-reflection technique for the examination of protein adsorption. *J. Biomed. Mater. Res.* **11,** 915–938.

24. Lok, B. K. (1981) PhD thesis, Protein adsorption onto cross-linked polydimethysiloxane using total internal reflection fluorescence. Stanford University, Stanford, CA.

25. DeLisi, C. and Metzger, H. (1976) Some physical chemical aspects of receptor-ligand interactions. *Immunol. Commun.* **5/5,** 417–436.

26. Eigen, M. (1974) Diffusion control in biochemical reactions, in *Quantum Statistical Mechanics in the Natural Sciences* (Mintz, S. L. and Wiedermayer, S. M., eds.), Plenum, New York.

27. Andrade, J. D. and Hlady, V. (1986) Protein adsorption and materials biocompatability: A tutorial review and suggested hypotheses. *Adv. Polymer Sci.* **70,** 1–63.

28. Sutherland, R. M. and Dähne, C. (1987) IRS devices for optical immunoassays, in *Biosensors: Fundamentals and Applications* (Turner, A. P. F., Karube, I., and Wilson, G. S., eds.), Oxford University Press, Oxford, UK.

29. Sutherland, R. M., Dähne, C., Bregnard, A., Hybl, E., and Maystre, J.-L. (1987) Evanescent wave immunoassays, in *Complementary Immunoassays* (Collins, W. P., ed.), Wiley, Chichester.

30. Kronick, M. N. and Little, W. A. (1975) A new immunoassay based on fluorescence excitation by internal reflection spectroscopy. *J. Immunol. Meth.* **8,** 235–242.

31. Andrade, J. D. and Van Wagenen, R. (1983) Process for conducting fluorescence immunoassays without added labels and employing attenuated internal reflection. US Patent 4,368,047.

32. Sutherland, R. M., Dähne, C., Place, J. F., and Ringrose, A. R. (1984) Optical detection of antibody–antigen reactions at a glass–liquid interface. *Clin. Chem.* **30,** 1533–1538.

33. Sutherland, R. M., Dähne, C., Place, J. F., and Ringrose, A. R. (1984) Immunoassays at a quartz–liquid interface: Theory, instrumentation and preliminary application to the fluorescent immunoassay of human immunoglobulin G. *J. Immunol. Meth.* **74,** 253–265.

34. Hirschfeld, T. E. (1984) Fluorescent immunoassay employing optical fiber in capillary tube. US Patent 4,447,546.

35. Sutherland, R. M., Dähne, C., Slovacek, R., and Bluestein, B. (1987) Interface immunoassays using the evanescent wave, in *Non-isotopic Immunoassays* (Ngo, T. T., ed.), Plenum, New York.

36. Badley, R. A., Drake, R. A. L., Shanks, I. A., Smith, A. M., and Stephenson,

P. R. (1986) Optical biosensors for immunoassays—the fluorescence capillary fill device. *Proc. R. Soc.*

37. Leidberg, B., Nylander, C., and Lundström, I. (1983) Surface plasmon resonance for gas detection and biosensing. *Sens. Act.* **4,** 299–304.

38. Layton, D. G., Pettigrew, R. M., Smith, A. M., Petty-Saphon, S., and Fisher, J. H. (1984) Qualitative and quantitative assays of biological samples etc. by using thin film coating on plate with optical observation of changes. EP Patent 0112721.

39. Giaever, I. (1976) Visual detection of CEA antigen on surfaces. *J. Immunol.* **116,** 766–771.

40. Rothen, A. (1974) Immunological reactions between films of antigen and antibody molecules. *J. Biol. Chem.* **168,** 75–97.

41. Hirschfeld, T. (1971) Total reflection fluorescence spectroscopy. US Patent 3,604,937.

42. Raether, H. (1977) Surface plasmon oscillations and their applications, in *Physics of Thin Films, Advances in Research and Development* (Haas, G. and Farnecombe, M. H., eds.), Academic, New York, pp. 145–261.

43. Harrick, N. J. (1967) *Internal Reflection Spectroscopy*, Interscience, New York.

44. Burghardt, T. P. and Axelrod, D. (1981) Measuring surface dynamics of biomolecules by total internal reflection fluorescence with photobleaching recovery and correlation spectroscopy. *Biophys. J.* **33,** 435–454.

45. Carniglia, C. K., Mandel, L., and Drexhage, H. (1972) Absorption and emission of evanescent photons. *J. Opt. Soc. Amer.* **62,** 479–486.

46. Lee, E. H., Benner, R. E., Fenn, J. B., and Chang, R. K. (1979) Angular distribution of fluorescence from liquids and monodispersed spheres by evanescent wave excitation. *App. Opt.* **18,** 862–870.

47. Ekins, R. P. (1985) Current concepts and future developments, in *Alternative Immunoassays* (Collins, W. P., ed.), Wiley, Chichester, UK, pp. 219–237.

48. Abelèls, F. (1972) *Optical Properties of Solids,* North-Holland, Amsterdam, p. 67.

49. Agranovic, V. M. (1982) *Surface Polaritons,* North-Holland, Amsterdam, pp. 243–251.

50. Raether, H. (1980) Excitation of plasmons and interband transitions by electrons. *Springer Tracts Mod. Phys.* **88,** pp. 1–196.

51. Flanagan, M. T. and Pantell, R. M. (1984) Surface plasmon resonance and immunosensors. *Electr. Lett.* **20,** 968–970.

52. North, J. B., Petty-Saphon, S., and Sawyers, C. G. (1986) Assay of biological, biochemical and chemical species by fluorescent measurement on a surface having a pre-formed relief profile. WO Patent 8601901.

Luminescence in Biosensor Design

Pierre R. Coulet and Loïc J. Blum

1. Introduction

The production of light during some biochemical reactions (bioluminescence) has been recognized as a powerful tool for biochemical and clinical analysis. The light emission can be measured with great sensitivity, and consequently analysis can be performed at very low detection levels.

The light-producing reaction of the firefly has been the most extensively studied bioluminescent system. The light emission occurs in the presence of the enzyme luciferase and requires luciferin (which is a specific substrate), Mg^{2+}, molecular oxygen, and ATP. The reaction has been used either to measure ATP directly or in coupled reactions that produce or consume ATP. Bioluminescence is also observed in marine bacteria. The bacterial bioluminescent system requires oxygen, a long-chain aliphatic aldehyde, $FMNH_2$, and a luciferase. An oxidoreductase activity, closely associated with the luciferase activity in crude extracts of luminescent bacteria, enables NADH or NADPH to be measured. This coupled bioluminescent system is also useful in the analysis of some NAD(P)-H dependent enzymes and their substrates.

Chemiluminescent reactions (chemical generation of light) also are of particular interest in biochemical analysis, especially those that require H_2O_2 for light emission. Hydrogen peroxide is a product of several enzymatic reactions, and these enzyme-catalyzed processes can be coupled to chemiluminescent detection.

The high sensitivity of luminescent analysis, coupled with the advantages provided by immobilized enzymes (reusability, enhancement of stability), led several groups to take an interest in the use of immobilized reagents for bio- and chemiluminescent analysis, including biosensor applications. These as well as other studies on the use of immobilized reagents in luminescent analysis are discussed in this chapter.

2. Bioluminescent Reactions

2.1. Firefly Bioluminescence

Since the discovery by McElroy of the absolute requirement of ATP in the bioluminescent reaction of firefly luciferase *(1)*, this reaction has been widely used for analytical purposes. In most of the studies using firefly bioluminescent reactions, the species *Photinus pyralis* is employed, but the mechanism of light emission is fundamentally the same in all species of fireflies *(2)*. The luciferin–luciferase system of *Photinus pyralis* has been reviewed by DeLuca and McElroy *(2–4)*. Briefly, the oxidation of luciferin by O_2, in the presence of ATP-Mg, and catalyzed by luciferase (EC 1.13.12.7), occurs in two steps. First, the activation by the enzyme (E) of D-luciferin (LH_2) with ATP-Mg leads to the formation of pyrophosphate (PPi) and enzyme-bound luciferyl-adenylate (E-LH_2-AMP) (Reaction [1]). This enzymatic complex next reacts with oxygen to produce CO_2, AMP, oxyluciferin, and light (Reaction [2]).

$$E + LH_2 + ATP + Mg^{2+} \longleftrightarrow E\text{-}LH_2\text{-}AMP + PPi \tag{1}$$

$$E\text{-}LH_2\text{-}AMP + O_2 \longrightarrow \text{oxyluciferin} + AMP + CO_2$$
$$+ h\nu \ (\lambda max = 560 \ nm) \tag{2}$$

If the concentrations of all the substrates except ATP are kept constant, the light intensity can be directly related to the ATP concentration. A third reaction catalyzed by luciferase is the formation of a nonproducing light complex, dehydroluciferyl-adenylate, from dehydroluciferin (L) and ATP-Mg (Reaction [3]).

$$E + L + ATP + Mg^{2+} \longleftrightarrow E\text{-}L\text{-}AMP + PPi \qquad (3)$$

Dehydroluciferin, present in crude extracts of firefly luciferase and in luciferin preparations from the same natural source, acts as a competitive inhibitor of luciferase *(5,6)*. However, this drawback is easily overcome with the use of crystalline firefly luciferase *(7)* and synthetic luciferin, the chemical synthesis of which was first achieved by White et al. *(8)*. Now, commercial preparations of luciferase (crude extracts or crystalline), synthetic luciferin, and luciferin–luciferase reagent kits are widely available *(9)*.

2.2. Bacterial Bioluminescence

The bacterial bioluminescent reaction is also catalyzed by a luciferase (EC 1.14.14.3) isolated from marine bacteria that can be grown in the laboratory *(10,11)*. The taxonomy of luminous bacteria has been a source of controversy until a definitive classification was adopted *(12,13)*. For clarity, actual genus and species appellations are used in this chapter, whatever the names appearing in the referenced papers. In the different luminescent bacteria, the same light-emission mechanism is involved and the luciferases are similar *(11)*. The light production catalyzed by the bacterial luciferase occurs according to Reaction 4, in which R-CHO is an aliphatic aldehyde with a chain length of eight or more carbon atoms.

$$FMNH_2 + R\text{-}CHO + O_2 \longrightarrow FMN + R\text{-}COOH$$
$$+ H_2O + h\nu \quad (\lambda max = 490 \text{ nm}) \qquad (4)$$

In addition, luminescent bacteria contain an NAD(P)H:FMN oxidoreductase (EC 1.6.8.1) that catalyses the following reaction:

$$NAD(P)H + H^+ + FMN \longrightarrow NAD(P)^+ + FMNH_2 \qquad (5)$$

This reductase was isolated from various strains of bioluminescent bacteria, such as *Vibrio fischeri*, *Photobacterium phosphoreum*, *Photobacterium leiognathi (14)*, and *Vibrio harveyi (15)*, as well as in several species of nonluminous aerobic and anaerobic bacteria *(14)*.

The two most useful light-emitting systems were isolated from *V. harveyi* and *V. fischeri*. Distinct FMN reductases for NADH and NADPH were identified in extracts of *V. harveyi*, whereas *V. fischeri* appears to have only one oxidoreductase *(16)*.

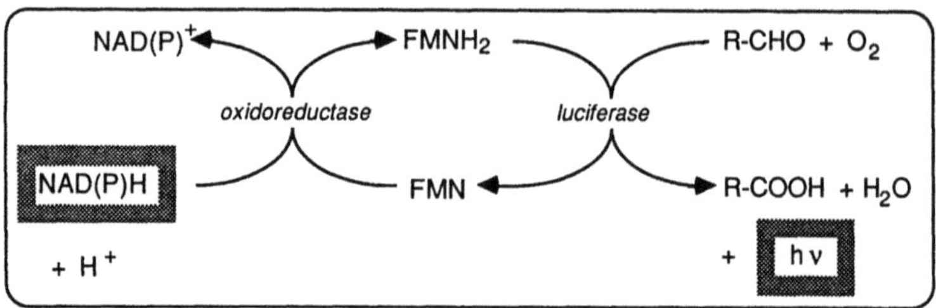

Fig. 1. Schematic representation of the coupled bacterial reaction with luciferase and oxidoreductase.

Thus, the bacterial bienzymatic system allows NAD(P)H to be assayed as shown in Fig. 1, provided that it is the limiting substrate. Decanal or tetradecanal are generally used as the aliphatic aldehyde. Moreover, the use of coupled reactions makes it easy to determine various substrates and enzymes involved in NAD(P)H-producing or - consuming reactions.

3. Immobilized Enzymes for Bioluminescent Analysis

The immobilization of bioluminescent reagents on solid supports allows their reusability, which significantly reduces cost. In addition, an increase in enzyme stability is often observed. Insoluble derivatives of bioluminescence enzymes were first prepared by Erlanger et al. by the reaction of luciferases from *V. fischeri* and *P. leiognathi* with the azide of polyacrylic acid *(17)*, and the properties of these immobilized enzyme preparations and their use in studies on the mechanism of bioluminescence were reported. Since then, various supports have been used for immobilizing firefly and bacterial luciferases, generally via covalent binding to preactivated matrices.

3.1. Analytical Systems with Immobilized Firefly Luciferase

Alkylamine glass beads glued to a glass rod were first used to immobilize firefly luciferase from *Photinus pyralis (18)*. The enzyme, covalently linked with glutaraldehyde, had a lower optimum pH (7.25) than soluble luciferase (7.75), and it emitted light with a major peak at 615 nm instead of at 562 nm. The bound enzyme exhibited only 0.07–0.16% of its activity compared to the soluble enzyme, but was stable for at least 3 wk when kept at 4°C in $0.1M$ phosphate buffer, pH 7.0, containing 1 mM β-mercaptoethanol. A linear relationship between peak light intensity and ATP concentration was observed in the range 1×10^{-8}–$1 \times 10^{-5}M$ using the immobilized preparation. The assays were performed in 0.5 mL of medium containing the reagents plus the glass rod, and disposed in an Aminco Chem-Glow photometer. Creatine kinase (EC 2.7.3.2) (CK) has also been assayed using immobilized luciferase by measuring the ATP formed from ADP and creatine phosphate (Reaction [6]).

$$\text{CK}$$
$$\text{ADP + creatine phosphate} \longleftrightarrow \text{ATP + creatine} \tag{6}$$

The increase in light intensity was directly proportional to the amount of CK in the range 1.27–9.5 IU. Later, the properties of firefly luciferase immobilized on Sepharose 4B and CL 6B were studied *(19)*. The recovered activities ranged from 44–84% of the initial value depending on the preparations. When stored in a glycerol- and dithiothreitol (DTT)-containing buffer, no loss in enzyme activity was observed after freezing at –196°C and thawing. Blocking reagents used in the coupling procedure did not affect the shape of the pH-activity profile or its optimum. A linear response of peak light intensity to ATP concentration was observed between 0.4×10^{-9} and $4 \times 10^{-6}M$. The intraassay precision was 3.2% at $1 \times 10^{-5}M$ ATP.

An automated flow system using luciferase immobilized on Sepharose CL 6B and packed into a small flow cell was described by Kricka et al. *(20)*. The design of the immobilized-enzyme-packed flow cell is shown in Fig. 2, with an Aminco Chem-Glow photometer used with a flow-cell adapter. The linear range for ATP assays was 3–3000 pmol. With a 20-s wash between samplings, the precision of the bioluminescent assay was acceptable (CV = 10.3%) but was significantly improved (CV = 3.6%) by increasing the time of the wash cycle to 80 s. The carry-over was then 5.9%. More than 100 assays could be performed with the same flow cell during a period of 3–4 wk. A problem affecting the analytical performance of such a system was product

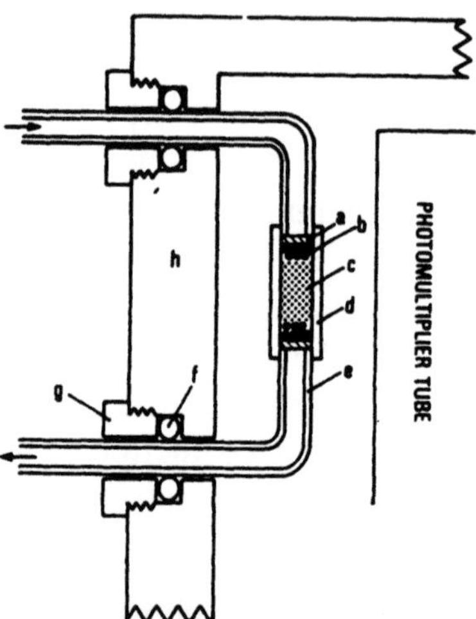

Fig. 2. Design of the immobilized enzyme-packed flow cell and its location in the flow-cell adapter: (a) polyethylene disk; (b) glass beads; (c) Sepharose-immobilized enzymes; (d) Tygon tubing—1/4-in. od, 3/16-in. id; (e) Tygon tubing—3/16-in. od, 1/8-in. id; (f) O ring; (g) knurled screw; (h) body of flow adapter *(20)*.

inhibition as a result of product accumulation at the enzyme after repeated use. This drawback was eliminated either by extensive washing (80 s) of the immobilized enzyme preparation or by including Triton X-100 in the wash buffer with a 20-s wash cycle. The effect of such a nonionic detergent is to promote removal of the product from the enzyme *(21)*. A larger problem was microbial growth in the packed bed after use for a long period at ambient temperature, which markedly enhanced the background light level despite the presence of the antibiotic gentamicin.

After isolation of luciferase from firefly *Luciola mingrelica*, Brovko and coworkers immobilized the enzyme on CNBr-Sepharose 4B and on cellophane films *(22,23)*. Bioluminescent intensity was measured and used to determine ATP in the range 50 nM–1 mM. In addition, the possibility of using this immobilized enzyme to determine the activity of ATPase and pyruvate kinase was demonstrated. The same sample of luciferase immobilized on cellophane film was used 40 times for ATP quantification with a precision of 7%. After 1 mo of storage at 4°C in a solution containing 6.5 mM DTT, luciferase immobilized on Sepharose 4B retained 60% of its initial activity, whereas after only 5 d of storage under the same conditions, the cellophane-bound enzyme retained only 50% of its initial activity. The mechanism of inactivation and reactivation of Sepharose-bound luciferase and the role of sulfhydryl groups in these processes were also investigated, and a kinetic scheme was proposed *(24)*. In further studies, luciferase from *Luciola mingrelica* was immobilized on various polysaccharide-based carriers (Sepharose, Ultrodex®, cellophane, and Ultrogel® activated by CNBr) and on albumin and polyacrylamide gel *(25,26)*. Only immobilization on cyanogen bromide activated polysaccharide carriers resulted in highly active immobilized luciferase. Better activity was obtained with Ultrodex-bound enzyme, whereas the Sepharose-bound luciferase exhibited better stability at pH 7.5 and 25°C. Kinetic properties of immobilized luciferase differed little from those of the soluble enzyme. For each polysaccharide-carrier-bound enzyme, the optimum pHs for activity and stability were determined. Cellulose films activated by cyanuric chloride or sodium periodate were also used *(27)*, and retained enzyme activity as high as that observed with CNBr-Sepharose luciferase. Light measurements were performed with an LKB 1250 luminometer and with enzymatic cellulose films, calibration curves for ATP determination ranged from 0.1×10^{-9} to $1 \times 10^{-6}M$. Using CNBr-Sepharose, a bioluminescent method was developed for creatine kinase assay, involving an immobilized firefly extract containing luciferase and adenylate kinase (EC 2.7.4.3) *(28,29)*. Instead of ADP, an AMP–ATP mixture yielding ADP by the adenylate kinase (AK) reaction (7) was used. A great excess of AMP with respect to ATP entailed a fairly complete conversion of ATP to ADP, which is a substrate of CK.

$$\text{AK}$$
$$\text{ATP} + \text{AMP} \longleftrightarrow 2\ \text{ADP} \tag{7}$$

Then, the amount of ATP formed per unit of time was measured by the increase in light intensity vs time. Despite inhibition of CK by luciferin and inhibition of luciferase by creatine phosphate (the other substrate of CK), the lowest detection limit for CK activity was 0.5 ±

0.2 U/L in the sample, and the linear range of the determined CK activities was 0.5–1000 U/L.

A continuous-flow system using luciferase (from *Photinus pyralis*) immobilized on a nylon coil placed in front of the photomultiplier window of an LKB 1250 luminometer was described by Carrea et al. *(30)*. The bioluminescent reactor consisted of a 0.5-m Nylon-6 tube (1 mm id) wound around a Plexiglas™ support. The flow system involved two streams; one a working bioluminescent solution and the other a continuous flow of air into which a 5-80-µL sample was injected. The assay was linear between 0.3 and 100 pmol of ATP, and 20 samples/h could be treated with no carry-over. The relative standard deviations for both intra- and interassays were 5–8%. Up to 900 ATP samples could be analyzed with a single reactor containing about 0.4 mg of covalently linked enzyme with a half-life of 15 d. More recently, the authors have reported the use of luciferase immobilized on epoxy methacrylate beads in a flow reactor for luminescent determination of ATP in platelets *(31)*. Enzymatic beads were packed in a small glass column (2 mm id, 3 cm in length) that was positioned inside the luminometer. The half-life of the immobilized luciferase was 3 d at 25°C. The sensitivity of the assay was 0.3 pmol of ATP, and the response was linear between 1 and 500 pmol.

Worsfold and Nabi *(32)* described a flow-injection analyzer for determination of ATP with firefly luciferase immobilized on Sepharose 4B and on controlled-porosity glass (CPG) beads. A glass coil (2.5 mm id × 60 mm length) containing immobilized enzyme was placed directly in front of an end-window photomultiplier tube. Simultaneously, ATP samples and luciferin were injected into separate buffered carrier streams using a rotary polytetrafluoroethylene valve. The Sepharose system had a detection limit three orders of magnitude better and a sensitivity two orders of magnitude greater than the CPG system. The dynamic range for ATP measurements was $1 \times 10^{-12} - 1 \times 10^{-7}M$ with a relative standard deviation <2% over the whole range. No loss of activity over a 2-mo period was shown by the Sepharose-bound enzyme when stored at 4°C, but a gradual loss of activity was observed when the immobilized enzyme was kept at room temperature in the flow-injection manifold. Creatine kinase and creatine

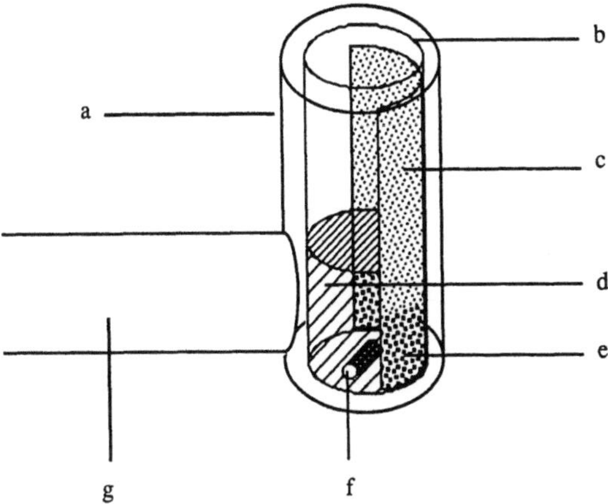

Fig. 3. Reaction cuvet for bioluminescence measurements with collagen-bound luciferase: (a) sample chamber of the luminometer; (b) reaction cuvet; (c) collagen strip; (d) reaction medium; (e) immobilized enzyme area; (f) stirring bar; (g) photomultiplier tube.

phosphate assays could also be performed with the same manifold used for the ATP assay. For the enzyme assay, the CK sample was injected into a carrier stream containing ADP and creatine phosphate, and ATP was produced *in situ*. Results for CK standards ranged from 10 to 400 U/L. Creatine phosphate measurements were performed by injecting the sample into a carrier stream containing CK, with good precision over the range 1×10^{-5}–$0.1M$.

In our laboratory, acyl-azide-activated collagen strips have been used for immobilization of firefly luciferase from *Photinus pyralis (33)*. Light measurements were performed with a Berthold Biolumat (Wildbad, FRG) LB 9500 luminometer. The enzyme was bound to a 1-cm^2 area at the tips of 1×4.5-cm collagen strips. The reaction cuvet (Fig. 3) contained 1 mL of buffered reagents, the luciferase strip, and a small magnetic stirring bar. This provided reproducible mixing of sample and reagent and a constant diffusion rate of substrates from the bulk solution to the enzymatic surface. Therefore the measurement precision was improved. If immobilization was performed in the presence of enzyme substrates, i.e., luciferin and ATP, the bound luciferase activity was twice as high as when immobilization was done without substrate. This was attributed to the luciferyl–adenylate complex formed during the immobilization process having a protective effect against binding to active site luciferase groups. The presence of certain amino acids at the active site of luciferase has been demonstrated *(34)*, and our coupling procedure *(35)* has resulted in covalent binding betwen enzyme amino groups and collagen-activated carboxyl groups. For ATP concentrations $\leq 1 \times 10^{-7}M$, stable light emission could be observed for at least 30 min, thus making possible activity measurements of ATP-producing or -consuming enzymes. The optimum pH for collagen-bound luciferase was 7.75, as it was for the soluble enzyme. Similarly, the activation energy of the immobilized enzyme was unchanged compared to the free-enzyme energy. ATP measurements could be performed with good precision (CV = 3%) in the range $1 \times 10^{-11} - 3 \times 10^{-6}M$, and ATP concentrations as low as 10 pM (10 fmol/ assay) could be determined. After extraction with boiling Tris-borate buffer, human blood ATP could be measured with such strips; the results were in good agreement with those obtained with soluble luciferase. The strips, when stored dehydrated at 4°C, retained 20% of their initial activity over at least 8 mo. Atypical kinetic properties of collagen-bound firefly luciferase were observed *(36)*.

3.2. Analytical Systems
with Immobilized Bacterial Luciferase

Since the first report *(37)* using immobilized bacterial bioluminescence enzymes for analytical purposes, several immobilized-enzyme systems have been described, including either batchwise or flow-analysis methods.

3.2.1. Batchwise Systems

In the first system, both reductases and luciferase from *Vibrio harveyi* were covalently attached to arylamine glass beads cemented on glass rods placed in front of the photomultiplier tube window *(37)*. The immobilized-enzyme properties were similar to those of the soluble forms regarding pH (6.5) and substrate optima (0.0005% decanal; 2.5 μM FMN and 0.2 mM NADH). Linearity in initial maximum light intensity as a function of the NADH or NADPH concentration ranged, respectively, from 1×10^{-8}–$5 \times 10^{-4} M$ and from 1×10^{-7}–$2 \times 10^{-3} M$. A single rod could be used for over 100 consecutive assays of NADH without any apparent loss of activity. Such rods have also been used to monitor reactions producing NADH or NADPH, and picomolar concentrations of malate dehydrogenase, alcohol dehydrogenase, glucose-6-phosphate dehydrogenase (G-6-PDH), lactate dehydrogenase, and hexokinase (HK) have been assayed *(38)*. Glucose determinations have been performed in the range 3×10^{-8} to $3 \times 10^{-6} M$ using soluble HK and G-6-PDH with immobilized luciferase–oxidoreductase enzymes. Using the same support and by coimmobilizing G-6-PDH *(39)*, it was possible to quantitate a G-6-P concentration of 2 n*M*. By coimmobilizing a fourth enzyme, HK, a 30 n*M* glucose concentration could be reproducibly detected.

With Sepharose 4B, immobilized-enzyme preparations of higher activity than the arylamine glass beads were obtained *(40)*. Coimmobilization of bacterial luciferase–oxidoreductase with 3α- or 3β-hydroxysteroid dehydrogenase on such a polysaccharide carrier enabled androsterone or testosterone to be determined at the picomolar level (0.8 pmol in a 0.5-mL assay mixture). In the same way, enzymes coimmobilized on Sepharose 4B with purified bacterial bioluminescence enzymes have been used to develop bioluminescent assays for D-glucose, L-lactate, 6-P-gluconate, L-malate, L-alanine, L-glutamate, NAD, and NADP *(41)*. Although the lower limits of detection were different for each metabolite, 10–100 pmol/assay could be measured at a reproducibility of 2.5–4.0%. Primary bile acids have also been assayed, using a bioluminescence technique involving the light-emitting system from *V. harveyi* coimmobilized on Sepharose 4B with 7α-hydroxysteroid dehydrogenase *(42)*. For glycine and taurine conjugates of 7α-hydroxycholic and chenodeoxycholic acids, as well as for the unconjugated acids, the assays were linear at 1.5–50 μ*M* with intra- and interassay precisions of 6–8 and 8–10%, respectively. This bioluminescence assay showed good agreement with gas–liquid chromatography, radioimmunoassay, or end-point enzymatic assays. In a further report *(19)*, activities after immobilization to Enzacryls® and CH-Sepharose 4B with a variety of reactive groups (aldehyde, diazonium salt, and *N*-hydroxy-succinimide ester) were only 0.01–0.1% of the activity of cyanogen bromide activated Sepharose 4B. The major limitation for detection of very small amounts of substrates or metabolites using either glass beads or Sepharose as support was a high background-light level associated with the immobilized enzymes.

A bioluminescent assay specific for 12α-hydroxy bile acids involving 12α-hydroxysteroid dehydrogenase coimmobilized on Sepharose 4B with bacterial luciferase and either NADPH:FMN oxidoreductase or bacterial diaphorase gave a lower detection limit of 4 pmol/0.5 mL assay *(43)* at an intraassay precision of about 8%. The preparation containing the oxidoreductase was stable over a 2-mo period at 4°C. That containing diaphorase lost considerable activity during the same period, although thiol compounds stabilized the diaphorase system. With a more complex coimmobilized system of 3α-hydroxysteroid dehydrogenase, diaphorase, and bacterial luciferase (isolated from *V. harveyi*), a bioluminescence assay for total 3α-hydroxy bile acids in serum has been described *(44)*. The system is linear over the range 0.4–100 μ*M* with 92–110% recovery of added standards and an intraassay precision of 6.2–8.2%. The determination of chenobiol bioequivalence has been achieved using an immobilized multienzyme bioluminescence technique *(45)*. The enzyme system, bound on Sepharose beads, included bacterial luciferase, diaphorase, and 7α-hydroxysteroid dehydrogenase, and involved the bioluminescence quantitation of unconjugated 7α-hydroxy bile acids.

Commercially available bacterial luciferase–oxidoreductase from *V. fischeri* has been immobilized in bovine serum albumin *(46)* and used to measure NADH or NADPH. It has also been coupled to an NADPH-producing system for creatine kinase activity determination. In this kinetic assay, a linear relationship between reaction rate and creatine kinase activity was obtained over the clinically important range, and the immobilized enzyme was usable for at least 100 assays.

CNBr-activated agarose has also been used for immobilization of bacterial extract or partially purified enzymes from *V. harveyi (47)*. With the partially purified preparation, NADH and FMN could be measured at 1×10^{-12}–$1 \times 10^{-7}M$ and 1×10^{-10}–$1 \times 10^{-7}M$, respectively. The coimmobilization of formate dehydrogenase with bacterial bioluminescence enzymes allowed the determination of NAD and formate, both at the picomolar level. Two other coimmobilized enzymatic systems enabled microassays of metabolites at the nanomolar level: one for glucose-6-phosphate using glucose-6-phosphate dehydrogenase and the *V. harveyi* enzymes; a second for glucose-1-phosphate, using phosphoglucomutase.

In our laboratory, determination of NADH at $1 \times 10^{-9} - 2 \times 10^{-5} M$ was successfully performed using the luciferase–oxidoreductase system from *V. fischeri* immobilized on collagen strips *(48)*. This support has higher mechanical strength and greater ease of handling, particularly for repeated use, compared to particulate or gel supports. The apparatus and reaction cuvet were identical to those used for the collagen-bound firefly luciferase *(33)* (Fig. 3). With the immobilized bacterial system, a steady-state luminescent signal was obtained within 2 min if 2 m*M* DTT was added to the reaction medium. The optimum conditions for NADH measurements were $1.3 \times 10^{-5} M$ FMN, 0.004% decanal and pH of 6.4. The precision was 5% at 10 n*M* NADH. After 2 wk of storage at 4°C in phosphate buffer, 70% of the initial activity remained, whereas complete loss of activity was observed with the same enzymes in solution.

3.2.2. Flow Systems

Kurkijärvi et al. have first established the feasibility of continuous-flow bioluminescent assays using a bioreactor packed with bacterial bioluminescence enzymes immobilized on Sepharose 4B *(49)*. The packed-glass column was placed in front of the photomultiplier tube of an LKB Wallac luminometer 1250. The luminescence responded linearly to NADH from 1 pmol to 10 nmol with 2–20 µL sample volumes. About 400 NADH assays could be performed with a single enzyme column without any change in sensitivity or accuracy. However, problems with packing or disruption of the matrix were encountered after 4 d of intensive use.

NADH, glucose-6-phosphate, and primary bile acids have been assayed with an automated flow system identical to that used for ATP determination *(20)*. Bacterial luciferase and NADH:FMN oxidoreductase (from *V. harveyi*) were coimmobilized on Sepharose 4B with either glucose-6-phosphate dehydrogenase or 7α-hydroxysteroid

                                                                      Coulet and Blum

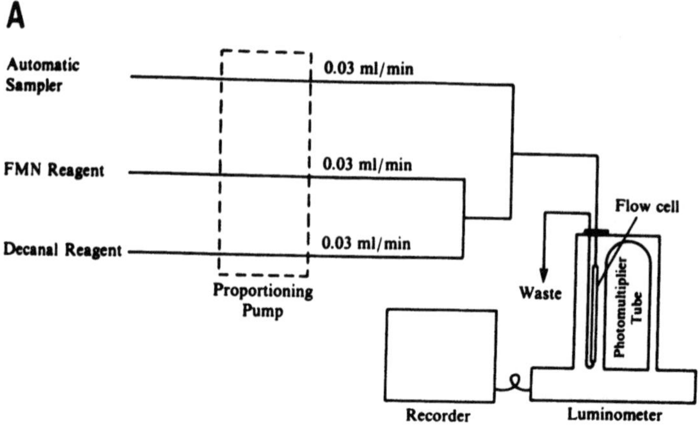

Fig. 4(A). The flow manifold.

dehydrogenase, depending on the analyte to be assayed. The Sepharose-bound enzymes were packed in small flow-cells placed in front of a photomultiplier tube (Fig. 2). The sensitivity of the coimmobilized bacterial luciferase system was limited by the variable level of background light associated with the NAD–FMN–decanal assay mixture. In general, measurements could be performed from several to several hundred picomoles. The coefficient of variation was 2–10%, depending on the concentration of the sample and the duration of the washing between samples (20- or 80-s wash). Thirty samples/h could be assayed and up to 700 consecutive measurements could be performed with the same packed flow cell. However, bacterial contamination of the Sepharose limited the continuous use of the immobilized preparation. Subsequently, the design of the flow cell was modified to reduce background light and improve sensitivity *(50)*. The bed volume of the flow cell, as well as the reagent flow rates, was reduced, and FMN and decanal were separated into individual reservoirs. These modifications resulted in lowering background light to about 0.1% of its previous value and in a 1000-fold increase in sensitivity for measurement of NADH. The flow-cell was a capillary tube containing Sepharose-bound enzymes (Fig. 4). The flow rate through the cell was 0.09 mL/min. The sampling time was 2 min and was followed by a 4-min wash cycle. With 60-μL samples, concentrations of NADH from 1×10^{-10}

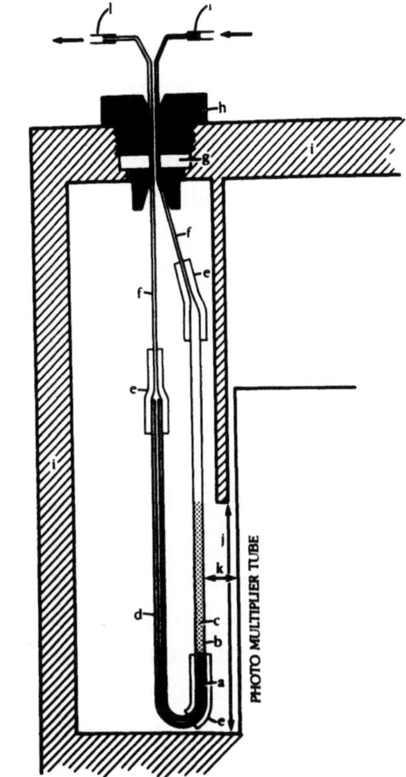

Fig. 4(B). Design of the enzyme-packed capillary flow cell: (a) glass-bead bed support; (b) capillary tube flow cell; (c) Sepharose-enzyme bed; (d) thick-wall capillary for outflow; (e) silicone rubber tubing; (f) 22-gage stainless steel hypodermic needle; (g) rubber septum; (h) knurled screw port; (i) body of sample chamber of LKB 1250 luminometer; (j) window of sample chamber (38 mm); (k) distance from flow cell to PMT, ~5 mm; (l) Teflon™ tubing *(50)*.

to $1 \times 10^{-6}M$ could be measured, with an intra-assay coefficient of variation of 2–10%. Using coimmobilized 6-phosphogluconate dehydrogenase, the sample concentration range for 6-phosphogluconate was $1 \times 10^{-9} - 1 \times 10^{-5}M$, but contamination of the Sepharose enzyme bed caused the background light emission to increase gradually after only a few hours of operation.

Roda and coworkers developed a continuous-flow bioluminescent method for primary bile acid analysis using the luminescence enzymes from *V. fischeri* coimmobilized with 7α-hydroxydehydrogenase in a nylon coil (1 m long, 1.0 mm id) *(51)*. The flow system involved two continuous streams: one containing the working solution and the other containing air into which a known volume of sample was intermittently injected. The nylon coil was placed inside an LKB Model 1250 luminometer in front of the window of the photomultiplier tube. The assay was highly specific for 7α-hydroxy bile acids, with a standard curve in the range 10–2500 pmol having satisfactory precision (CV 5–10%). Using the same flow system with only the bioluminescence enzymes, NADH measurements could be performed in the range 1–2500 pmol *(52)* with intra- and interassay precision of 5–10% at a rate of more than 20 samples/h with no carry-over. The continuous-flow determination of ethanol, glycerol, aldehyde, and 3α-, 7α-, and 12α-hydroxy bile acids has also been performed using bioluminescence enzymes and suitable dehydrogenases immobilized on separate nylon coils *(53)*. The dehydrogenase reactors were on the outside and the luciferase-oxidoreductase reactor on the inside of the luminometer (Fig. 5). The flow system involved four streams: the first contained the working bioluminescent solution, the second and third supplied the immobilized dehydrogenase with NAD solution and sodium pyrophosphate buffer (pH 9), and the fourth was air in which samples were injected. Assays at the picomolar level could be conducted with an acceptable precision. Recently, a bioluminescent continuous-flow sensor for l-alanine determination in serum and urine was described with an alanine dehydrogenase reactor separate from the bioluminescent reactor *(54)*. Urine samples had to be deproteinized and diluted before use and serum samples simply diluted after filtration. The bioluminescent response was linear from 50–1500 pmol. About 25–30 measurements could be performed per hour. After 2 mo of use at 50 samples/d, the residual activity of the enzymatic system was about 50% of its initial value. The enzyme coils were used at room temperature and stored at 4°C. Using a similar device with leucine dehydrogenase, branched-chain l-amino acids analyses in serum and urine have been performed in the range 20–2000 pmol *(55)*. The catalytic activity of

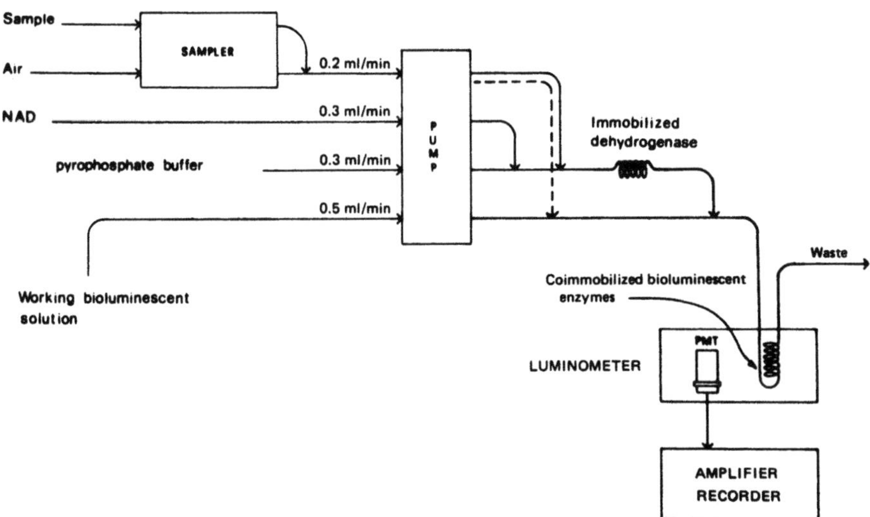

Fig. 5. Manifold for bioluminescent continuous-flow asssay *(53)*.

serum l-lactate dehydrogenase was determined by monitoring the NADH produced by this enzyme with the bioluminescent flow system using nylon-immobilized bacterial enzymes *(56)*. The dehydrogenase reaction of L-lactate with NAD took place in a flow-through mixing coil preceding the bioluminescent detector coil. At 37°C, a linear response was obtained from 1 to 5000 IU/L, and at 25°C, the linear range was from 3 to 2000 IU/L.

A flow-injection method for determination of NADH, NADPH, FMN, and $FMNH_2$ has been described by Nabi and Worsfold *(57)*. A partially purified extract from *V. harveyi* containing bacterial luciferase and oxidoreductase was used for immobilization on cyanogen bromide activated Sepharose 4B. The bound enzyme was placed in a glass coil in front of a photomultiplier tube. For NADPH measurements, the sample and decanal were simultaneously injected into separate carrier streams both containing FMN and DTT. FMN determination was performed by simultaneously injecting FMN and NADH into separate carrier streams containing decanal and DTT. For the $FMNH_2$ determination, the streams contained only DTT, and $FMNH_2$ and decanal were injected. For bioluminescent determination of each analyte, calibration data ranged between 1×10^{-9} and $1 \times 10^{-4}M$, with relative standard deviations over the calibration range of 1.2–4.1 %, 2.3–3.4%, and 1.5–6.6% for NADH, NADPH, and FMN or $FMNH_2$, respectively. Enzyme activity was maintained for several days at room temperature within the flow-injection analysis manifold, but the purity of the extracted enzymes was an important factor in determining the limit of detection by this technique. In order to perform ethanol assays, alcohol dehydrogenase was immobilized on CNBr-activated Sepharose 4B and incorporated upstream of the coimmobilized bacterial light-emitting system in the reaction coil *(58)*. The detection limit was $1 \times 10^{-6}M$ and the relative standard deviation 1.8% ($n = 5$) over the range 1×10^{-6}–$1 \times 10^{-1}M$. Alcohol dehydrogenase could also be assayed in the range 0.03–30 pmol.

4. Chemiluminescent Systems for Hydrogen Peroxide Detection

Several chemiluminescence reactions that require hydrogen peroxide are available as an alternative colorimetric determination of H_2O_2 with the advantage of lower detection limits.

4.1. Chemiluminescent Reactions

Luminol and related hydrazides *(59)* undergo chemiluminescence in aqueous solution (Fig. 6a). Light is also produced at alkaline conditions in the presence of an oxidizing agent and a catalyst *(60,61)*. Peroxidase (EC 1.11.1.7) and ferricyanide are the most common catalysts used, with peroxidase allowing the reaction to proceed at near-neutral pH *(62)*.

Lucigenin requires only alkaline hydrogen peroxide to produce light, and the oxidation proceeds without catalyst. The main reaction pathway *(63)* is shown in Fig. 6b, although reductants can produce light in alkaline conditions with lucigenin in the absence of H_2O_2 *(64,65)*.

In peroxyoxalate chemiluminescence, esters of oxalic acid react with hydrogen peroxide to produce a high-energy intermediate, 1,2-dioxetane, that transfers its energy to a fluorescer. The excited fluorescer then emits light by returning to the ground state (Fig. 6c). Oxalate esters, such as *bis*(2,4,6-trichlorophenyl)oxalate (TCPO) and *bis*(2,4,5-trichloro-6-pentoxycarbonyl)oxalate (CPPO), are often used with

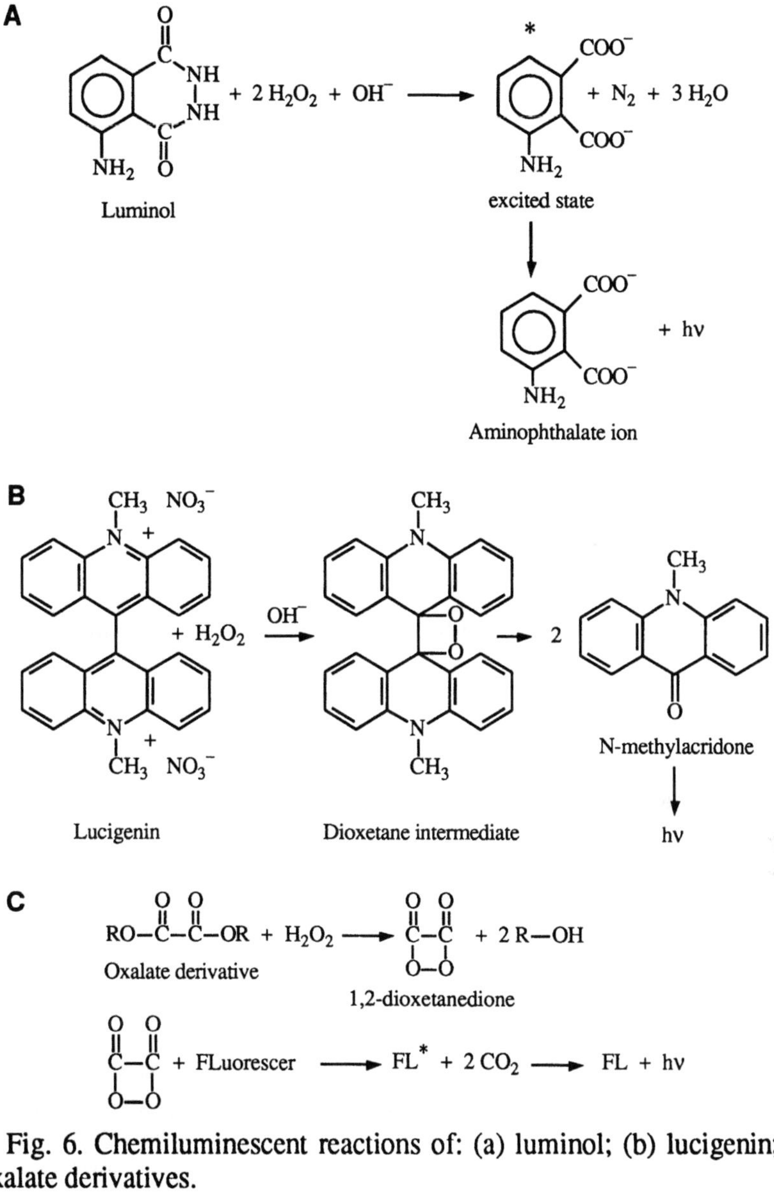

Fig. 6. Chemiluminescent reactions of: (a) luminol; (b) lucigenin; (c) oxalate derivatives.

perylene or diphenylanthracene as a fluorescer. The structural effects of the oxalate derivatives, the fluorescer effects, and the mechanism of the luminescent reaction have recently been reviewed (66,67). The peroxyoxalate chemiluminescence reaction can proceed within a greater pH range than luminol and lucigenin, but the solvent system

Table 1
Examples of Chemiluminescent Determination
of Enzymes and Metabolites with Soluble Reagents

Analyte	Chemiluminescent reaction	Reference
Glucose	luminol/Fe(CN)$_6^{3-}$	68,69
Glucose, glucose oxidase	luminol/peroxidase	70
Hydrogen peroxide	luminol/microperoxidase	71
Hydrogen peroxide	CPPO[a]/perylene	72
Hypoxanthine, xanthine, inosine, adenosine	luminol/peroxidase	73
NADH, lactate dehydrogenase	methylene blue + TCPO[b]/perylene	74
Organic reductants	lucigenin	75
Superoxide dismutase	luminol/peroxidase	76
	lucigenin	77
Uric acid, ascorbate	luminol/ferrihaem	78

[a] CPPO, bis (2,4,5-trichloro-6-pentoxycarbonyl) oxalate;

[b] TCPO, bis (2,4,6-trichlorophenyl) oxalate.

presents a major drawback since the efficient oxalate derivatives are soluble only in organic solvents, such as dimethoxy ethane, dioxane, or ethyl acetate.

4.2. Immobilized Reagents in Chemiluminescence Analysis

The design of heterogeneous-phase chemiluminescence assays has included immobilization of the catalyst for either a coupled reaction leading to H_2O_2, for the chemiluminescence reaction itself, or immobilization of the luminogenic substrate involved in the light-emission reaction.

4.2.1. Luminol Reaction

Blood glucose was measured using Sepharose-bound glucose oxidase and a mixed luminol–ferricyanide reagent to produce chemi-

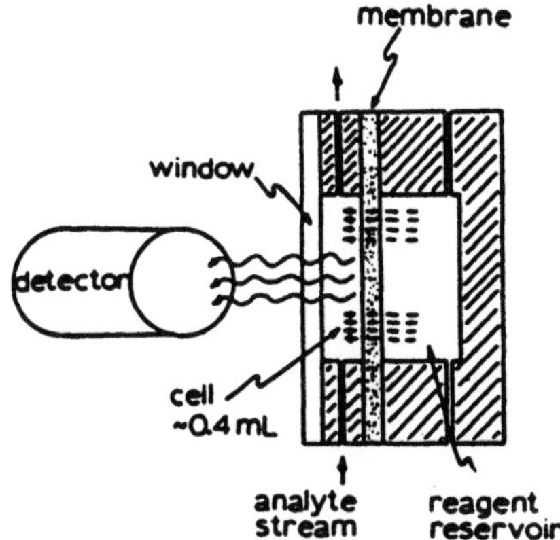

Fig. 7. Schematic diagram of the microporous-membrane flow cell. The dashed arrows indicate the direction in which the reagent flows through the membrane *(72)*.

luminescence proportional to glucose concentration *(68)*. Also, Murachi and coworkers developed a flow-injection analysis of glucose and uric acid in serum with immobilized glucose oxidase and urate oxidase, respectively, in combination with chemiluminescence detection *(69)*.

Another enzymatic assay by flow-injection analysis with detection by luminol–ferricyanide chemiluminescence has been described by Petersson et al. *(70)* for determination of glucose, creatinine, free cholesterol, and lactic acid using specific oxidases as H_2O_2-generating enzymes immobilized on CPG.

Horseradish peroxidase, immobilized on new polyamide membranes supplied in a preactivated form *(71)*, was placed in a cylindrical flat-bottomed cuvet in the sample chamber of a Berthold Biolumat LB 9500 luminometer. With $1 \times 10^{-4}M$ luminol, H_2O_2 could be detected between 1×10^{-8} and $1 \times 10^{-4}M$ by measuring the maximum intensity of the emitted light. Adding soluble cholesterol oxidase allowed chemiluminescent determination of cholesterol. Two modes of detection were used. In one, chemiluminescent signal integration during 150 s was linear with cholesterol in the range 2.5×10^{-8}–$2.5 \times 10^{-5}M$. In the second method, the measurement of maximum light intensity gave a linear calibration graph from 1×10^{-8}–$2.5 \times 10^{-4}M$, with a response time dependent on the concentration. This chemiluminescent technique has been adapted to free-cholesterol analysis in serum, with a coefficient variation of 8%. The operational stability of polyamide-bound peroxidase was fairly good, since 65% of the initial activity remained after 8 d, and the residual activity was still 50% of its initial value after 20 d.

A novel approach to chemiluminescence analysis in a flowing stream was accomplished by Nieman and coworkers using a microporous-membrane flow cell (Fig. 7) separating a reagent reservoir from an analyte stream *(72,73)*. The membrane flow cell created a stable pH gradient, ensuring a coupled reaction and the chemiluminescent reaction to proceed at their respective optimum pH. Using nonimmobilized enzymes, the chemiluminescence determination of glucose *(74)* and cholesterol *(75)* has been achieved.

4.2.2. Peroxyoxalate Reaction

TCPO chemiluminescence in the presence of perylene was used for determining hydrogen peroxide *(76)* and coupled to a flow-through column of immobilized aldehyde oxidase to determine formaldehyde and formic acid in natural waters *(77)*. Using immobilized oxalate oxidase and immobilized uricase (urate oxidase), oxalic acid and uric acid have been assayed by this technique in the range $2 \times 10^{-8} - 8 \times 10^{-4}M$ *(78)*, and L-amino acid oxidase on porous glass beads and the chemiluminescent system TCPO/9,10-diphenylanthracene have been used to determine L-amino acids *(79)*.

The peroxyoxalate chemiluminescence reaction also has been applied to the determination of acetylcholine (ACh) and choline (Ch) by high-performance liquid chromatography and enzyme generation of H_2O_2 *(80)*.

4.2.3. Immobilized Luminogenic Substrates

A chemiluminescent thiol derivative of luminol (mercaptoacetyl-3-aminophthalic acid hydrazide) has been bound on activated thiopropyl–Sepharose 6B, used to exchange with a thiol in a sample, and chemiluminescence of the displaced thiol derivative of luminol measured by reaction with hydrogen peroxide and ferricyanide *(81)*. Five thiol standards were tested (L-cysteine, reduced bovine insulin, reduced urease, dihydrolipoic acid, and dithiothreitol) with a detection limit of 5 pmol. Proteinases also have been assayed using immobilized isoluminol (6-amino-2,3-dihydro-1,4 phthalazinedione) coupled to peptides *(82)*.

Luminol has also been immobilized with glutaraldehyde on silanized particles of CPG and silica *(83)*. The immobilized chemiluminescent reagent was packed into a flow cell and used in a flow-injection system. Immobilized luminol was hydrolyzed from the support prior to or during reaction so that the light emission occurred in solution. Hydrogen peroxide determination was done from 1×10^{-6}–$1 \times 10^{-3}M$ with hemin as the catalyst.

5. Bio- and Chemiluminescence Analysis with Fiberoptic Sensors

Only a few fiberoptic-based biosensors involving biolumines-cence or chemiluminescence reactions have been reported so far. A new approach in luminescence measurements was proposed by Free-man and Seitz, who immobilized peroxidase in polyacrylamide gel on the end of a fiberoptic *(84)*. When the probe was immersed in a solu-tion of hydrogen peroxide and luminol, light was emitted and trans-mitted through the fiber to a photomultiplier tube. The detection limit was close to $1 \times 10^{-6}M$ peroxide and the time required for chemilumi-nescence to reach steady state was about 4 s at $1 \times 10^{-5}M$ H_2O_2.

Abdel-Latif and Guilbault *(85)* demonstrated the potential of mi-celles to enhance the chemiluminescent intensity of TCPO with hydrogen peroxide in the presence of perylene. With cetyltrimethyl-ammonium bromide as a surfactant, the problems of mixing of or-ganic and aqueous solvents were eliminated, and reproducibility was improved. Incorporation of fiberoptics with this chemiluminescent system allowed hydrogen peroxide to be measured in the range 8×10^{-4}–$8 \times 10^{-9}M$ with a coefficient of variation (5 measurements) of 0.3% at $1 \times 10^{-7}M$ H_2O_2. Using glucose oxidase immobilized on a polyamide membrane, glucose could be assayed, with the fiberoptic sensor, in the range 3×10^{-2}–$3 \times 10^{-6}M$ with a coefficient of variation of 0.3% for 5 measurements at $1 \times 10^{-4}M$.

Novel fiberoptic biosensors based on bioluminescence reactions have been reported by our laboratory *(86,87)* for NADH or ATP mea-surements. The fiberoptic sensor (Fig. 8) consists of an enzymatic

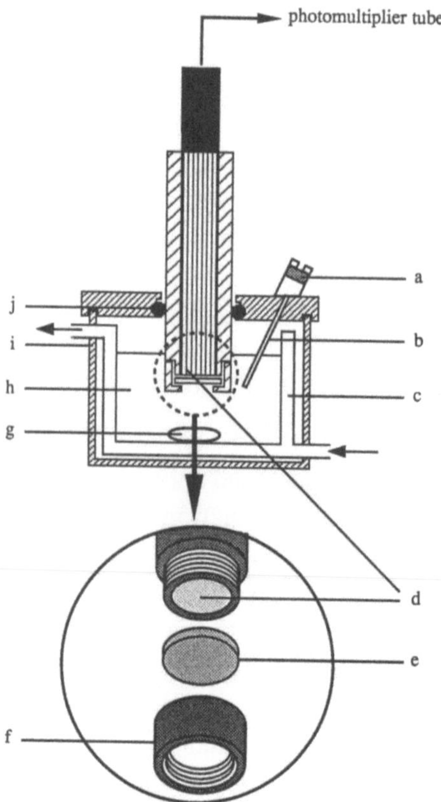

Fig. 8. Fiberoptic biosensor setup. (a) septum; (b) needle guide; (c) thermostatted reaction vessel; (d) fiber bundle; (e) enzymatic membrane; (f) screw cap; (g) stirring bar; (h) reaction medium; (i) black PVC jacket; (j) O ring (reprinted from Ref. *(86)*, p. 721, by courtesy of Marcel Dekker, Inc).

membrane maintained in close contact with the tip of a glass-fiber bundle by a screw cap. The other end of the bundle is placed close to the window of the photomultiplier tube of a Berthold LB 9500 luminometer. The biosensor is immersed in a stirred and thermostatted medium (4.5 mL final volume), and samples are introduced by injection through a septum. Luciferase and oxidoreductase from *V. fischeri* are coimmobilized on Pall® membranes and NADH measurements performed in the range 3×10^{-10}–$3 \times 10^{-6} M$. Using firefly luciferase from *Photinus pyralis*, the standard curve for ATP assays is linear from 2.8 $\times 10^{-10}$ to $1.6 \times 10^{-6} M$. The peroxidase-mediated chemiluminescence

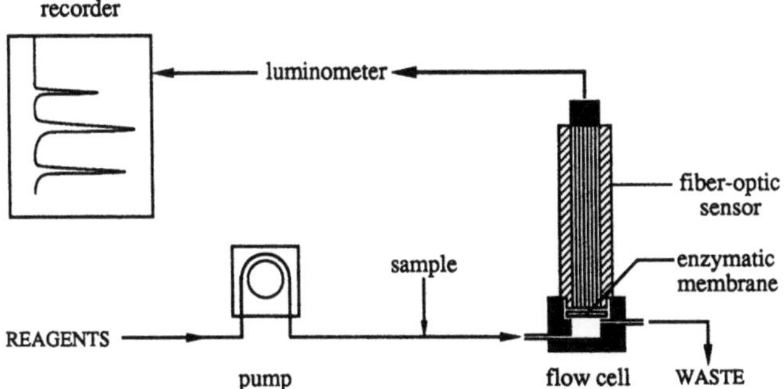

Fig. 9. Continuous-flow fiberoptic sensor for the bioluminescent determination of NADH.

reaction of luminol has been also used to perform H_2O_2 measurements with the fiberoptic sensor. For this purpose, horseradish peroxidase was immobilized on a preactivated nylon membrane and hydrogen peroxide measured from 2×10^{-8} to $2 \times 10^{-5} M$.

The performance of our bioluminescence-based fiberoptic biosensor was optimized using the bacterial bioluminescence system, giving special attention to the storage and operational stability of the immobilized bacterial enzymes from *V. fischeri (88)* and *V. harveyi (89)*. The potentialities of this NADH fiberoptic sensor have been extended by coimmobilizing suitable dehydrogenases for the microdetermination of ethanol, sorbitol, and oxaloacetate *(90)*. The alcohol dehydrogenase and sorbitol dehydrogenase reactions were directly coupled to the bioluminescent system in the presence of NAD^+ in excess, allowing the rate of NADH production to be measured by the peak light intensity. In contrast, the assay of oxaloacetate must be conducted in a sequential manner, because it involves the simultaneous consumption of NADH by both malate dehydrogenase and bacterial oxidoreductase (*see* Reaction [5]). Thus, NADH was injected first and, after a constant light emission was obtained, the oxaloacetate sample was injected, leading to a light decrease. Either the variation of light intensity or the rate of light decrease could be linearly related to the oxaloacetate concentration.

We have also reported a new approach in the design of biosensors by developing a multifunctional fiberoptic sensor for the microdetermination of ATP and NADH *(91)*. Firefly luciferase from *P. pyralis* was coimmobilized with the bacterial bioluminescent system from *V. fischeri* on a preactivated polyamide membrane, allowing alternate determination of the two analytes without changing the sensing element of the biosensor by changing the coreactants in the medium. When using such a bioactive matrix associated with the fiberoptic sensor, the alternate measurement of ATP or NADH could be performed with the same probe in the linear ranges of 1×10^{-10}–$1 \times 10^{-6}M$ and 1×10^{-9}–$1 \times 10^{-6}M$, respectively. For 10 measurements, the relative standard deviation was 6% at $9 \times 10^{-9}M$ ATP or $4 \times 10^{-8}M$ NADH.

Besides the batch analysis mentioned above, performed with our fiberoptic biosensor, a continuous-flow method for bioluminescent determination of NADH has been described *(92)*. The bioluminescent flow sensor (Fig. 9) consisted of a specially designed flow cell adapted to the fiberoptic probe. Twenty-five samples/h could be measured with no carry-over and no loss of activity after 150 assays over a 3-d period. The assay was linear from 2×10^{-12} to 1×10^{-9}mol NADH with a relative standard deviation of 3.4% at 1×10^{-10}mol.

6. Conclusion and Future Trends

Research efforts with biosensors have proceeded with various kinds of transducers, especially electrochemical probes. Expected improvements in biosensor design must include lower detection limits, avoidance of interference, miniaturized electronics, and matching for disposable sensing elements. Light measurement appears to be a key parameter for exploitation in the design of novel biosensors to suit these needs. Results already obtained in our group with fiberoptic sensors, as well as those reported by a few other laboratories involved in this field, appear very promising and may lead to a new generation of very attractive biosensors.

References

1. McElroy, W. D. (1947) The energy source for bioluminescence in an isolated system. *Proc. Nat. Acad. Sci. USA* **33,** 342–345.
2. DeLuca, M. (1976) Firefly luciferase, in *Advances in Enzymology*, vol. 44 (Meister, A., ed.), Wiley, New York, pp. 37–68.

3. DeLuca, M. and McElroy, W. D. (1978) Purification and properties of firefly luciferase, in *Methods in Enzymology*, vol. 57 (DeLuca, M. A., ed.), Academic, New York, pp. 3–15.

4. McElroy, W. D. and DeLuca, M. (1985) Firefly luminescence, in *Chemi- and Bioluminescence* (Burr, J. G., ed.), Marcel Dekker, New York, pp. 387–399.

5. Denburg, J. L., Lee, R. T., and McElroy, W. D. (1969) Substrate-binding properties of firefly luciferase. I. Luciferin-binding site. *Arch. Biochem. Biophys.* **134,** 381–394.

6. Lemasters, J. L. and Hackenbrook, C. R. (1977) Kinetics of product inhibition during firefly luciferase luminescence. *Biochemistry* **16,** 445–447.

7. Green, A. A. and McElroy, W. D. (1956) Crystalline firefly luciferase. *Biochim. Biophys. Acta* **20,** 170–176.

8. White, E. H., McCapra, F., Field, G. F., and McElroy, W. D. (1961) The structure and synthesis of firefly luciferin. *J. Am. Chem. Soc.* **83,** 2402,2403.

9. Leach, R. L. and Webster, J. J. (1986) Commercially available firefly luciferase reagents, in *Methods in Enzymology*, vol. 133 (DeLuca, M. A. and McElroy, W. D., eds.), Academic, New York, pp. 51–70.

10. Hastings, J. W., Baldwin, O. T., and Nicoli, M. Z. (1978) Bacterial luciferase: Assay, purification, and properties, in *Methods in Enzymology*, vol. 57, (DeLuca, M. A., ed.), Academic, New York, pp. 135–152.

11. Nealson, K. H. (1978) Isolation, identification, and manipulation of luminous bacteria, in *Methods in Enzymology*, vol. 57 (DeLuca, M. A., ed.), Academic, New York, pp. 153–166.

12. Baumann, P., Furniss, A. L., and Lee, J. V. (1984) Genus I. *Vibrio.*, in *Bergey's Manual of Systematic Bacteriology*, vol. 1 (Krieg, N. R., ed.), William & Willkins, Baltimore, pp. 518–538.

13. Baumann, P. and Baumann, L. (1984) Genus II. *Photobacterium.*, in *Bergey's Manual of Systematic Bacteriology*, vol. 1 (Krieg, N. R., ed.), Williams & Willkins, Baltimore, pp. 539–545.

14. Puget, K. and Michelson, A. M. (1972) Studies in bioluminescence. VII. Bacterial NADH: Flavin mononucleotide oxidoreductase. *Biochimie* **54,** 1197–1204.

15. Duane, W. and Hastings, J. W. (1975) Flavin mononucleotide reductase in luminous bacteria. *Mol. Cell. Biochem.* **6,** 53–64.

16. Gerlo, E. and Charlier, J. (1975) Identification of NADH-specific and NADPH-specific FMN reductases in *Beneckea harveyi. Eur. J. Biochem.* **57,** 461–467.

17. Erlanger, B. F., Isambert, M. F., and Michelson, A.M. (1970) Insoluble bacterial luciferases: A new approach to some problems in bioluminescence. *Biochem. Biophys. Res. Commun.* **40,** 70–76.

18. Lee, Y., Jablonski I., and DeLuca M. (1977) Immobilization of firefly luciferase on glass rods. Properties of immobilized enzyme. *Anal. Biochem.* **80,** 496–501.

19. Wienhausen, G. K., Kricka, L. J., Hinkley, J. E., and DeLuca, M. (1982) Properties of bacterial luciferase/NADH:FMN oxidoreductase and firefly luciferase immobilized onto Sepharose. *Applied Biochem. Biotechnol.* **7**, 463–473.

20. Kricka, L. J., Wienhausen, G. K., Hinkley, J. E., and DeLuca, M. (1983) Automated bioluminescent assays for NADH, glucose–6-phosphate, primary bile acids and ATP. *Anal. Biochem.* **129**, 392–397.

21. Kricka, L. J. and DeLuca, M. (1982) Effect of solvents on the catalytic activity of firefly luciferase. *Arch. Biochem. Biophys.* **217**, 674–681.

22. Brovko, L. Yu., Ugarova, N. N., Vasileva, T. E., Dombrovski, V. A., and Berezin, I.V. (1978) Use of immobilized firefly luciferase for quantitative determination of ATP and enzymes that synthesize and destroy ATP. *Biochemistry USSR* **43**, 633–639.

23. Ugarova, N. N., Brovko, L. Yu., and Berezin, I. V. (1980) Immobilized firefly luciferase and its use in analysis. *Anal. Lett.* **13**, 881–892.

24. Brovko, L. Yu and Ugarova N. N. (1980) Kinetics and mechanism of the inactivation and reactivation of immobilized luciferase of fireflies *Luciola mingrelica* and the role of sulfhydryl groups in these processes. *Biochemistry USSR* **45**, 607–613.

25. Brovko, L. Yu., Kost, N. V., and Ugarova, N. N. (1980) Immobilized luciferase from the fireflies *Luciola mingrelica*. Change in the pH dependence of the catalytic activity and stability of the enzyme after immobilization on various polysaccharide carriers. *Biochemistry USSR* **45**, 1199–1204.

26. Ugarova, N. N., Brovko, L. Y., and Kost N. V. (1982) Immobilization of luciferase from the firefly *Luciola mingrelica*. Catalytic properties and stability of the immobilized enzyme. *Enzyme Microbial Technol.* **4**, 224–228.

27. Ugarova, N. N., Brovko, L. Y., and Beliaieva, E. I. (1983) Immobilization of luciferase from the firefly *Luciola mingrelica*: Catalytic properties and thermostability of the enzyme immobilized on cellulose films. *Enzyme Microbial Technol.* **5**, 60–64.

28. Ivanova, L. V., Brovko, L. Yu., Shekhovtsova, T. N., Ugarova, N. N., and Dolmanova, I. F. (1986) Bioluminescent method of determining the creatine phosphokinase activity using immobilized firefly luciferase. *J. Anal. Chem. USSR* **41**, 593–599.

29. Ugarova, N. N., Brovko, L. Yu., Ivanova, L. V., Shekhovtsova, T. N., and Dolmanova, I. F. (1986) Bioluminescent assay of creatine kinase activity using immobilized firefly extract. *Anal. Biochem.* **158**, 1–5.

30. Carrea, G., Bovara, R., Mazzola, G., Girotti, S., Roda, A., and Ghini,S. (1986) Bioluminescent continuous-flow assay of adenosine 5'-triphosphate using firefly luciferase immobilized on nylon tubes. *Anal. Chem.* **58**, 331–333.

31. Carrea, G., Bovara, R., Girotti, S., Ferri, E., Ghini,S., and Roda, A. (1989) Continuous-flow bioluminescent determination of ATP in platelets using firefly luciferase immobilized on epoxy methacrylate. *J. Biolum Chemilum* **3**, 7–11.

32. Worsfold, P. J. and Nabi, A. (1986) Bioluminescent assays with immobilized firefly luciferase based on flow injection analysis. *Anal. Chim. Acta* **179**, 307–313.

33. Blum, L. J., Coulet, P. R., and Gautheron, D. C. (1985) Collagen strip with immobilized luciferase for ATP bioluminescent determination. *Biotechnol. Bioeng.* **27,** 232–237.

34. Travis, J. and McElroy, W. D. (1966) Isolation and sequence of an essential sulfhydryl peptide at the active site of firefly luciferase. *Biochemistry* **5,** 2170–2176.

35. Coulet, P. R., Julliard, J. H., and Gautheron, D. C. (1974) A mild method of general use for covalent coupling of enzymes to chemically activated collagen films. *Biotechnol. Bioeng.* **16,** 1055–1068.

36. Blum, L. J. and Coulet, P. R. (1986) Atypical kinetics of immobilized firefly luciferase. *Biotechnol. Bioeng.* **28,** 1154–1158.

37. Jablonski, E. and DeLuca, M. (1976) Immobilization of bacterial luciferase and FMN reductase on glass rods. *Proc. Natl. Acad. Sci. USA* **73,** 3848–3851.

38. Haggerty, C., Jablonski, E., Stav, L., and DeLuca, M. (1978) Continuous monitoring of reactions that produce NADH and NADPH using immobilized luciferase and oxidoreductases from *Beneckea harveyi. Anal. Biochem.* **88,** 162–173.

39. Jablonski, E. and DeLuca, M. (1979) Properties and uses of immobilized light-emitting enzyme systems from *Beneckea harveyi. Clin. Chem.* **25,** 1622–1627.

40. Ford, J. and DeLuca, M. (1981) A new assay for picomole levels of androsterone and testosterone using co-immobilized luciferase, oxidoreductase and steroid dehydrogenase. *Anal. Biochem.* **110,** 43–48.

41. Wienhausen, G. and DeLuca, M. (1982) Bioluminescent assays of picomole levels of various metabolites using immobilized enzymes. *Anal. Biochem.* **127,** 380–388.

42. Roda, A., Kricka, L. J., DeLuca, M., and Hofmann, A. F. (1982) Bioluminescence measurement of primary bile acids using immobilized 7α-hydroxysteroid dehydrogenase: Application to serum bile acids. *J. Lipid Res.* **23,** 1354–1361.

43. Schoelmerich, J., Hinkley, J. E., MacDonald, I. A., Hofmann, A. F., and DeLuca, M. (1983) A bioluminescent assay for 12-a-hydroxy bile acids using immobilized enzymes. *Anal. Biochem.* **133,** 244–250.

44. Schoelmerich, J., van Berge Henegouwen, G. P., Hofmann, A. F., and DeLuca, M. (1984) A bioluminescence assay for total 3α-hydroxy bile acids in serum using immobilized enzymes. *Clin. Chim. Acta* **137,** 21–32.

45. Rossi, S. S., Clayton, L. M., and Hofmann, A. F. (1986) Determination of chenodiol bioequivalence using an immobilized multi-enzyme bioluminescence technique. *J. Pharm. Sci.* **75,** 288–290.

46. Rodriguez, O. and Guilbault, G. G. (1981) Immobilized bacterial luciferase for microscale analysis of creatine kinase activity. *Enzyme Microb. Technol.* **3,** 69–72.

47. Ugarova, N. N., Lebedeva, O. V., and Frumkina, I. G. (1988) Bioluminescent microassay of various metabolites using bacterial luciferase co-immobilized with multienzyme systems. *Anal. Biochem.* **173,** 221–227.

48. Blum, L. J. and Coulet, P. R. (1984) Bioluminescent determination of reduced nicotinamide adenine dinucleotide with immobilized bacterial lucif-

erase and flavin mononucleotide oxidoreductase on collagen film. *Anal. Chim. Acta* **161,** 355–358.

49. Kurkijärvi, K., Raunio, R., and Korpela, T. (1982) Packed-bed reactor of immobilized bacterial bioluminescence enzymes: A potential high-sensitivity detector for automated analyzers. *Anal. Biochem.* **125,** 415–419.

50. Vellom, D. C. and Kricka, L. J. (1986) Continuous-flow bioluminescent assays employing Sepharose-immobilized enzymes, in *Methods in Enzymology,* vol. 133 (DeLuca, M. A. and McElroy, W. D., eds.), Academic, New York, pp. 229–237.

51. Roda, A., Girotti, S., Ghini, S., Grigolo, B., Carrea, G., and Bovara, R. (1984) Continuous–flow determination of primary bile acids, by bioluminescence, with use of nylon-immobilized bacterial enzymes. *Clin. Chem.* **30,** 206–210.

52. Girotti, S., Roda, A., Ghini, S., Grigolo, B., Carrea, G., and Bovara, R. (1984) Continuous flow analyses of NADH using bacterial bioluminescent enzymes immobilized on nylon. *Anal. Lett.* **17,** 1–12.

53. Roda, A., Girotti, S., and Carrea, G. (1986) Flow systems utilizing nylon-immobilized enzymes, in *Methods in Enzymology,* vol. 133 (DeLuca, M. A. and McElroy, W. D., eds.), Academic, New York, pp. 238–248.

54. Girotti, S., Roda, A., Piazzi, S., Carrea, G., Piacentini, A. L., Angelloti, M. A., Bovara, R., and Ghini, S. (1987) Bioluminescent flow sensors: L-Alanine determination in serum and urine. *Anal. Lett.* **20,** 1315–1330.

55. Girotti, S., Roda, A., Angelloti, M. A., Ghini, S., Carrea, G., Bovara, R., Piazzi, S., and Merighi, R. (1988) Bioluminescence flow system for determination of branched-chain L-amino acids in serum and urine. *Anal. Chim. Acta* **205,** 229–237.

56. Girotti, S., Bassoli, C., Cascione, M. L., Ghini, S., Carrea, G., Bovara, R., Roda, A., Motta, R., and Petilino, R. (1989) Bioluminescent flow sensors: L-Lactate dehydrogenase activity determination in serum. *J. Biolum. Chemilum.* **3,** 41–45.

57. Nabi, A. and Worsfold, P. J. (1986) Bioluminescence assays with immobilised bacterial luciferase using flow injection analysis. *Analyst* **111,** 1321–1324.

58. Nabi, A. and Worsfold, P. J. (1987) Flow injection procedures for the determination of ethanol and alcohol dehydrogenase using co-immobilised bacterial luciferase and oxidoreductase. *Analyst* **112,** 531–533.

59. Roswell, D. H. and White, E. H. (1978) The chemiluminescence of luminol and related hydrazides, in *Methods in Enzymology,* vol. 57 (DeLuca, M. A., ed.), Academic, New York, pp. 409-423.

60. White, E. H., Zafiriou, O., Kägi, H. H., and Hill, J. H. M. (1964) Chemiluminescence of luminol: the chemical reaction. *J. Am. Chem. Soc.* **86,** 940–941.

61. White, E. H. and Bursey, M. M. (1964) Chemiluminescence of luminol and related hydrazides. The light emission step. *J. Am. Chem. Soc.* **86,** 941,942.

62. Cormier, M. J. and Prichard, P. M. (1968) An investigation of the luminescent peroxidation of luminol by stopped flow techniques. *J. Biol. Chem.* **243,** 4706–4714.

63. Maskiewicz, R., Sogah, D., and Bruice, T.C. (1979) Chemiluminescent reaction of lucigenin. I. Reactions of lucigenin with H_2O_2. *J. Am. Chem. Soc.* **101,** 5347–5354.

64. Isacsson, U. and Wettermark, G. (1974) Chemiluminescence in analytical chemistry. *Anal. Chim. Acta* **68,** 339–362.

65. Totter, J. R. (1975) Light production in alkaline mixtures of reducing agents and dimethylbiacridylium nitrate. *Photochem. Photobiol.* **22,** 203–211.

66. Mohan, A. G. (1985) Peroxyoxalate chemiluminescence, in *Chemi- and Bioluminescence* (Burr, J. G., ed.), Marcel Dekker, New York, pp. 245–258.

67. Imai, K., Miyaguchi, K., and Honda, K. (1985) High-performance liquid chromatography—chemiluminescence reaction detection system of fluorescent compounds using TCPO and hydrogen peroxide, in *Bioluminescence and Chemiluminescence: Instruments and Applications*, vol. 2 (Van Dyke, K., ed.), CRC, Boca Raton, pp. 65–75.

68. Bostick, D. T. and Hercules, D. M. (1975) Quantitative determination of blood glucose using enzyme induced chemiluminescence of luminol. *Anal. Chem.* **47,** 447–452.

69. Tabata, M., Fukunaga, C., Ohyabu, M., and Murachi, T. (1984) Highly sensitive flow injection analysis of glucose and uric acid in serum using an immobilized enzyme column and chemiluminescence. *J. Appl. Biochem.* **6,** 251–258.

70. Petersson, B. A., Hansen, E. H., and Ruzicka, J. (1986) Enzymatic assay by flow-injection analysis with detection by chemiluminescence: Determination of glucose, creatinine, free cholesterol and lactic acid using an integrated FIA microconduit. *Anal. Lett.* **19,** 649–665.

71. Blum, L. J., Plaza, J. M., and Coulet, P. R. (1987) Chemiluminescent analyte microdetection based on the luminol H_2O_2 reaction using peroxidase immobilized on new synthetic membranes. *Anal. Lett.* **20,** 317–326.

72. Nau, V. and Nieman, T. A. (1979) Application of microporous membranes to chemiluminescence analysis. *Anal. Chem.* **51,** 424–428.

73. Pilosof, D. and Nieman, T. A. (1980) Localization of light emission in microporous membrane chemiluminescence cells. *Anal. Chem.* **52,** 662–666.

74. Pilosof, D. and Nieman, T. A. (1982) Microporous membrane flow-cell with nonimmobilized enzyme for chemiluminescent determination of glucose. *Anal. Chem.* **54,** 1698–1701.

75. Malavolti, N. L., Pilosof, D., and Nieman, T. A. (1985) Determination of cholesterol with a microporous membrane chemiluminescence cell with cholesterol oxidase in solution. *Anal. Chim. Acta* **170,** 199–207.

76. Williams D. C. III, Huff, G. F., and Seitz, W. R. (1976) Evaluation of peroxyoxalate chemiluminescence for determination of enzyme generated peroxide. *Anal. Chem.* **48,** 1003–1006.

77. Rigin, V. I. (1981) Determination of formaldehyde and formic acid in natural water by applying an immobilized enzyme and a chemiluminescence finish. *J. Anal. Chem. USSR* **36,** 1111–1115.

78. Rigin, V. I. (1982) Chemiluminescence determination of microamounts of oxalate and urate by means of flow-through columns containing immobilized enzymes. *J. Anal. Chem. USSR* **37,** 1302–1306.

79. Rigin, V. I. (1983) Determination of microamounts of L-amino acids by enzymatic oxidation reaction. *J. Anal. Chem. USSR* **38,** 1328–1330.

80. Honda, K., Miyaguchi, K., Nishino, H., Tanaka, H., Yao, T., and Imai, K. (1986) High-performance liquid chromatography followed by peroxyoxalate chemiluminescence detection of acetylcholine and choline utilizing immobilized enzymes. *Anal. Biochem.* **153,** 50–53.

81. Lippman, R. D. (1980) Sensitive solid-phase chemiluminescence microassay of thiols. *Anal. Chim. Acta* **116,** 181–184.

82. Branchini, B. R., Salituro, F. G., Hermes, J. D., and Post, N. J. (1980) Highly sensitive assays for proteinases using immobilized luminogenic substrates. *Biochem. Biophys. Res. Commun.* **97,** 334–339.

83. Hool, K. and Nieman, T. A. (1987) Chemiluminescence analysis in flowing streams with luminol immobilized on silica and controlled-pore glass. *Anal. Chem.* **59,** 869–872.

84. Freeman, T. M. and Seitz, W. R. (1978) Chemiluminescence fiber optic probe for hydrogen peroxide based on the luminol reaction. *Anal. Chem.* **50,** 1242–1246.

85. Abdel-Latif, M. S. and Guilbault, G. G. (1988) Fiber-optic sensor for the determination of glucose using micellar enhanced chemiluminescence of the peroxyoxalate reaction. *Anal. Chem.* **60,** 2671–2674.

86. Blum, L. J., Gautier, S. M., and Coulet, P. R. (1988) Luminescence fiber-optic biosensor. *Anal. Lett.* **21,** 717–726.

87. Blum, L. J., Gautier, S. M., and Coulet, P. R. (1989) Design of luminescence photobiosensors. *J. Biolum. Chemilum.* **4,** 543-550.

88. Gautier, S. M., Blum, L. J., and Coulet, P. R. (1989) Fibre-optic sensor with co-immobilised bacterial bioluminescence enzymes. *Biosensors* **4,** 181–194.

89. Blum, L. J., Gautier, S. M., and Coulet, P. R. (1989) Highly stable bioluminescence-based fiber-optic sensor using immobilized enzymes from *Vibrio harveyi*. *Anal. Lett.* **22,** 2211–2222.

90. Gautier, S. M., Blum, L. J., and Coulet, P. R. (1990) Fibre-optic biosensor based on luminescence and immobilized enzymes: Microdetermination of sorbitol, ethanol and oxaloacetate. *J. Biolum. Chemilum.* **5,** 57–63.

91. Gautier, S. M., Blum, L. J., and Coulet, P. R. (1990) Alternate determination of ATP and NADH with a single bioluminescence-based fiber-optic sensor, in *Sensors and Actuators*, B1, 580–584.

92. Blum, L. J., Gautier, S. M., and Coulet, P. R. (1989) Continuous flow bioluminescent assay of NADH using a fiber-optic sensor. *Anal. Chim. Acta* **226,** 331–336.

In Vivo Applications of Fiberoptic Chemical Sensors

Amos Gottlieb, Skip Divers,
and Henry K. Hui

1. Introduction

As stated at the beginning of this volume, the term "biosensor" refers to sensors that use biomolecules in the molecular recognition or transduction processes. Although there have been many proposals to use fiberoptic biosensors in vivo, almost all the work to date has been in vitro. In the more general class of fiberoptic chemical sensors, in vivo applications have progressed further. Intravascular fiberoptic blood-gas chemical sensors have been developed and are currently undergoing clinical evaluation.

Frequent measurement of oxygen partial pressure (P_{O_2}), carbon dioxide partial pressure (P_{CO_2}), and pH of arterial blood of critically ill patients in operating rooms and intensive care units is essential to clinical diagnosis and management. Normally these data are collected by manually withdrawing a sample of blood and analyzing it with a benchtop blood-gas analyzer. This instrument utilizes a series of electrodes, each of which measures the concentration of a single parameter. It has been suggested that patient management could be improved if blood gases could be monitored continuously. During the last decade, numerous efforts have been made to develop fiberoptic sensors capable of continous in vivo blood-gas monitoring.

Progress in the development of fiberoptic chemical sensors for continuous blood-gas monitoring and other in vivo applications is discussed in this chapter. Critical issues common to the development of both in vivo fiberoptic biosensors and chemical sensors are presented first, followed by a discussion of the progress that has been made in developing specific sensors for oxygen, carbon dioxide, and pH. Design considerations that affect the development of in vivo fiberoptic chemical sensors should provide insight into those to be faced in the development of in vivo fiberoptic biosensors.

2. Critical Issues and Design Criteria

In vivo sensor development requires consideration of factors related to the inability to control or prepare the sample and the need to minimize the impact of the sensor's presence on the host organism. These factors include biocompatibility, sensor materials design, selectivity and sensitivity, optical compensation schemes, and calibration.

2.1. Biocompatibility

The problem of biocompatibility has been succinctly stated *(1)*: "There is a dual aspect to biocompatibility—the sensor must not adversely affect the body and the body must not adversely affect the sensor." Sensors for intravascular use must be engineered to be sterile, nontoxic, nonpyrogenic, and as blood-compatible as possible.

2.1.1. Blood Compatibility

Intravascular fiberoptic sensors need to be nonthrombogenic and resistant to platelet and protein deposition. Although optical sensors are less sensitive to fouling than electrochemical sensors, their function may be adversely affected by the presence of biological deposits. Deposition of protein on a pH sensor could buffer the local pH and thus affect the accuracy of measurement *(2)*. Although it has been reported that thrombus buildup can affect the time-response of fiberoptic blood-gas sensors but not their equilibrium accuracy *(3)*, this need not always be true *(2)*. Deposition of metabolically active material can affect sensor accuracy by changing the local environment but, again, exceptions have been noted *(4)*.

The thrombogenicity of a sensor depends on its shape, chemical and physical surface properties, method of manufacture, and placement in the vessel. Several reviews *(5–7)* discuss the development and preparation of materials having improved blood compatibility. These materials include those based on immobilized heparin or antiplatelet agents and those that reduce the thermodynamic driving potential for platelet and protein deposition. Other approaches *(8,9)* use endogenous materials, such as albumin, to provide a protective coating.

Anticoagulants are generally used in conjunction with sensors deployed in blood *(4,10)*. Typically, the sensors are introduced via a catheter, and a slow heparin flush is maintained until they are removed. The Cardiovascular Devices (CDI, 3M Health Care) System 1000 is the only integrated intravascular blood-gas sensor whose construction has been discussed in the open literature *(11)*. This device apparently achieves blood compatibility by synergistic interaction of the heparin flush, a smooth probe surface, an overcoat, and heparin covalently attached to all exposed surfaces.

2.1.2. Sterilization

Devices for temporary or permanent implantation are normally sterilized by heat, ionizing radiation, or ethylene oxide, but all of these techniques have potential drawbacks. The sensors in the CDI System 1000 intravascular blood-gas sensor assembly are sterilized by single-use steam sterilization *(12)*. Hewlett-Packard has reported a sensor that can be sterilized by gamma radiation *(13)*.

Several potential problems can be encountered when sensors are subject to heat sterilization. Markedly different thermal coefficients of expansion can lead to various parts moving with respect to one another. For instance, optical fibers potted into a standard fiberoptic connector with epoxy may protrude from the connector following heat sterilization, a process referred to as "pistoning," resulting in the face of the fiber no longer being aligned with the end of the connector. This may have serious consequences on the overall performance of the sensor. Other problems arise if sensor materials or indicators are thermally or hydrolytically unstable. Heat sterilization is also incompatible with plastic fibers having glass transition temperatures near or below the autoclave temperature.

Radiation sterilization is convenient and has the advantage of not subjecting the materials to the temperature changes that accompany heat sterilization. Unfortunately, radiation sterilization is not compatible with many glass fibers *(14)* and fluorescent indicators *(2)*. Glass fibers may become colored upon exposure to gamma radiation, and, unless the resulting color disappears as the material relaxes, the performance of the sensor may be radically altered.

Chemical sterilization with ethylene oxide has the potential disadvantage of an undesirable chemical reaction occurring. In addition, if the sensor is "wet-packed," some ethylene oxide may be hydrolyzed, with the residue affecting the selectivity and sensitivity of the device *(2)*.

2.2. Sensor Design

2.2.1. Mechanical Design

Intravascular blood-gas sensors are normally placed in a standard 20-gage catheter, with allowance for a space between the sensor and the wall of the catheter. The resulting lumen should be large enough to permit the withdrawal of blood samples, introduction of a continuous heparin flush, and measurement of blood-pressure waveform. Probe packages with a total diameter of 620 μm are reported to meet this requirement *(11,15)*. In most situations, an integrated package containing sensors for all three blood-gases is preferred.

Two geometries have been considered for fiberoptic blood-gas sensors. In one case, the matrix containing the indicator is attached to the end of an optical fiber; in the other, the indicator is embedded in the cladding (Fig. 1). Although the former approach is widely used, the latter approach has been the subject of theoretical *(16)* and experimental *(17–19)* work and is favored for evanescent wave sensors.

Most fiberoptic chemical sensors employ either a single- or double-fiber configuration, as described elsewhere in this volume. In the double-fiber configuration, separate optical fibers are used to carry light to and from the sensor tip. In a single-fiber configuration, all the

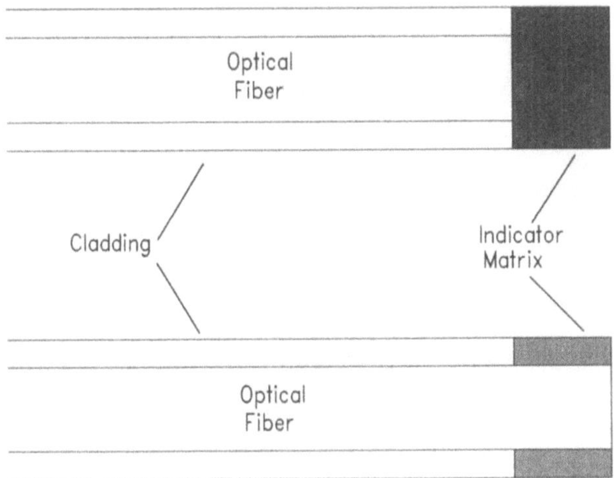

Fig. 1. Two geometries for fiberoptic blood-gas sensors.

light travels in one fiber. Although single-fiber sensors are typically of smaller diameter than the two-fiber counterpart, absorbance-based single-fiber sensors may require more complicated monitoring instrumentation. An experimental comparison of single- and double-fiber configurations for fluorescence-based sensors has recently been reported *(20)*.

In blood-gas sensing, it is desirable to have a single sensor package capable of measuring temperature, pH, P_{CO_2}, and P_{O_2}. The simplest approach is one in which each analyte is measured with a separate optical-fiber sensor and the separate sensors are integrated into a single unit. The CDI sensor package of this type has an overall diameter of 620 μm (Fig. 2) *(11)*. This design requires the monitor to have a separate optical train for each analyte.

Another approach is to incorporate sensing chemistry for two or more analytes onto a single-fiber. Proposals have appeared for sensors measuring CO_2 and O_2 *(21)*, but to date these sensors have been demonstrated only in vitro. An integrated sensor system with multiple sensing chemistries on a single fiber has been described in the patent literature *(15)*. Both approaches present substantial challenges in terms of large-volume production.

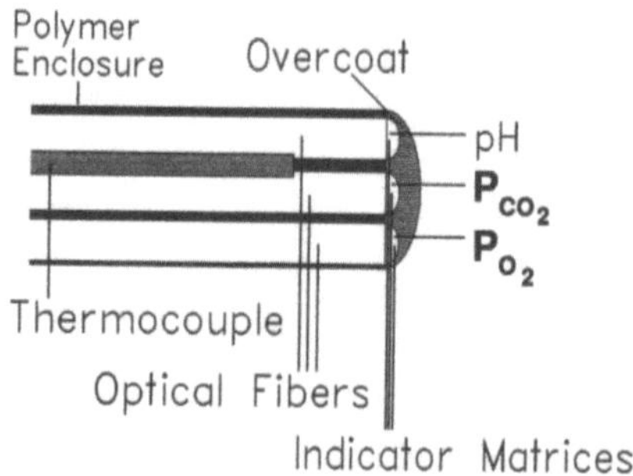

Fig. 2. Schematic diagram of the CDI integrated fiberoptic blood-gas sensor.

2.2.2. Temperature Compensation

At least two factors make temperature compensation of fiberoptic blood-gas sensors necessary. First, almost all sensor chemistries involve equilibria that are temperature-dependent. Second, and more important, the pH, P_{CO_2}, and P_{O_2} of biological fluids, such as blood, are strongly temperature-dependent *(22,23)*. Blood-gas analyzers normally measure samples at a fixed temperature of 37°C, and corrections must be made for comparisons with data at other temperatures. Correction nomograms *(24)* and algorithms *(25)* are known, but it has recently been suggested that further refinements are necessary *(25)*.

2.3. Selectivity and Sensitivity

Although fiberoptic sensors for in vivo use cannot rely on general analytical chemistry techniques, selectivity can be enhanced by the use of membranes. One design for a CO_2 sensor *(26)*, modeled after the Severinghouse electrode *(27)*, involves pH measurement of an indicator solution isolated from the sample by a gas-permeable membrane. Carbon dioxide can pass through the membrane, but hydronium ions cannot. Thus, the internal pH indicator is immune to changes in sample pH, but sensitive to changes in sample P_{CO_2}. The selectivity that can be achieved with a membrane must be carefully evaluated, since any gas capable of entering into acid-base chemistry can affect the output from the sensor. Thus, a CO_2 sensor might be sensitive to gases such as ammonia, and some ammonia sensors *(28)* might be sensitive to variations in P_{CO_2}.

These problems notwithstanding, membranes are widely used in fiberoptic sensors, and their judicious application has aided the construction of sophisticated sensors, including one capable of measuring (in vitro) both oxygen and halothane (a general anesthetic agent) *(29)*.

2.4. Optical Compensation Schemes

Many fiberoptic chemical sensors are intensity-based, i.e., the intensity of light from the sensor is related to the concentration of analyte. Therefore it is important that sensors be designed so that artifacts from fiber bending or indicator photodegradation do not compromise the integrity of the measurement. This is commonly achieved by using a ratio approach, in which the concentration of analyte is related to the ratio of light intensities at two different wavelengths *(30,31)*. As long as both wavelengths are similarly affected by the artifacts and dissimilarly affected by analyte concentration, the resulting measurement is immune to the presence of artifacts.

In certain cases, immunity to artifacts can be achieved by measuring the time-dependent intensity of light, rather than its steady-state intensity. In quenching-based oxygen sensors (*see* Section 3.1.), measurement of the fluorescence lifetime rather than the steady-state fluorescence intensity can be of use *(32,33)*. Similarly, balanced-bridge arrangements have been proposed to achieve immunity with simple intensity-based sensors *(34)*. In any case, it is imperative that sensors used for in vivo sensing incorporate a suitable optical compensation scheme.

2.5. Calibration

Since the in vivo environment is one that cannot be duplicated in the laboratory, it is unclear what standards and methods should be used to calibrate in vivo sensors and, more important, what references can be used to evaluate the accuracy of these probes. Yafuso and coworkers *(3)* have suggested that the accuracy of P_{O_2} and P_{CO_2} measurements can be verified by tonometry of blood with primary gas standards, but that pH must be determined relative to standard pH buffer solutions. They, as well as others *(17)*, have pointed out that blood-gas values of identical samples may differ substantially when determined with different blood-gas analyzers. It has become common practice to evaluate intravascular blood-gas sensors by comparison to benchtop blood-gas analyzers *(4,10,17)*. Although this may seem arbitrary from a scientific standpoint, it is pragmatic in light of the fact that current medical diagnosis and treatment are based on data collected in this fashion. The method of Bland and Altman is the most suitable for comparison of data from in vivo fiberoptic sensors with that from blood-gas analyzers *(35)*.

3. Oxygen Sensors

A number of chemical or physical phenomena can be monitored optically to determine the amount of oxygen in solution. Among these are luminescence quenching by oxygen *(36,37)*, oxygen-induced chemiluminescence *(38)*, oxygen-mediated redox equilibria, and reversible complexation of oxygen to organometallics *(39)*. Although there have been attempts to develop fiberoptic oxygen sensors based on each of these phenomena, the most successful to date has been based on luminescence quenching. Sensors based on this phenomenon have been successfully used in both animal and human trials, and the discussion here is limited to this approach.

3.1. Principle of Operation

The reemission of absorbed light by a molecule is referred to as fluorescence or phosphorescence, depending on the details of the photophysics involved. The term "luminescence" includes both fluorescence and phosphorescence, as discussed earlier in this volume.

The intensity of light emitted from a solution containing an oxygen-quenchable luminescent material is inversely related to the concentration of oxygen in the solution. The theoretical relationship *(36,40–42)* between luminescence intensity and oxygen concentration is

$$L_0/L = K\,P_{O_2} + 1 \tag{1}$$

where L_0 is the intensity of luminescence in the absence of oxygen, L is the intensity in the presence of oxygen at a partial pressure of P_{O_2}, and K is a constant. Equation (1), is a variation of the Stern-Volmer ex-

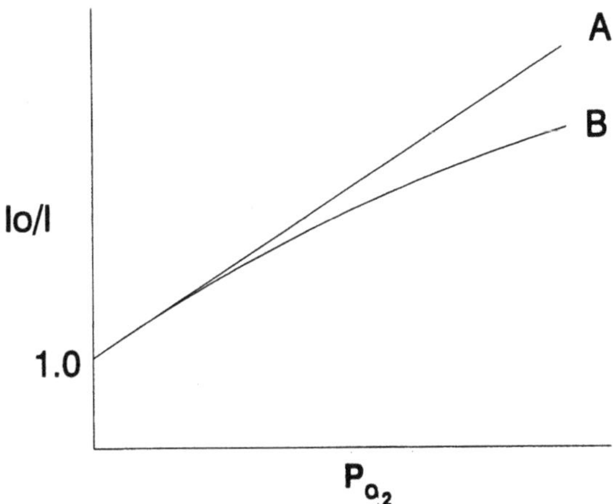

Fig. 3. L_0/L as a function of P$_{O_2}$. (A) Theoretical. (B) Experimental.

pression, is valid for a fiberoptic probe prepared by immobilizing an indicator in a matrix at the end of a fiber. Equation (1) can be rearranged to give Eq. (2):

$$P_{O_2} = [(L_0/L) - 1]/K \tag{2}$$

Empirically, the relationship between oxygen concentration and the reciprocal of the luminescence intensity is usually nonlinear. For many sensors, plots of L_0/L vs P$_{O_2}$ show a negative curvature (Fig. 3). Although the physical basis of the curvature is not fully understood, modified algorithms (Eq. [3]) using the term L_x to account for stray light and unquenchable luminescence have been proposed to correct for it (*21,43*).

$$P_{O_2} = \frac{\dfrac{(L_0 - L_x)}{(L - L_x)} - 1}{K} \tag{3}$$

The corresponding plots of $(L_0 - L_x)/(L - L_x)$ vs P$_{O_2}$ are reported to be quite linear, and similar corrections have been used by other groups (*44*). Alternative approaches (*45*) use an exponential expression, either as in Eq. (4) or with *n* equal to unity (*46*).

$$P_{O_2} = K[(L_0/L)^m - 1]^n \tag{4}$$

For an ideal solution, the relationship of P_{O_2} to luminescence lifetime is analogous to that for luminescence intensity,

$$P_{O_2} = [(\tau_0/\tau) - 1]/K \tag{5}$$

where τ_0 is the luminescence lifetime in the absence of oxygen and τ is the luminescence lifetime in the presence of an oxygen partial pressure equal to P_{O_2}. Luminescence lifetimes are typically determined by two methods *(42)*: pulse lifetime measurements, and phase and modulation measurements. As with intensity, a nonlinear relationship between reciprocal lifetime and oxygen partial pressure is usually observed experimentally *(32)*. Approaches to overcoming this problem using the pulse lifetime method have been presented *(33)*.

Luminescence-quenching devices for measuring oxygen tension have been known for many years *(47,48)*, but Hesse was the first to propose a fiberoptic oxygen sensor for in vivo use *(49)*. In an East German patent filed in 1973, he described a system in which a substrate, such as a polymer containing an oxygen quenchable fluorescent material, is attached to the end of an optical fiber through which the exciting and emitting light is transmitted. He recognized that the excitation and emission light could be separated by differences in wavelength or, if modulated, by differences in phase. He also described the use of a second fiber as a reference to correct for effects of temperature or interfering phenomena.

Lubbers and Opitz independently proposed a similar sensor, and reported on response time and boundary effects *(41, 50, 51)*. Their patents are the basis for an extracorporeal blood-gas monitor used in conjunction with heart-lung machines *(52)* and for an integrated intravascular blood-gas monitor currently under development *(11)*. In both of these sensor systems, the indicator is located on the distal face of the optical fiber. Sensors having the luminescent material in a sheath or cladding, rather than at the terminus, have also been reported *(17, 53, 54)* and allow interaction via the evanescent wave. In 1982, patent applications were filed that considered both lifetime and intensity measurements for fiberoptic oxygen sensors *(45, 55)*. Lifetime-based sensors

Fig. 4. Oxygen-sensor configurations.

have since been reported that use pulse *(33)* and phase modulation *(32)* techniques.

3.2. Design Considerations and Current Methodology

One design *(46)* (Fig. 4A) consists of two plastic fibers (250 μm diameter) sealed into a section of porous hydrophobic tubing filled with microspheres. The oxygen indicator is adsorbed onto the surface of the microspheres. Excitation light enters through one fiber, and emission light, together with some scattered light, exits through the other fiber.

Other designs include a rod of controlled-pore glass welded to the end of a single 140-μm-diameter glass fiber (Fig. 4B), and a polymer containing an indicator deposited on the end *(50)* (Fig. 4C) or side *(17)* (Fig. 4D) of an optical fiber.

The performance of a luminescence-quenching oxygen sensor is largely determined by the slope, K (Eq. [2]), of the modified Stern-Volmer equation. This term is the product of three factors: the quenching rate constant, the oxygen solubility coefficient of the matrix, and the luminescence lifetime of the indicator. Analyses of the optimum value for K are given elsewhere *(46,51)*.

The compounds that have been proposed for use as indicators fall into two classes: polycyclic aromatics and organometallics and are discussed elsewhere in this vol and *(43,46,56–60)*. Considerable work *(37,55,61,62)* has been done with platinum-group metal complexes, including ruthenium α-diimine complexes. More recently, reports involving the use of fluorinated porphyrins *(33)* and lanthanide complexes *(63)* have appeared. These materials are particularly attractive because of their relative photostability. With lanthanides, a terbium complex is used as the oxygen-sensitive indicator and a europium complex as an oxygen-insensitive reference.

Both porous and nonporous matrices are used for containing oxygen indicators. Porous materials have been widely investigated *(46,64,65)* because of the large slope usually obtained with their use. Typical porous substrates include silica gel, controlled-pore glass, and porous polystyrene. Unfortunately, problems occur regarding shelf life and sensitivity to species other than oxygen resulting from surface adsorption.

Silicone polymers are the most important nonporous matrices because of their relatively high oxygen permeability. A broad range of materials less permeable to oxygen have also been used, but these require indicators with higher sensitivities. Plasticizers have been used to increase the sensitivity to oxygen of some indicator–matrix pairs *(66)*. One problem in the preparation of stable, high-concentration indicator–matrix compositions has been caused by differences in the solubility parameter of the indicator and matrix. This problem has been addressed by derivatizing the dye to change its solubility parameter *(57,58)* and by covalently bonding the dye to the polymer *(59,60)*. For sensors using water-soluble organometallic indicators, it is possible to incorporate the indicator into the polymer matrix as an emulsion *(67)*.

As mentioned previously, quenching-based oxygen sensors operate on steady-state intensity or luminescence-lifetime measurements. In theory, sensors based on lifetime measurements have a distinct advantage over those that measure steady-state intensities. Because lifetime-based sensors measure the time-response of intensity rather than simple intensity, they are insensitive to many variations in in-tensity unrelated to oxygen concentration, including indicator photobleaching, fiber bending, and connector misalignment. Sensors based on measurements of steady-state intensity need a reference to correct for these aberrations. Luminescence-lifetime systems based on pulse-lifetime measurements *(33)* and phase-shift techniques *(32)* have been reported.

Several reference schemes have been suggested for intensity-based sensors, including a separate oxygen-insensitive dye to absorb and emit light in regions of the spectrum not used by the indicator dye *(30,55,68)*, and compounds capable of dual emission *(30)*.

3.3. *In Vivo Studies and Results*

Fiberoptic oxygen probes have been used to measure oxygen partial pressures (P_{O_2}) in arterial blood of sheep, swine, dogs, and humans, and intraocular P_{O_2} in dogs. Both porous and polymeric matrices have been used. In the following sections, the term "bias" is defined as the mean difference between the fiberoptic P_{O_2} and the P_{O_2} as determined by a bench-top blood-gas analyzer, whereas "precision" is the standard deviation of the difference *(35)*. In general, fiberoptic oxygen sensors show better performance in vitro than in vivo.

3.3.1. *Sensors with Porous Matrices*

The first in vivo evaluation of a fiberoptic oxygen probe was carried out in the carotid artery of a sheep, using a 0.6-mm-diameter probe containing perylene dibutyrate dispersed onto porous polystyrene beads contained within a polypropylene membrane *(46)*. The fiberoptic sensor roughly tracked the blood-oxygen tension.

A modified version of the above probe has been used to measure intraocular oxygen tension in normal and diabetic dogs *(69)*. Measurements of anterior-chamber oxygen tensions, taken simultaneously with a fiberoptic probe and a polargraphic electrode in the same eye, gave similar response times. In other experiments, a fiberoptic probe and a polarographic electrode were placed in the preretinal vitreous of contralateral eyes. With the dogs breathing 21% oxygen, the fiberoptic probe measured preretinal oxygen tension as 26 ± 5 mm Hg (mean $\pm$ SD, $n = 5$), whereas the electrode gave 23 ± 7 mm Hg (mean $\pm$ SD, $n = 5$). The difference between these measurements is statistically insignificant.

In 1985 and 1986, workers at Kelsius Inc. tested two designs of fiberoptic oxygen sensors in dogs. In one design, the optical fiber was

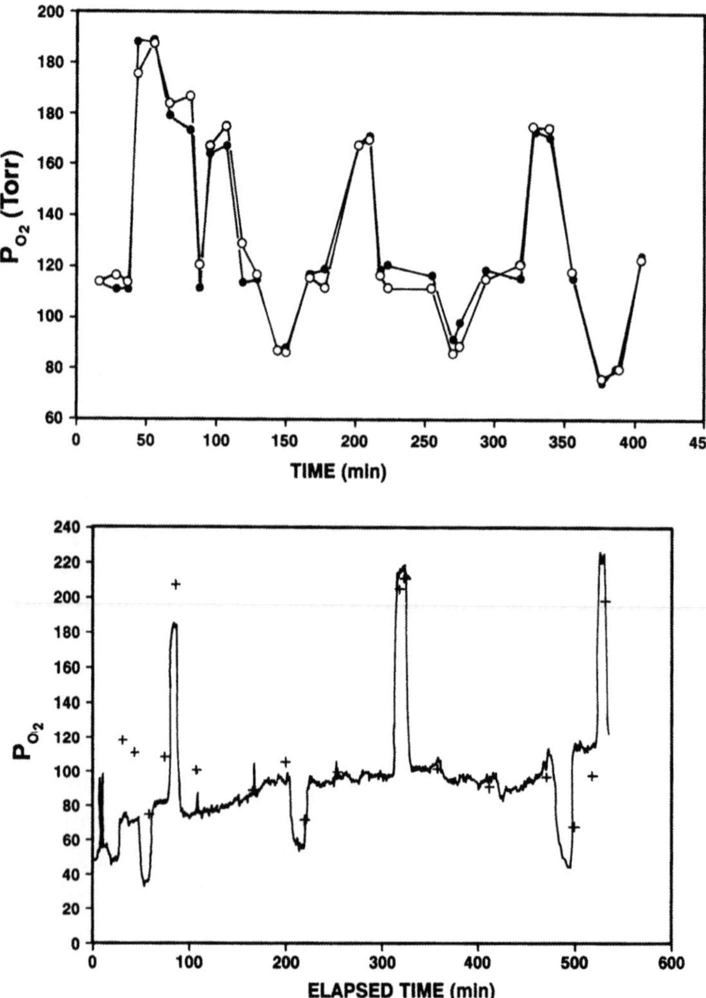

Fig. 5. In vivo oxygen-sensor performance in a dog. (Top) Single-unit oxygen sensor: o in vivo; •, standard blood-gas analyzer. (Bottom) Oxygen data from an integrated (pH, P_{CO_2}, P_{O_2}) fiberoptic blood-gas sensor. (Reproduced courtesy of Kelsius, Inc.)

tipped with a porous matrix; in the second, a polymer tip was used. In the porous-matrix design, the indicator was covalently bound to the surface of the matrix. Octadecyl substituents were also attached to the surface to provide blood compatibility *(8,9)*. Probes of this design were tested in vivo, both as a single unit (Fig. 5, top) and as part of an integrated sensor containing P_{O_2}, P_{CO_2}, and pH probes (Fig. 5, bottom). The single-unit device was calibrated in vivo. Relative to the bench-top blood-gas results, the device showed a bias of 1 mm Hg and a precision of 8 mm Hg for 31 blood-gas measurements over the range 75–208 mm Hg.

Figure 5 (bottom) shows data from the integrated unit that was calibrated prior to insertion. The device showed a bias of 7 mm Hg and a precision of 21 mm Hg relative to the analysis obtained with a benchtop blood-gas instrument. Similar results were found using the polymer-based sensor.

3.3.2. Sensors with Polymeric Substrates

Intravascular oxygen sensors have also been developed at Puritan-Bennett, CDI (3M Health Care), and American Baxter. Puritan-Bennett porcine tests *(17)* involved a single-unit fiberoptic sensor evaluated by comparison with results obtained from two conventional blood-gas analyzers over the P_{O_2} range 50–300 mm Hg. Comparative analysis was performed by drawing intermittent arterial blood samples and analyzing them concomitantly on a Corning Model 170 and an Instrumentation Laboratory (IL) Model 1303 (Fig. 6). The intravascular sensor showed a bias of 4 mm Hg and a precision of 9 mm Hg over the entire range when compared to the IL 1303. The results were similar to those obtained by comparison of the two conventional analyzers: A bias of 7 mm Hg and a precision of 8 mm Hg was observed when the Corning 170 was correlated with the IL 1303.

Evaluation of an integrated device containing both P_{O_2} and P_{CO_2} sensors in seven human subjects (Fig. 7) *(70)* demonstrated a bias of –1 mm Hg and a precision of 10 mm Hg for the intravascular sensor over the P_{O_2} range of 67–600 mm Hg when compared to a Corning Model 178 blood-gas analyzer. Data obtained through analysis between an IL 1306 blood-gas analyzer and the Corning 178 for the same samples over the identical range revealed a bias of –3 mm Hg and a precision of 20 mm Hg. When the data were limited to those values below 150 mm Hg, the intravascular sensor bias was –1 mm Hg and the precision was 5 mm Hg. These results compared favorably to those obtained for the IL 1306 for the range below 150 mm Hg which showed a bias of 1 mm Hg and a precision of 7 mm Hg.

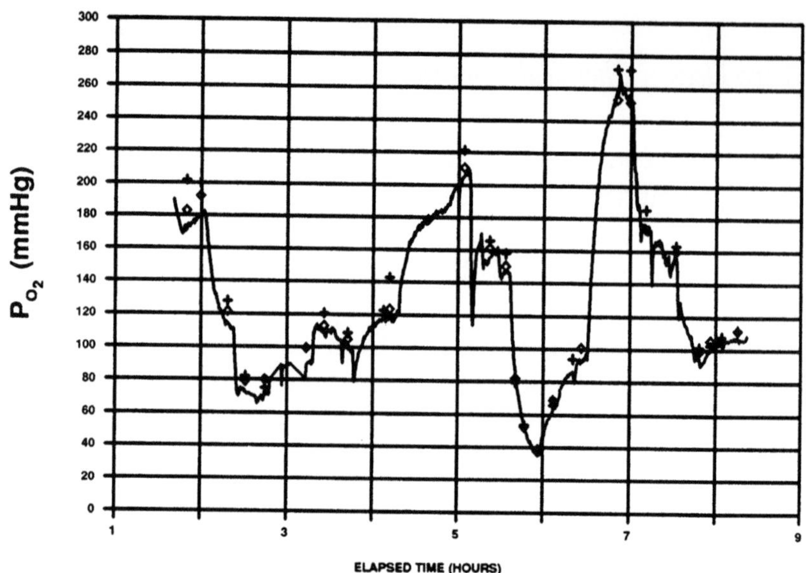

Fig. 6. In vivo performance of a single-unit oxygen sensor from Puritan-Bennett Corp. in a pig: —, in vivo sensor; +, Corning Model 178 blood-gas analyzer; ◊, Instrumentation Laboratories Model 1303 blood-gas analyzer (reproduced courtesy of the Puritan-Bennett Corp.).

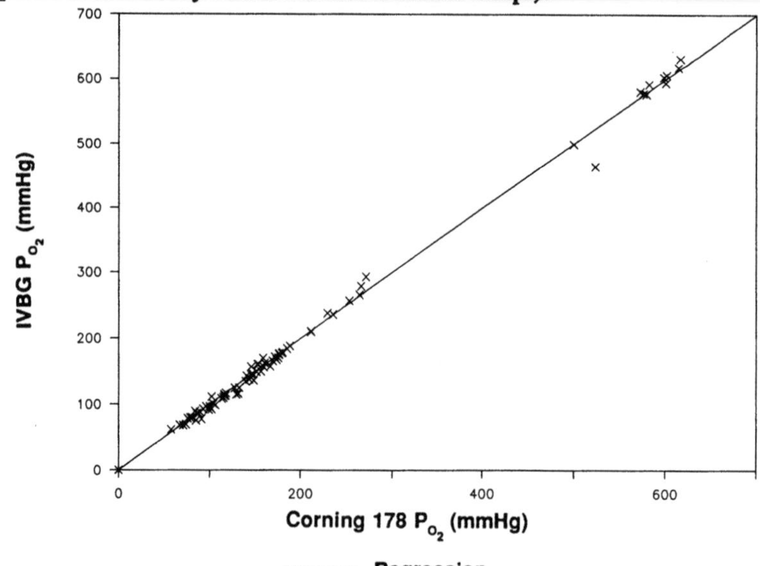

Fig. 7. In vivo oxygen-sensor performance: pooled data from a series of integrated Puritan-Bennett sensors (CO_2 and O_2) in humans (reproduced courtesy of the Puritan-Bennett Corp.).

340

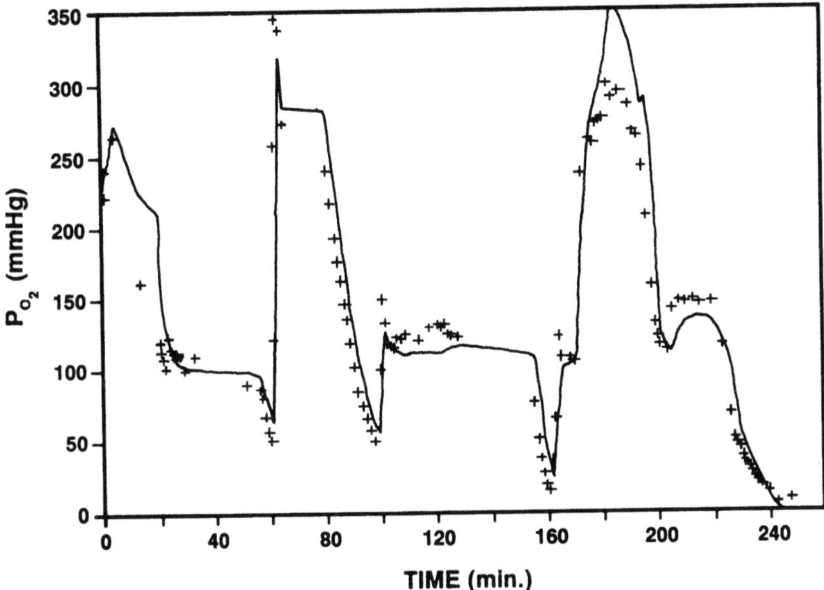

Fig. 8. In vivo oxygen-sensor performance for CDI intravascular blood gas sensor in a dog: —, in vivo; +, samples analyzed on blood-gas analyzer (reproduced with permission of American Society of Clinical Pathologists, *Lab. Med.* **19(10),** 629–635, 1988).

The intravascular blood-gas sensor being developed by CDI (3M Health Care) *(11)* has been evaluated in both dogs and humans *(10,71,72)*. A plot of the data from a continuous fiberoptic sensor and from a standard blood-gas monitor is shown in Fig. 8 *(73)*. In one study *(10)* with six dogs, each monitored for six h, sensors were calibrated before insertion and showed a bias of –17 mm Hg and a precision of 46 mm Hg. When the comparison was limited to samples with P_{O_2} values <150 mm Hg (172 samples), the performance of the fiberoptic sensors improved substantially, with a bias of 4 mm Hg and a precision of 13 mm Hg.

In a related set of clinical tests *(10)* in 12 patients, sensors were recalibrated after introduction into the radial artery. The bias was –1 mm Hg, with a precision of 9 mm Hg. When analysis was performed on a

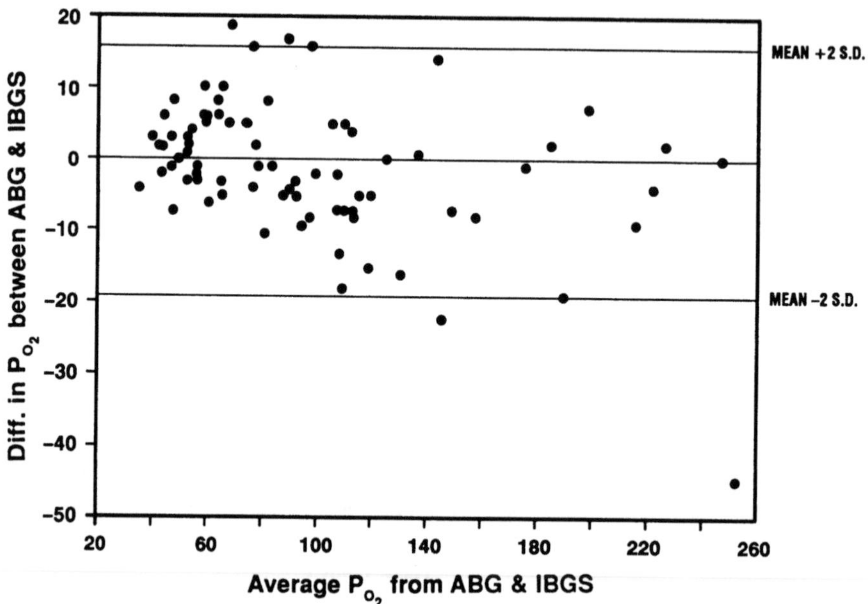

Fig. 9. In vivo oxygen-sensor performance for several CDI intravascular blood-gas sensors in human subjects. Horizontal lines show mean ±2 SD. ABG = arterial blood gas; IBGS = intravascular blood-gas system (reproduced with permission of the copyright owner, B. A. Shapiro et al., "Preliminary evaluation of an intra-arterial blood gas system in dogs and humans," *Crit. Care Med.* **17**(5), 455–460. © by Williams and Wilkins, 1989).

restricted set of data with P_{O_2} values <150, the bias was 0.5 mm Hg and the precision was 8 mm Hg. Visual inspection of the data from this experiment (*see* Fig. 9) suggests that the scatter is not correlated with P_{O_2}.

In a more recent study *(72)*, data from 10 patients undergoing surgery were obtained with sensors calibrated prior to insertion via an 18-gage cannula. For 68 samples collected, an intraarterial sensor and laboratory analyzer agreed with a bias of –3 mm Hg and precision of 24 mm Hg.

A single-unit oxygen sensor developed at American Baxter and tested in dogs *(4,74)* and humans *(75)* differs from the Puritan-Bennett and CDI sensors in terms of calibration and biocompatibility. The Baxter device is calibrated at one point, simply by exposing the probe to air. The Puritan-Bennett and CDI probes require two-point calibration, carried out with tonometered gases. The Baxter probe uses a benzalkonium heparin coating; the CDI unit uses covalently bonded heparin. Although the indicator and matrix were not discussed in the cited reference, a patent describing the preparation of an oxygen indicator in which a quenching indicator is covalently bound to a silicone polymer has recently been issued to American Baxter *(59)*. The results in dogs are similar to those found with other devices: a bias of 2.7 mm Hg and a precision of 37.3 mm Hg, based on a total of 290 blood samples from four dogs. If the analysis is restricted to samples with a P_{O_2} <200 mm Hg, the bias becomes 2 mm Hg and the precision becomes 27 mm Hg. A similar trend is seen in the human data, which shows a bias of −1 mm Hg and a precision of 19 mm Hg. The precision improves to 12 mm Hg when the data are restricted to P_{O_2} values <150 mm Hg.

4. pH Sensors

4.1. Principle of Operation

Optical measurements of pH are based on the determination of absorbance, reflectance, or fluorescence to measure the concentration of an indicator involved in a pH-mediated equilibrium *(76,77)*.

If the acidic dissociation constant for the indicator, K_a, is defined as

$$K_a = \frac{[H^+][A^-]}{[HA]} \tag{6}$$

where HA is the acidic form of the indicator and $HA \rightleftharpoons H^+ + A^-$, then the pH of a solution can be determined from the concentration of the acid and base forms of the indicator according to the Henderson-Hasselbalch equation

$$pH = pK_a - \log\,[HA]/[A^-] \tag{7}$$

In sensor applications, C, the total concentration of indicator ([HA] + [A$^-$]), will normally be constant, and the pH can be determined by measuring the concentration of the base form, A$^-$:

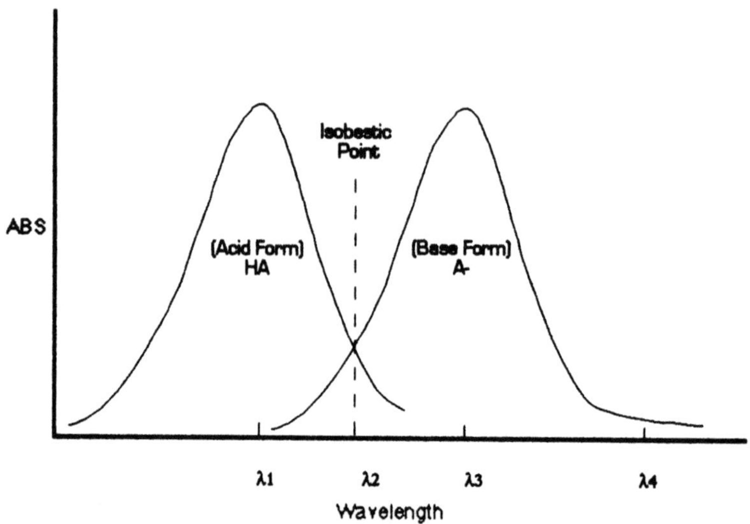

Fig. 10. Absorbance spectra of a pH indicator.

$$pH + pK_a - \log\left(\frac{C}{[A^-]} - 1\right) \qquad (8)$$

A number of pH sensors have been developed using matrices containing suitable pH indicators attached to the terminus of optical fibers. The absorbance, reflectance, or fluorescence of these matrices are then monitored via the fiber.

4.1.1. Absorbance Sensors

Figure 10 shows the absorbance spectrum typical of an indicator matrix. The acid (HA) and base (A$^-$) forms of the indicator absorb light in different spectral regions. At wavelength λ_3, the absorbance (*Abs*) of the matrix is proportional to the concentration of A$^-$, in accordance with Beer's Law

$$Abs = \varepsilon l\, [A^-] = \log I_0/I \qquad (9)$$

with the molar extinction coefficient (ε), the path length (l), and the concentration of A$^-$ as the variables. I_0 and I are the intensities of transmitted light when A$^-$ is absent and present, respectively. The pH of the matrix, and presumably of the surrounding solution, can be determined from the intensity of transmitted light *(78)*

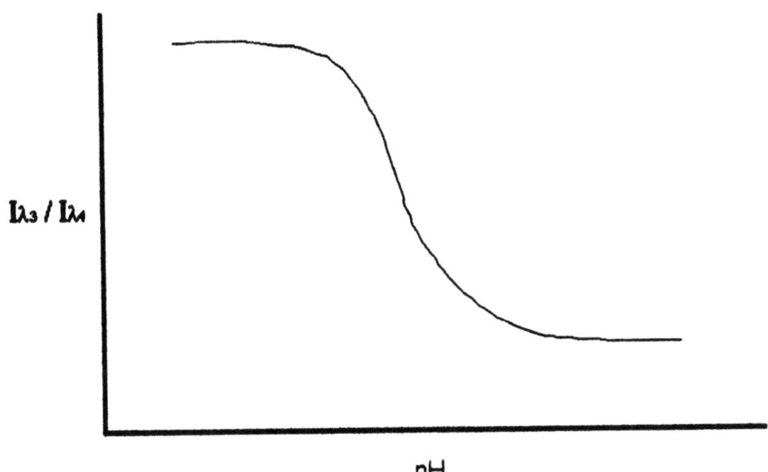

Fig. 11. Plot of pH vs $I_{\lambda 3}/I_{\lambda 4}$.

$$\text{pH} + pK_{a} - \log\left(\frac{C\varepsilon l}{\log(I_{0}/I)} - 1\right) \tag{10}$$

In many situations it may be impractical to measure I_{0}, but measuring the intensity at a second wavelength (λ_{4}) proportional to I_{0} is an alternative *(78)*. The intensity of this reference must be independent of pH. Thus its wavelength can be in a region away from the absorption peaks, as shown in Fig. 10, or it can be at the isosbestic point (λ_{2}). The pH can then be determined from the ratio of intensities at the different wavelengths

$$\text{pH} + pK_{a} - \log\left(\frac{K_{1}}{K_{2} + \log(I_{\lambda 3}/I_{\lambda 4})} - 1\right) \tag{11}$$

where K_{1} is the product of the total indicator concentration, the extinction coefficient, and the path length and K_{2} is the log of the ratio of I_{0} to $I_{\lambda 4}$. A plot of pH vs $I_{\lambda 3}/I_{\lambda 4}$ is shown in Fig. 11.

Over a 0.4-pH-unit range, the central region of the $I_{\lambda 3}/I_{\lambda 4}$-vs-pH curve can be considered linear and the pH expressed as

$$\text{pH} = m(I_{\lambda 3}/I_{\lambda 4}) + b \tag{12}$$

where the slope, *m*, and the intercept, *b*, are determined experimentally by using buffer solutions for a two-point calibration.

4.1.2. Fluorescence Sensors

The quantitative relationship between fluorescence intensity and concentration is *(79)*

$$F = I_0 \phi_F K[1 - \exp(-\varepsilon lc)] \tag{13}$$

where F is the fluorescence intensity; I_0, the incident intensity; ϕ_F, the fluorescence quantum yield; K, the optical efficiency of the measuring instrument; ε, the extinction coefficient; l, the pathlength; and c the concentration of the species of interest. For systems in which the absorbance is <0.05, this expression can be simplified to

$$F = 2.3 I_0 \phi_F K(\varepsilon lc) \tag{14}$$

For a pH indicator matrix in which the concentration of the base form of the indicator is being measured, the pH can be expressed as

$$pH + pK_a - \log\left(\frac{D}{F_{A^-}} - 1\right) \tag{15}$$

where D is a constant and F_{A^-} is the fluorescence intensity.

For an indicator where the acid and base forms fluoresce at different wavelengths, the pH may be expressed in terms of a constant, a, and the ratio of intensities:

$$pH = pK_a - \log a(F_{HA}/F_{A^-}) \tag{16}$$

In the case of indicators such as 8-hydroxy-1,3,6-pyrenetrisulfonic acid (HPTS), the acid and base forms are excited at different wavelengths, but emit at the same wavelength (Fig. 12) *(80)*. It has been suggested that the pH can be determined from the ratio of the fluorescence intensities resulting from sequential excitation of the acid (405 nm) and base (470 nm), using the following algorithm *(81)*

$$pH = pK_a - m \log[k_1(F_{405}/F_{460}) + k_2] \tag{17}$$

An alternative algorithm, which considers the contribution to the 405-nm excitation from both forms of HPTS, can be derived from the work of Zhujun and Seitz *(31)*

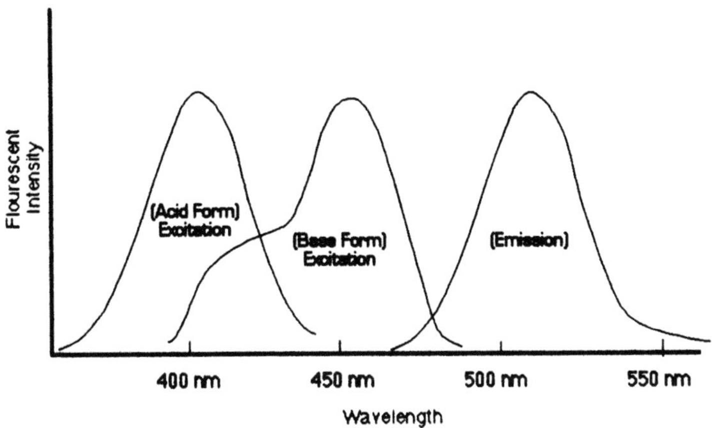

Fig. 12. Excitation and emission spectra for 8-hydroxy-1,3,6-pyrene-trisulfonic acid (HPTS).

$$pH = -\log[k_1(F_{405}/F_{470}) + k_2] \qquad (18)$$

In some situations, it may be possible to use the fluorescence that results from excitation at the isosbestic wavelength as an internal reference. However, if high concentrations of indicator are used, these expressions are not accurate. Zhujun and Seitz *(31)* have presented an analysis of the corrections that must be applied to account for the effects of concentration.

Lubbers and Opitz were among the first to suggest using membranes in conjunction with fluorescent indicators and optical fibers to prepare optical pH sensors for in vivo use *(50)*. Their work is the basis for a commercial extracorporeal blood-gas monitor and for an intravascular blood-gas monitor currently under development *(11)*.

Peterson, Goldstein, and Fitzgerald were the first to test an in vivo fiberoptic pH sensor *(78,82,83)*, and their absorbance-based sensor has been used in physiological studies *(84–90)*. In recent years, the development of absorbance- and reflectance-based fiberoptic pH sensors has been the focus of work in both the academic and the industrial communities *(91)*.

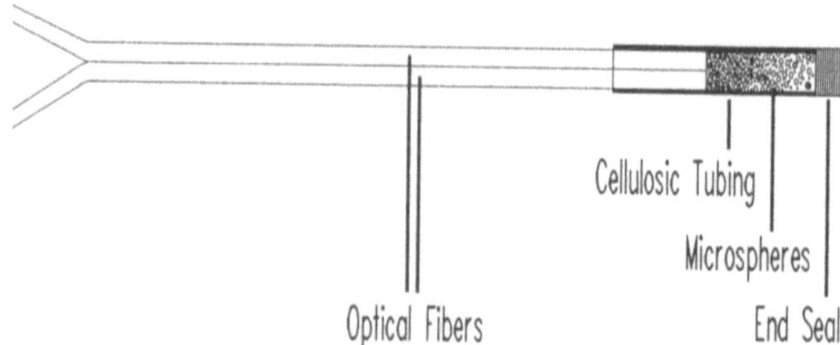

Fig. 13. Schematic diagram of an early fiberoptic pH sensor *(78)*.

4.2. Design Considerations and Current Methodology
4.2.1. Mechanical Design

The sensor developed by Peterson et al. *(78,82,83)* is shown schematically in Fig. 13. The device consists of two plastic fibers sealed into a section of cellulosic dialysis tubing filled with two types of microspheres: 5–10-µm-diameter polyacrylamide spheres containing Phenol Red as an indicator, and 1-µm-diameter polystyrene spheres to scatter light. The tubing holds the microspheres and also allows for diffusion of the aqueous portion of the sample while excluding large molecules. Green light (558 nm) and red light (600 nm) from a tungsten source are launched alternately into the input fiber. A portion of the green light is absorbed by the indicator, and a portion is scattered into the output fiber. The pH of the solution determines the amount absorbed. Red light is not absorbed by Phenol Red, and a constant amount is scattered into the output fiber. The optical signal from the output fiber is measured, and the pH is calculated from the ratio of green to red light, according to Eq. (11).

Markle and coworkers *(92)* tried to use the probes described above to measure the pH of beating canine hearts, but found the design too flexible. The sensor was redesigned to fit into a 25-gage stainless steel needle and used in a number of studies. Other redesigns *(84,85)* incorporated the probe into a myocardial transducer capable of measuring both pH and temperature. Another design, employing a 22-gage needle, has been used to monitor the pH of arterial blood *(93)*. This design uses a fiber bent back on itself. A different bent-fiber design has recently been described in the patent literature *(94)*. Other designs include a noninvasive conjunctival sensor *(93)* and one for monitoring intrapartum fetal tissue, in which an optical pH sensor is integrated with an ECG electrode *(95,96)*.

Additional developments include using light-emitting diodes (LEDs) as illumination sources *(97–100)*, sensors that measure reflectance rather than absorbance *(91)*, and designs that use a single-fiber to carry light to and from the sample *(101)*. A recent analysis describes the use of index-matching materials and modulation and phase-sensitive detection techniques to reduce the effects of stray light in single-fiber sensors *(102)*.

To date, the only fluorescence-based pH sensor suitable for in vivo use and described in the open literature is that of Cardiovascular Devices, Inc. (CDI 3M Health Care) *(11)*. This sensor uses a fluorescent matrix (probably prepared by covalently bonding HPTS to cellulose *[103]*) located at the end of an optical fiber. The excitation and emission are transmitted between the instrument and the sensor by a single fiber. This pH sensor is integrated with sensors for oxygen, carbon dioxide, and temperature into a single unit for use in human radial arteries.

4.2.2. Indicators

A number of pH indicators have been evaluated for use in fiberoptic pH sensors, including phenol red and rosolic acid dyes that apparently become permanently fixed to the polyacrylamide during polymerization *(78,83)*. The pK_a of phenol red shifts from 7.9 to 7.57 upon immobilization.

Other groups have also reported indicators that are affected by immobilization. Bacci et al. *(104)* studied the spectral properties of bromothymol blue, phenol red, chlorophenol red, and alizarin. Upon deposition onto a styrene/divinylbenzene copolymer, three of these compounds showed changes in their absorbance spectra. The spectra of the alkaline species undergo a red shift and broaden substantially, whereas the absorbance of the protonated species becomes much less sensitive to pH. Sensors using Congo red *(105)* and cresol red *(106)* as indicators have also been reported.

A large number of fluorescent pH indicators are known *(107)*, but, of these, only 4-methylumbelliferone *(108,109)*, 8-hydroxy-1,3,6-pyrenetrisulfonic acid (HPTS) *(31,80,110,111)*, and derivatives of fluorescein *(112,113)* have been widely used with fiberoptic sensors. HPTS is particularly popular because of its relative photostability, ability to be ratioed, and low toxicity in mammals *(114)*. Two long-wavelength indicators have recently become available: SNARF™ (seminaphthorhodafluor) and SNAFL™ (seminaphthofluorescein) *(115)*. These hold considerable promise because they can be excited between 500–600 nm (i.e., they are potentially excitable with LEDs), and they can be ratioed.

Opitz and Lubbers have predicted that maximum sensitivity should occur when the pK_a of the dye is equal to the pH of the solution, but that maximum resolution should result when the pK_a is about 0.3 U greater than the pH *(116)*.

4.2.3. Indicator Matrices

A variety of polymeric materials have been used as substrates to immobilize pH indicators for use in fiberoptic pH sensors. Peterson and coworkers found that phenol red becomes permanently fixed to the polymer during acrylamide polymerization and cannot be washed out either by water or by treatment with acid or base *(78,83)*. The dyed polymer is, nonetheless, sufficiently permeable that sensors made with this material have acceptable response times.

Polyacrylamides have also been used as substrates for fluorescent indicators. Munkholm et al. *(113)* have reported pH sensors prepared by dipping the end of a glass fiber into aqueous acrylamide, *N,N'*-methylene-*bis*-acrylamide, and acryloylfluorescein. During polymerization, the end of the fiber becomes covered with a thin layer of the indicator matrix consisting of covalently bound dye and polymer. Since the glass surface is activated by plasma or silane treatment prior to polymerization, the polymer is also covalently bonded to the glass. Because of the thin, concentrated indicator layer, these sensors showed a fast (9-s) response time and a high signal-to-noise ratio.

Indicators have also been covalently bonded to other materials. Crosslinked polyvinyl alcohol has been used for covalent immobilization of dyes such as fluoresceinamine *(117)*. HPTS has been covalently bonded to porous glass *(110)* and electrostatically immobilized on an ion-exchange membrane *(31)*. Yafuso and Hui *(103)* have prepared sensors that use an indicator matrix of HPTS covalently bonded to aminoalkylated cellulose. Other cellulose-based sensors include one using immobilized fluoresceinamine *(112)* and another employing Congo red immobilized on a porous cellulosic polymer film *(105)*. The latter sensor shows fast response times in consequence of the porous character of the matrix.

Dye immobilization has also been achieved by entrapping a polymer-bound dye in a macromolecular network *(118)* and by adsorbing dye onto a substrate, such as a styrene/divinylbenzene copolymer *(91,104)* or an ion-exchange resin.

4.2.4. Selectivity and Sensitivity

Temperature, ionic strength, and interfering contaminants can affect the performance of fiberoptic pH sensors. Corrections for temperature can be made with an empirical temperature coefficient *(78)*.

It is well known that the pK_a of pH indicators is sensitive to ionic strength *(77)*. Consequently, the accuracy of fiberoptic pH sensors that use these indicators may be affected by changes in ionic strength *(119–122)*, with some showing a change of 0.01 pH units per 11% change in ionic strength *(78)*. Zhujun and Seitz *(31,123)* have reported that, although the fluorescence intensity of HPTS immobilized on an ion-exchange membrane is sensitive to changes in ionic strength, the pH calculated from a ratio of fluorescence intensities is not.

One approach to decreasing the sensitivity to changes in ionic strength is to immobilize the indicator in a highly charged environment *(120)*. A second approach is to use two indicators having different sensitivities to ionic strength and pH *(116,119)*. The true pH can be calculated from the measured values and the ratio of indicator ionic-strength dependencies.

Most fluorescent pH indicators are subject to quenching by such species as oxygen. Fortunately, for dyes such as HPTS, this effect is essentially inconsequential under normal physiological conditions.

4.3. In Vivo Studies and Results

Fiberoptic pH sensors have been used to measure the pH of blood in sheep, dogs, and humans, and to measure the pH in cardiac, fetal, and conjunctival tissues. Most of these studies have involved absorbance-based sensors, but recently results from fluorescence-based sensors have appeared. In general, fiberoptic pH sensors perform better in vitro than in vivo.

4.3.1. Blood Sensors

In the first report on the use of an intravascular fiberoptic pH sensor, the probe was calibrated using pH 7.00 and 7.40 buffers and inserted through an 18-gage needle into the right jugular vein of a ewe sheep systemically treated with heparin. A reference pH microelectrode was placed in the left jugular vein. Over a 3-h period, fair correspondence was observed between the intravascular probes and discrete samples measured with a blood-gas analyzer *(78,82)*.

Abraham and coworkers *(93)*, using a sensor designed to fit into a 22-gage stainless steel needle, evaluated the sensor in eight mongrel dogs. In each experiment the sensor was calibrated using three buffer solutions (pH 6.425, 6.967, and 7.592) and placed in the carotid artery of a dog systemically treated with heparin. For a range of pH values from 6.500 to 7.770, the mean difference between the fiberoptic sensor and a blood-gas analyzer was 0.060 pH units. At the end of each experiment, the sensors were checked for drift by measuring the pH of the calibration solutions. The greatest drift for an individual sensor was 0.155 pH units but the mean drift was never >0.042 pH U over more than 6 h.

The intravascular blood-gas sensor system being developed by CDI uses a fluorescence-based pH sensor *(11)* and has been evaluated in dogs and humans *(10,71,72)*. A plot of data from this continuous fiberoptic sensor and a blood-gas analyzer is shown in Fig. 14. In one study *(10)* involving six dogs, each monitored for 6 h, 420 blood-gas samples were used for comparison of the fiberoptic sensor with a blood-gas analyzer. The sensors used in this study underwent a two-point calibration before insertion and showed a bias of –0.02 pH units and a precision of 0.03 pH units.

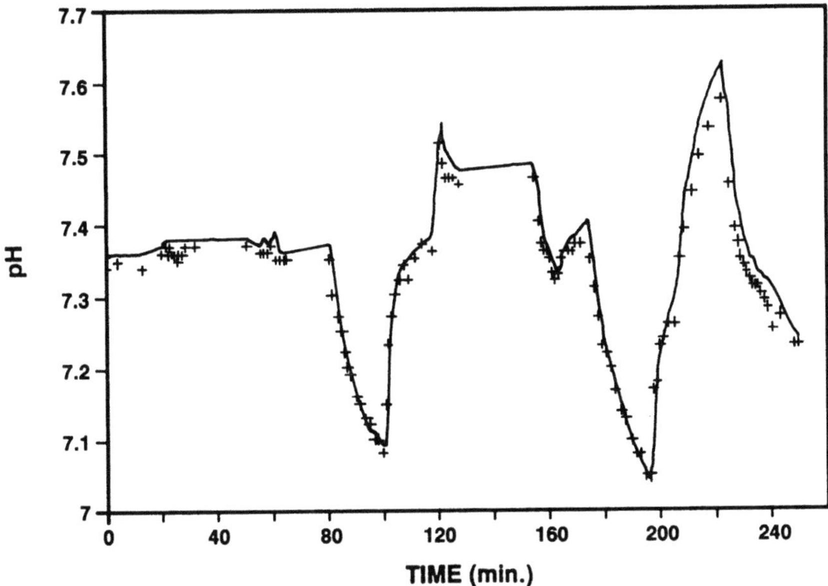

Fig. 14. In vivo pH performance of a CDI intravascular blood-gas sensor in a dog: —, fiberoptic sensor; +, samples analyzed by blood-gas analyzer (reproduced with permission of the copyright owner, M. W. Miller et al., "Continuous in vivo monitoring of blood gases," *Lab. Med.* **19(10)** 629–635, © by American Society of Clinical Pathologists, 1988).

In a related study using human subjects *(10)*, sensors were introduced into 12 patients, and 79 blood-gas samples were drawn for comparison. In this study, the sensors were subject to a one-point recalibration after insertion into the radial artery. The bias was 0.002 pH units with a precision of 0.02 pH units.

In a more recent study *(72)*, sensors were tested in 10 patients undergoing surgery, with the sensors calibrated prior to insertion via an 18-gage cannula. For the 68 samples that were collected, the intraarterial sensor and laboratory blood-gas analyzer were in agreement with a bias of –0.039 pH units and a precision of 0.057 pH units.

4.3.2. Tissue Sensors

Modified designs of the pH sensor first proposed by Peterson have been used in studies of myocardial ischemia *(84–90)* and in detection of

                                                      Gottlieb, Divers, and Hui

acute rejection after cardiac transplantation *(124)*. Still other designs have been presented for monitoring the pH of the conjunctiva *(125)* and for continuous intrapartum monitoring of the pH of fetal scalp tissue *(95)*. In the latter application, the use of these sensors to monitor over 300 term or postdate high-risk fetuses has recently been reported *(96)*.

5. Carbon Dioxide Sensors

5.1. Principle of Operation

Carbon dioxide partial pressure, P_{CO_2}, is usually measured by monitoring the pH of a bicarbonate buffer solution that is in equilibrium with a sample, but separated from it by a gas-permeable membrane. The P_{CO_2} is related to the pH of the buffer through the interplay of a series of chemical equilibria.

$$CO_2\,(g) \rightleftharpoons P_{CO_2}\,(s)$$

$$CO_2 + H_2O \rightleftharpoons H_2CO_3 \overset{K_1}{\rightleftharpoons} H^+ + HCO_3^- \tag{19}$$

$$HCO_3^- \overset{K_2}{\rightleftharpoons} H^+ + CO_3^{-2}$$

Severinghaus and Bradley *(26)* suggested that Eq. (19) can be used when the concentration of sodium bicarbonate in the buffer is between 0.001 and 0.1*M*.

$$\alpha P_{CO_2} = ([H^+][Na^+]/K_1) \tag{20}$$

where K_1 is the first dissociation constant for carbonic acid and $[Na^+]$ is the concentration of sodium bicarbonate. Measurement of $[H^+]$ thus permits the measurement of P_{CO_2}. An alternative approach, in which the water content of a matrix is used to measure P_{CO_2}, has recently been proposed *(126)*.

In 1958, Severinghaus and Bradley *(26)* reported on a P_{CO_2} electrode prepared from a pH electrode, a bicarbonate buffer solution, and a gas-permeable membrane. Lubbers and Opitz *(50,127)* developed an optical sensor on the same principle, but replaced the pH electrode with a fluorescent pH-indicator matrix at the end of an optical fiber. In 1983 Vurek et al. described a system that uses an absorptive dye *(27)*.

5.2. Design Considerations

The sensor of Vurek and coworkers *(27)* has a design similar to that of the pH sensor constructed by Peterson *(78,83)*. The sensor consists of a piece of silicone rubber tubing 0.6 mm in diameter filled with a solution of phenol red, potassium carbonate, and potassium chloride; sealed at one end with white silicone adhesive and at the other with a pair of optical fibers.

The design proposed by Zhujun and Seitz *(128)* uses a pair of membranes separated from a bifurcated optical fiber by a cavity filled with bicarbonate buffer. The external membrane is of silicone and the internal membrane of HPTS (12.6 $\mu g/cm^2$) immobilized on an ion-exchange membrane. This arrangement reportly allows for use of a relatively large volume of buffer solution without compromising the sensor's response time.

Recently, Munkholm et al. *(129)* reported a compact design that uses a single fiber for transmitting both excitation and emission. This device consists of a pH sensor (prepared by copolymerizing fluoroscein and either acrylamide or 2-hydroxyethyl methacrylate (HEMA) onto a fiber) soaked in bicarbonate buffer and overcoated with a silicone jacket.

These sensors are not specific for carbon dioxide, but respond to any volatile material capable of acid–base chemistry. Fortunately, carbon dioxide is the only such species present in most physiological situations. Interference may occur, however, because of materials such as acetic acid, that are produced during the curing of some silicone polymers *(27)*.

5.3. In Vivo Studies and Results

Fiberoptic CO_2 sensors have been used to measure the P_{CO_2} of blood in dogs and humans and to measure conjunctival P_{CO_2}.

Evaluation of an integrated sensor produced by the Puritan-Bennett Corporation for continuous monitoring of arterial P_{O_2} and P_{CO_2} in humans has recently appeared *(70)*. The sensors were introduced through a 20-gage cannula into the radial arteries of seven patients undergoing neurosurgical procedures. The sensors, in place for 4–15 h, were monitored during surgery and during the recovery phase in an inten-

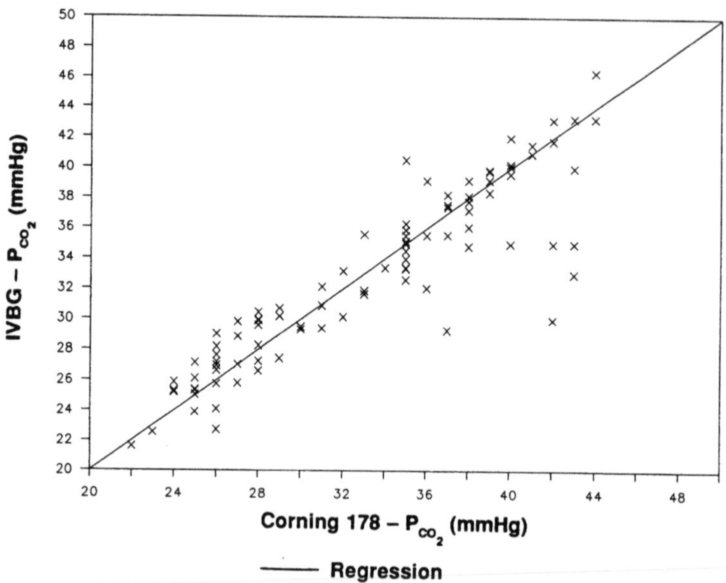

Fig. 15. In vivo P_{CO_2} pooled data with integrated PuritanBennett sensors (CO_2 and O_2) in humans (reproduced courtesy of the Puritan-Bennett Corp).

sive care unit. Calibration was undertaken in vitro, and 93 blood samples were collected for comparison assay on a blood-gas analyzer (Fig. 15). These sensors showed a bias of –0.3 mm Hg and a precision of 3 mm Hg relative to the blood-gas analyzer.

The intravascular blood-gas sensor system being developed by CDI (3M Health Care) *(10,71,72)* has been evaluated in both dogs and humans. In one study involving six dogs *(71)* (Figs. 16 and 17), with the sensors calibrated in vitro, the bias was 1 mm Hg and the precision was 4 mm Hg. In a related set of human clinical tests *(71)*, sensors were introduced into 12 patients, and 79 blood-gas samples were drawn. The fiberoptic sensors, recalibrated after introduction into the radial artery, showed a bias of 0.4 mm Hg and a precision of 3 mm Hg.

In a more recent study, data collected from 10 patients undergoing surgery were reported *(72)*. The sensors were calibrated in vitro and introduced into the radial artery through an 18-gage cannula. A bias of

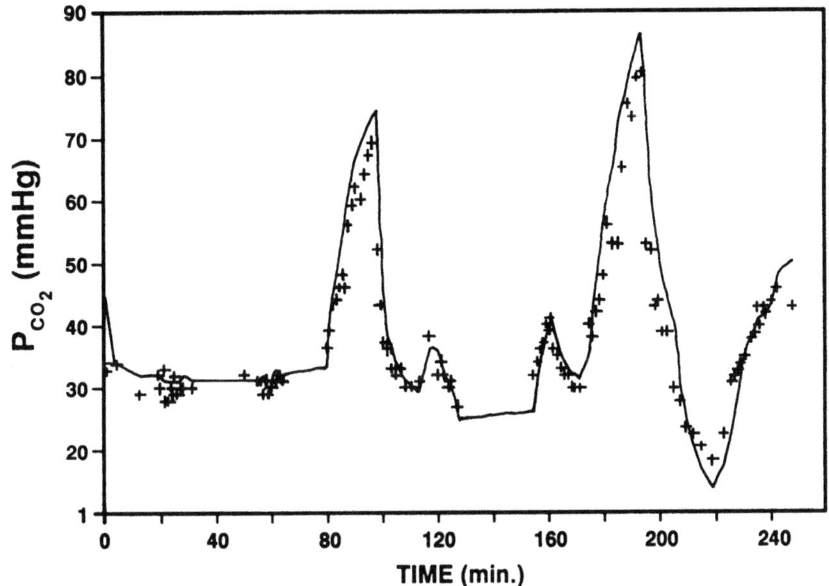

Fig. 16. In vivo P_{CO_2} results with CDI intravascular blood-gas sensor in a dog (reproduced with permission of the copyright owner, M. W. Miller et al., "Continuous in vivo monitoring of blood gases," *Lab. Med.* **19(10)** 629–635, © by American Society of Clinical Pathologists, 1988).

–3 mm Hg and a precision of 5 mm Hg were observed for the 68 blood samples collected.

Evaluation of a conjunctival P_{CO_2} sensor in dogs *(130,131)* has shown that under normal conditions and those of respiratory alkalosis and acidosis, conjunctival and arterial P_{CO_2} values are well correlated. Under conditions of metabolic acidosis and alkalosis, the correlation is significantly weaker *(131)*.

6. Conclusions

In vivo fiberoptic sensors for measurement of pH, P_{O_2}, and P_{CO_2} have been demonstrated and evaluated in animals and in human subjects. It is anticipated that these devices will find widespread clinical application during the next few years. As this technology matures, the

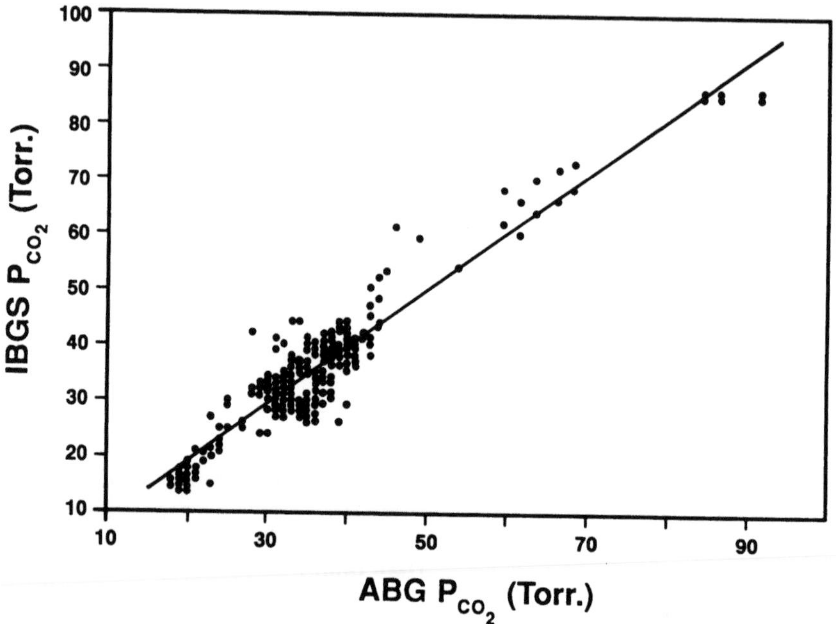

Fig. 17. Pooled data on in vivo P_{CO_2} with CDI intravascular blood-gas sensors in dogs (reproduced with permission of the copyright owner, B. A. Shapiro et al., "Preliminary evaluation of an intra-arterial blood gas system in dogs and humans," *Crit. Care Med.* **17**(5), 455–460, © by Williams and Wilkins, 1989).

lessons learned during its development should be of aid in the development of in vivo fiberoptic biosensors.

References

1. Thompson, M. and Vandenberg, E. T. (1986) In vivo probes: Problems and perspectives. *Clin. Biochem.* **19**, 255–261.
2. Regnault, W. F. and Picciolo, G. L. (1987) Review of medical biosensors and associated materials problems. *J. Biomed. Nater. Res.: Appl. Biomater.* **21**(A2), 163–180.
3. Yafuso, M., Arick, S. A., Hansmann, D., Holody, M., Miller, W. W., and Yan, C. F. (1989) Optical pH measurements in blood. *Proc. SPIE—Int. Soc. Opt. Eng.* **1067** (Optical Fibers in Medicine IV), 37–43.
4. Barker, S. J., Tremper, K. K., Hyatt, J., Zaccari, J., Heitzmann, H. A., Holman,

B. M., Pike, K., Ring, L. S., Teope, M., and Thaure, T. B. (1987) Continuous fiberoptic arterial oxygen tension measurements in dogs. *J. Clin. Monit.* **3**, 48–52.

5. Kim, S. W. and Feijen, J. (1985) Surface modification of polymers for improved blood compatibility. *CRC Crit. Rev. Biocompat.* **1**(3), 229–260.

6. Williams, D. F., ed. (1987) *Blood Compatibility*, vols. 1 and 2 (CRC Press, Boca Raton, FL).

7. Eberhart, R. C. (1985) Indwelling blood compatible chemical sensors. *Surg. Clin. North Am.* **65**(4), 1025–1040.

8. Eberhart, R. C., Munro, M. S., Williams, G. B., Kulkarni, P. V., Shannon, W. A. Jr., and Brink, B. E. (1987) Albumin adsorption and retension on alkyl derivatized polyurethane vascular grafts. *Artif. Organs* **11**, 375–382.

9. Munro, M. S., Quattrone, A. J., Ellsworth, S. R., and Eberhart, R. C. (1985) Nonthrombogenic articles having enhanced albumin affinity. US Patent 4,530,974.

10. Shapiro, B. A., Cane, R. D., Chomka, C. M., Bandala, L. E., and Peruzzi, W. T. (1989) Preliminary evaluation of an intra-arterial blood gas system in dogs and humans. *Crit. Care Med.* **17**(5), 455–460.

11. Gehrich, J. L., Lubbers, D. W., Opitz, N., Hansmann, D. R., Miller, W. W., Tusa, J. K., and Yafuso, M. (1986) Optical fluorescence and its application to an intravascular blood gas monitoring system. *IEEE Trans. Biomed. Eng.* **BME-33**(2), 117–132.

12. Miller, W. W., Gehrich, J. L., Hansmann, D. R., and Yafuso, M. (1988) Continuous in vivo monitoring of blood gases. *Lab. Med.* **19**(10), 629–635.

13. Gunther, M. and Rupp, L. (1990) Method for manufacturing a measuring probe. US Patent 4,900,381.

14. Friebele, E. J. (1979) Optical fiber waveguides in radiation environments. *Opt. Eng.* **18**(6), 552–561.

15. Boiarski, A. A. (1989) Integrated optic system for monitoring blood gases. US Patent 4,854,321.

16. Marcuse, D. (1988) Launching light into fiber cores from sources located in the cladding. *J. Lightwave Tech.* **6**(8), 1273–1279.

17. Hui, H. K., Divers, S., Lumsden, T., Wallner, T., and Weir, S. (1990) An accurate, low-cost, easily-manufacturable oxygen sensor. *Proc. SPIE—Int. Soc. Opt. Eng.* **1172** (Chemical, Biochemical, and Environmental Fiber Sensors), 233–238.

18. Blyler, L. L., Jr., Lieberman, R. A., Cohen, L. G., Ferrara, J. A., and Macchesney, J. B. (1989) Optical fiber chemical sensors utilizing dye-doped silicone polymer claddings. *Polym. Eng. Sci.* **29**(17), 1215–1218.

19. David, D. J., Willson, M. C., and Ruffin, D. S. (1976) Direct measurement of ammonia in ambient air. *Anal. Lett.* **9**(4), 389–404.

20. Louch, J. and Ingle, J. D., Jr. (1988) Experimental comparison of single- and

double-fiber configurations for remote fiber-optic fluorescence sensing. *Anal. Chem.* **60,** 2537–2540.

21. Wolfbeis, O. S., Weis, L. J., Leiner, M. J. P., and Ziegler, W. E. (1988) Fiber-optic fluorosensor for oxygen and carbon dioxide. *Anal. Chem.* **60,** 2028–2030.

22. Rahn, H. and Prakash, O., eds. (1985) *Acid-Base Regulation and Body Temperature (Developments in Critical Care Medicine and Anaesthesiology,* vol. 10) (Kluwer Academic, Boston).

23. Siggaard-Andersen, O., Wimberley, P. D., Gothgen, I. H., Fogh-Andersen, N., and Rasmussen, J. P. (1988) Variability of the temperature coefficients for pH, pCO_2, and pO_2 in blood. *Scand. J. Clin. Lab. Invest.* **48,** 85–88.

24. Kelman, G. R. and Nunn, J. F. (1966) Nomograms for correction of blood pO_2, pCO_2, pH, and base excess for time and temperature. *J. Appl. Physiol.* **21(5),** 1484–1490.

25. Burnett, R. W., Christiansen, T. F., Durst, R. A., Evenson, R., Fallon, K., Komjathy, Z. L., Ladenson, J. H., Moran, R. F., Pulwer, E., Weisberg, H. F., and Zee, D. (1982) Tentative standard for definitions of quantities and conventions related to blood pH and gas analysis. *National Committee for Clinical Laboratory Standards* **2(10),** 329–361.

26. Severinghaus, J. W. and Bradley, A. F. (1958) Electrodes for blood pO_2 and pCO_2 determination. *J. Appl. Physiol.* **13,** 515–520.

27. Vurek, G. G., Feustel, P. J., and Severinghaus, J. W. (1984) A fiber optic pCO_2 sensor. *Ann. Biomed. Eng.* **11,** 499–510.

28. Arnold, M. A. and Ostler, T. J. (1986) Fiber optic ammonia gas sensing probe. *Anal. Chem.* **58,** 1137–1140.

29. Wolfbeis, O. S., Posch, H. E., and Kroneis, H. W. (1985) Fiber optical fluorosensor for determination of halothane and/or oxygen. *Anal. Chem.* **57,** 2556–2561.

30. Lee, E. D., Werner, T. C., and Seitz, W. R. (1987) Luminescence Ratio Indicators for Oxygen. *Anal. Chem.* **59,** 279–283.

31. Zhujun, Z. and Seitz, W. R. (1984) A fluorescence sensor for quantifying pH in the range from 6.5 to 8.5. *Anal. Chim. Acta* **160,** 47–55.

32. Lippitsch, M. E. and Wolfbeis, O. S. (1988) Fibre-optic oxygen sensor with the fluorescence decay time as the information carrier. *Anal. Chim. Acta* **205,** 1–6.

33. Khalil, Gamal-E., Gouterman, M. P., and Green, E. (1989) Method for measuring oxygen concentration. US Patent 4,810,655.

34. Culshaw, B., Foley, J., and Giles, I. P. (1984) A balancing technique for optical fibre intensity modulated transducers. *Proc. SPIE—Int. Soc. Opt. Eng.* **574** (Proc. 2nd Int. Conf. Fibre Optic Sensors, Stuttgart), 117–120.

35. Bland, J. M. and Altman, D. G. (1986) Statistical methods for assessing agreement between two methods of clinical measurement. *Lancet* **i,** 307–310.

36. Vaughan, W. M. and Weber, G. (1970) Oxygen quenching of pyrenebutyric

acid fluorescence in water. A dynamic probe of the microenvironment. *Biochemistry* **9**(3), 464–473.

37. Wolfbeis, O. S. and Leiner, M. J. P. (1988) Recent progress in optical oxygen sensing. *Proc. SPIE—Int. Soc. Opt. Eng.* **906** (Optical Fibers in Medicine), 42–48.

38. Freeman, T. M. and Seitz, W. R. (1980) Oxygen probe based on tetrakis-(alkylamino)ethylene chemiluminescence. *Anal. Chem.* **53**(1), 98–102.

39. Zhujun, Z. and Seitz, W. R. (1985) Optical sensor for oxygen based on immobilized hemoglobin. *Anal. Chem.* **58**, 220–222.

40. Wolfbeis, O. S., Offenbacher, H., Kroneis, H., and Marsoner, H. (1984) A fast responding fluorescence sensor of oxygen. *Mikrochim. Acta* **I**, 153–158.

41. Lubbers, D. W. and Opitz, N. (1983) Optical fluorescence sensors for continuous measurement of chemical concentrations in biological systems. *Sens. Actuators* **4**, 641–654.

42. Lakowitz, J. R. (1983) Principles of Fluorescence Spectroscopy (Plenum, New York).

43. Kroneis, H. W. and Marsoner, H. J. (1983) A fluorescence-based sterilizable oxygen probe for use in bioreactors. *Sens. Actuators* **4**, 587–592.

44. Barnikol, W. K. R., Gaertner, T., Weiler, N., and Burkhard, O. (1988) Microdetector for rapid changes of oxygen partial pressure (pO_2) during the respiratory cycle in small animals. *Rev. Sci. Instrum.* **59**(7), 1204–1208.

45. Peterson, J. I. and Fitzgerald, R. V. (1984) Fiber optic pO_2 probe. US Patent 4,476,870.

46. Peterson, J. I., Fitzgerald, R. V., and Buckhold, D. K. (1984) Fiber-optic probe for in vivo measurement of oxygen partial pressure. *Anal. Chem.* **56**, 62–67.

47. Bergman, I. (1968) Improvements in or relating to gas detectors. UK Patent 1,190,583.

48. Stevens, B. (1971) Instrument for determining oxygen quantities by measuring oxygen quenching of fluorescent radiation. US Patent 3,612,866.

49. Hesse, Hans-C. (1974) Measuring Probe. GDR Patent 106,086.

50. Lubbers, D. W. and Opitz, N. (1985) Method and arrangement for measuring the concentration of gases. US Patent Re. 31,879.

51. Opitz, N. and Lubbers, D. W. (1987) Theory and development of fluorescence-based optochemical oxygen sensors: Oxygen optodes. *Int. Anesth. Clin.* **25**(3), 177–179.

52. Siggaard-Andersen, O., Gothgen, I. H., Wimberley, P. D., Rasmussen, J. P., and Fogh-Andersen, N. (1988) Evaluation of the Gas-STAT fluorescence sensors for continuous measurement of pH, pCO_2, and pO_2 during cardiopulmonary bypass and hypothermia. *Scand. J. Clin. Lab. Invest.* **48**, 77–84.

53. Buckles, R. G. (1982) Method for quantitative analysis using optical fibers. US Patent 4,321,057.

54. Buckles, R. G. (1983) Optical fiber apparatus for quantitative analysis. US Patent 4,399,099.

55. Bacon, J. R. and Demas, J. N. (1984) Method and apparatus for determining the presence of oxygen. UK Patent GB 2,132,348 A (Application).

56. Cox, M. E. and Dunn, B. (1985) Detection of oxygen by fluorescence quenching. *Appl. Opt.* **24(14)**, 2114–2120.

57. Yafuso, M., Yan, C. F., Hui, H. K., and Miller, W. W. (1989) Optical sensor. US Patent 4,849,172.

58. Marsoner, H., Kroneis, H., and Wolfbeis, O. (1987) Sensor element for determining the oxygen content and a method of preparing the same. US Patent 4,657,736.

59. Hsu, L. and Heitzmann, H. (1987) Dye containing silicone polymer composition. US Patent 4,712,865.

60. Klainer, S. M., Walt, D. R., and Gottlieb, A. J. (1988) Fibre optic sensing device and new polymer—useful as pH, oxygen, electrolyte or blood gas sensor. World Patent WO 8805533 A (Application).

61. Bacon, J. R. and Demas, J. N. (1987) Determination of oxygen concentration by luminescence quenching of a polymer–immobilized transition–metal complex. *Anal. Chem.* **59**, 2780–2785.

62. Murray, R. C. and Lefkowitz, S. M. (1988) Optical sensor for monitoring the partial pressure of oxygen. US Patent 4,752,115.

63. Nestor, J. R., Schiff, J. D., and Priest, B. H. (1990) Excitation and detection apparatus for remote sensor connected by optical fiber. US Patent 4,900,933.

64. Li, P. Y. F. and Narayanaswamy, R. (1989) Oxygen-sensitive reagent matrices for the development of optical fiber chemical transducers. *Analyst* **114**, 663–666.

65. Surgi, M. R. (1989) Design and evaluation of a reversible fiber optic sensor for determination of oxygen, in *Applied Biosensors* (Wise, D. L., ed.), Butterworths, Boston, pp. 249–290.

66. Marsoner, H. and Kroneis, H. (1986) Measuring device for deteriming the O_2 content of a sample. US Patent 4,587,101.

67. Wolfbeis, O. S., Leiner, M. J. P., and Posch, H. E. (1986) A new sensing material for optical oxygen measurement, with the indicator embedded in an aqueous phase. *Mikrochim. Acta 1986* **III(5–6)**, 359–366.

68. Lubbers, D. W. and Opitz, N. (1981) Photometer including auxiliary indicator means. US Patent 4,255,053.

69. Stefansson, E., Peterson, J. I., and Wang, Y. H. (1989) Intraocular oxygen tension measured with a fiber-optic sensor in normal and diabetic dogs. *Am. J. Physiol.* **256**, H1127–H1133.

70. Larson, C. P., Jr., Riccitelli, S. D., Divers, S., Hui, H. K., Wallner, T. G., Boyles, J. V. C., and Lumsden, T. J. (1990) Evaluation of a continuous, in vivo blood gas monitoring system in patients. *Abstracts of the Association of*

University Anesthetists Annual Meeting (Seattle, WA, May 3–5, 1990) (in press).

71. Shapiro, B., Cane, R., Chomka, C., and Gehrich, J. (1987) Evaluation of a new intra-arterial blood gas system in dogs and humans. *Anesthesiology* **67(3A)**, A640.

72. Barker, S. J., Hyatt, J., Tremper, K. K., Gehrich, J. L., Arick, S. M., Gerschultz, S., and Safdari, K. (1989) Fiberoptic intraarterial pHa, PaO_2, and $PaCO_2$ in the operating room. *Anesth. Analg.* **68**, S16.

73. Miller, W. W., Yafuso, M., Yan, C. F., Hui, H. K., and Arick, S. (1987) Performance of an in-vivo continuous blood-gas monitor with disposable probe. *Clin. Chem.* **33(9)**, 1538–1542.

74. Barker, S. J., Tremper, K. K., and Heitzmann, H. A. (1987) Continuous fiberoptic arterial oxygen tension in dogs. *Crit. Care Med.* **15**, 403.

75. Barker, S. J., Tremper, K. K., and Heitzmann, H. A. (1987) A clinical study of fiber-optic arterial oxygen tension. *Crit. Care Med.* **15**, 403.

76. Kolthoff, I. M. and Laitinen, H. A. (1941) *pH and Electro Titrations. The Colorimetric and Potentiometric Determination of pH*, 2nd ed. (Wiley, New York).

77. Bates, R. G. (1973) *Determination of pH. Theory and Practice*, 2nd ed. (Wiley-Interscience, New York).

78. Peterson, J. I., Goldstein, S. R., Fitzgerald, R. V., and Buckhold, D. K. (1980) Fiber optic pH probe for physiological use. *Anal. Chem.* **52**, 864–869.

79. Willard, H. H., Merritt, L. L., Dean, J. A., and Settle, F. A., Jr. (1981). *Instrumental Methods of Analysis*, 6th ed. (Wadsworth, Belmont, CA).

80. Wolfbeis, O. S., Furlinger, E., Kroneis, H., and Marsoner, H. (1983) Fluorimetric Analysis. I. A study on fluorescent indicators for measuring near neutral (physiological) pH-values. *Fresenius Z. Anal. Chem.* **314**, 119–124.

81. Junker, B. H., Wang, D. I. C., and Hatton, T. A. (1988) Fluorescence sensing of fermentation parameters using fiber optics. *Biotech. Bioeng.* **32**, 55–63.

82. Goldstein, S. R., Peterson, J. I., and Fitzgerald, R. V. (1980) A miniature fiber optic pH sensor for physiological use. *J. Biomech. Eng.* **102**, 141–146.

83. Peterson, J. I. and Goldstein, S. R. (1980) Fiber optic pH probe. US Patent 4,200,110.

84. Tait, G. A., Young, R. B., Wilson, G. J., Steward, D. J., and MacGregor, D. C. (1982) Myocardial pH during regional ischemia: Evaluation of a fiber-optic photometric probe. *Am. J. Physiol.* **243**, H1027–H1031.

85. Takach, T. J., Glassman, L. R., Ribakove, G. H., and Clark, R. E. (1986) Continuous measurement of intramyocardial pH: Correlation to functional recovery following normothermic and hypothermic global ischemia. *Ann. Thorac. Surg.* **42**, 31–36.

86. Watson, R. M., Markle, D. R., Ro, Y. M., Goldstein, S. R., McGuire, D. A., Peterson, J. I., and Patterson, R. E. (1984) Transmural pH gradient in canine myocardial isechmia. *Am. J. Physiol.* **246**, H232–238.

87. Watson, R. M., Markle, D. R., McGuire, D. A., Vitale, D., Epstein, S. E., and Patterson, R. E. (1985) Effect of verapamil on pH of ischemic canine myocardium. *J. Am. Coll. Cardiol.* **5(6),** 1347–1354.

88. Takach, T. J., Glassman, L. R., Milewicz, A. L., and Clark, R. E. (1986) Continuous measurement of intramyocardial pH: Relative importance of hypothermia and cardioplegic perfusion pressure and temperature. *Ann. Thorac. Surg.* **42,** 365–371.

89. Maturi, M. F., Greene, R., Speir, E., Burrus, C., Dorsey, L. M. A., Markle, D. R., Maxwell, M., Schmidt, W., Goldstein, S. R., and Patterson, R. E. (1989) Neuropeptide-Y. A peptide found in human coronary arteries constricts primarily small coronary arteries to produce myocardial ischemia in dogs. *J. Clin. Invest.* **83,** 1217–1224.

90. Ro, Y. M., Markle, D. R., Goldstein, S. R., Speir, E., Greene, R., Steadman, K., Aamodt, R., Epstein, S. E., and Patterson, R. E. (1989) Contrasting effects of verapamil and nifedipine on pH of ischemic myocardium in the dog. *J. Pharmacol. Exp. Ther.* **248(2),** 654–660.

91. Kirkbright, G. F., Narayanaswamy, R., and Welti, N. A. (1984) Fibre-optic pH probe based on the use of an immobilised colorimetric indicator. *Analyst* **109,** 1025–1028.

92. Markle, D. R., McGuire, D. A., Goldstein, S. R., Patterson, R. E., and Watson, R. M. (1981) A pH measurement system for use in tissue and blood employing miniature fiber optic probes, in *Advances in Bioengineering* (Viano, D., ed.) American Society of Mechanical Engineering, New York, pp. 123–126.

93. Abraham, E., Markle, D. R., Fink, S., Ehrlich, H., Tsang, M., Smith, M., and Meyer, A. (1985) Continuous measurement of intravascular pH with a fiberoptic sensor. *Anesth. Analg.* **64,** 731–736.

94. Costello, D. (1987) Fiber optic probe for quantification of colorimetric reactions. US Patent 4,682,895.

95. Chatterjee, M. S., Hetzel, F., and Kaminetzky, H. K. (1984) Fetal tissue pH—continuous monitoring. *Int. J. Gynaecol. Obstet.* **22(1),** 41–46.

96. Hochberg, H. M., Roby, P. V., Snell, H. M., Smith, W. D., and Chatterjee, M. S. (1988) Continuous intrapartum fetal scalp tissue pH and ECG monitoring by a fiberoptic probe. *J. Perinat. Med.* **16,** 71–86.

97. Grattan, K. T. V., Mouaziz, Z., and Palmer, A. W. (1987) Dual wavelength optical fiber sensor for pH measurement. *Biosensors* 17–25.

98. Guthrie, A. J., Narayanaswamy, R., and Welti, N. A. (1988) Solid-state instrumentation for use with optical-fibre chemical-sensors. *Talanta* **35(2),** 157–159.

99. Boisde, G. and Perez, J. J. (1987) Miniature chemical optical fiber sensors for pH measurements. *Proc. SPIE—Int. Soc. Opt. Eng.* **798** (Fiber optic Sensors II), 238–245.

100. Besar, S. S. A., Kelly, S. W., and Greenhalgh, P. A. (1989) Simple fibre optic spectrophotometric cell for pH determination. *J. Biomed. Eng.* **11,** 151–156.

101. Coleman, J. T., Eastham, J. F., and Sepaniak, M. J. (1984) Fiber optic based sensor for bioanalytical absorbance measurements. *Anal. Chem.* **56**, 2246–2249.
102. Skogerboe, K. J. and Yeung, E. S. (1987) Stray light rejection in fiber-optic probes. *Anal. Chem.* **59**, 1812–1815.
103. Yasuso, M. and Hui, H. K. (1989) Micro Sensor. US Patent 4,798,738.
104. Bacci, M., Baldini, F., and Scheggi, A. M. (1988) Spectophotometric investigations on immobilized acid-base indicators. *Anal. Chim. Acta* **207**, 343–348.
105. Jones, T. P. and Porter, M. D. (1988) Optical pH sensor based on the chemical modification of a porous polymer film. *Anal. Chem.* **60**, 404–406.
106. Moreno, M. C., Marinez, A., Millan, P., and Camara, C. (1986) Study of a pH sensitive optical fibre sensor based on the use of cresol red. *J. Mol. Struct.* **143**, 553–556.
107. Guilbault, G. G. (1973) *Practical Fluorescence* (Marcel Dekker, New York).
108. Lubbers, D. W., Opitz, N., Speiser, P. P., and Bisson, H. J. (1977) Nanoencapsulated fluorescence indicator molecules measuring pH and pO_2 down to submicroscopical regions on the basis of the optode-principle. *Z. Naturforsch.* **32c**, 133–134.
109. Lubbers, D. W. and Opitz, N. (1983) Blood gas analysis with fluorescence dyes as an example of their usefulness as quantitative chemical sensors, in *Proceedings of the International Meeting on Chemical Sensors, Analytical Chemistry Symposia, 17* (Seiyama, T., Fueki, K., Shiokawa, J., and Suzuki, S., eds.), Elsevier, New York, pp. 609–619.
110. Offenbacher, H., Wolfbeis, O. S., and Furlinger, E. (1986) Fluorescence optical sensors for continuous determination of near-neutral pH values. *Sens. Actuators* **9**, 73–84.
111. Seitz, W. R. and Zhujun, Z. (1985) Fluorescent fluid determination method and apparatus. US Patent 4,548,907.
112. Saari, L. A. and Seitz, W. R. (1982) pH sensor based on immobilized fluoresceinamine. *Anal. Chem.* **54**, 821–823.
113. Munkholm, C., Walt, D. R., Milanovich, F. P., and Klainer, S. M. (1986) Polymer modification of fiber optic chemical sensors as a method of enhancing fluorescence signal for pH measurement. *Anal. Chem.* **58**, 1427–1430.
114. Lutty, G. A. (1978) The acute intravenous toxicity of biological stains, dyes, and other fluorescent substances. *Toxicol. Appl. Pharmacol.* **44**, 225–249.
115. Haugland, R. P. (1989) *Handbook of Fluorescent Probes and Research Chemicals* (Molecular Probes, Eugene, OR).
116. Opitz, N. and Lubbers, D. W. (1985) The applicability of fluorescence indicators to measure hydrogen ion activities by optimizing accuracy and minimizing the influence of ionic strength, in *Ion Measurements in Physiology and Medicine* (Kessler, M.,Harrison, D. K., and Hoper, J., eds.), Springer-Verlag, Berlin, Heidelberg, pp. 122–127.
117. Zhujun, Z., Zhang, Y., Wangbai, M., Russell, R., Shakhsher, Z. M., Grant, C.

L., Seitz, W. R., and Sundberg, D. C. (1989) Poly(vinyl alcohol) as a substrate for indicator immobilization for fiberoptic chemical sensors. *Anal. Chem.* **61,** 202–205.

118. Wolfbeis, O., Kroneis, H., and Offenbacher, H. (1986) Sensor element for fluorescence-optical measurement. US Patent 4,568,518.
119. Opitz, N. and Lubbers, D. W. (1983) New fluorescence photometrical techniques for simultaneous and continuous measurements of ionic strength and hydrogen ion activities. *Sens. Actuators* **4,** 473–479.
120. Wolfbeis, O. S. and Offenbacher, H. (1986) Fluorescence sensor for monitoring ionic strength and physiological pH values. *Sens. Actuators* **9,** 85–91.
121. Janata, J. (1987) Do optical sensors really measure pH? *Anal. Chem.* **59,** 1351–1356.
122. Edmonds, T. E., Flatters, N. J., Jones, C. F., and Miller, J. N. (1988) Determination of pH with acid-base indicators: Implications for optical fibre probes. *Talanta* **25(2),** 103–107.
123. Seitz, W. R. (1987) Optical sensors based on immobilized reagents, in *Biosensors. Fundamentals and Applications* (Turner, A. P. F., Karube, I., and Wilson, G. S., eds.) Oxford University Press, New York, pp. 599–617.
124. Takach, T. J., Glassman, L. R., Rodriguez, E. R., Falcone, J. T., Ferrans, V. J., and Clark, R. E. (1986) Acute rejection after cardiac transplantation: Detection by interstitial myocardial pH. *Ann. Thorac. Surg.* **42,** 619–626.
125. Abraham, E., Fink, S. E., Markle, D. R., Pinholster, G., and Tsang, M. (1985) Continuous monitoring of tissue pH with a fiberoptic conjuctival sensor. *Ann. Emerg. Med.* **14(9),** 840–844.
126. Leader, M. J. and Kamiya, T. (1989) Sensor system. US Patent 4,833,091.
127. Lubbers, D. W. and Opitz, N. (1975) The pCO_2-/pO_2-Optode: A new probe for measurement of pCO_2 or pO_2 in fluids and gases. *Z. Naturforsch.* **30c,** 532–533.
128. Zhujun, Z. and Seitz, W. R. (1984) A carbon dioxide sensor based on fluorescence. *Anal. Chim. Acta* **160,** 305–309.
129. Munkholm, C., Walt, D. R., and Milanovich, F. P. (1988) A Fiber-optic sensor for CO_2 measurement. *Talanta* **35(2),** 109–112.
130. Abraham, E., Markle, D. R., Pinholster, G., and Fink, S. (1986) Noninvasive measurement of conjunctival pCO_2 with a fiberoptic sensor. *Crit. Care Med.* **14 (2),** 138–141.
131. Kram, H. B., Fink, S., Tsang, M., Markle, D., Appel, P. L., and Shoemaker, W. (1988) Noninvasive measurement of tissue carbon dioxide tension using a fiberoptic conjunctival sensor: Effects of respiratory and metabolic alkalosis and acidosis. *Crit. Care Med.* **16(3),** 280–284.

Index

MIX
Papier aus verantwortungsvollen Quellen
Paper from responsible sources
FSC® C105338

If you have any concerns about our products,
you can contact us on
ProductSafety@springernature.com

In case Publisher is established outside the EU,
the EU authorized representative is:
Springer Nature Customer Service Center GmbH
Europaplatz 3, 69115 Heidelberg, Germany

Printed by Libri Plureos GmbH
in Hamburg, Germany